Advances of Chitin and Chitosan-Based Materials: Preparation and Applications

Advances of Chitin and Chitosan-Based Materials: Preparation and Applications

Guest Editors

Dawei Zhang
Xianzhi Kong

MDPI

Basel • Beijing • Wuhan • Barcelona • Belgrade • Novi Sad • Cluj • Manchester

Guest Editors
Dawei Zhang
Material Science and Engineering College
Northeast Forestry University
Harbin
China

Xianzhi Kong
Institute of Petrochemistry
Heilongjiang Academy of Sciences
Harbin
China

Editorial Office
MDPI AG
Grosspeteranlage 5
4052 Basel, Switzerland

This is a reprint of the Special Issue, published open access by the journal *Polymers* (ISSN 2073-4360), freely accessible at: https://www.mdpi.com/journal/polymers/special_issues/7QD6X7I8GB.

For citation purposes, cite each article independently as indicated on the article page online and as indicated below:

Lastname, A.A.; Lastname, B.B. Article Title. *Journal Name* **Year**, *Volume Number*, Page Range.

ISBN 978-3-7258-5453-0 (Hbk)
ISBN 978-3-7258-5454-7 (PDF)
https://doi.org/10.3390/books978-3-7258-5454-7

Contents

polymers

Editorial

Closing Editorial: Advances in Chitin and Chitosan-Based Materials: Preparation and Applications

Xianzhi Kong [1,*] and Dawei Zhang [2,*]

[1] Institute of Petrochemistry, Heilongjiang Academy of Sciences, Harbin 150040, China
[2] State Key Laboratory of Woody Oil Resources Utilization, Northeast Forestry University, Harbin 150040, China
* Correspondence: kongxianzhi2023@hipc.org (X.K.); zhangdawei@nefu.edu.cn (D.Z.); Tel.: +86-0451-82623695 (X.K.)

1. Introduction

Chitin and chitosan-based materials are widely used and researched in healthcare, pharmaceutical, biomedical engineering, and related fields due to their biological activity. This Special Issue focuses on the preparation and application of materials based on chitin and chitosan, with the aim of presenting the latest research advances in biomedical applications and highlighting the potential for further innovative research. This Special Issue showcases the potential of chitin and chitosan-based materials by examining innovative approaches to maximize their structural, physical–chemical, and bioactive qualities.

2. An Overview of the Published Articles

The Special Issue highlights the potential uses of chitin and chitosan-based materials in biomedical, pharmaceutical, and sustainable practices and related fields. Liao et al. prepared a class of multifunctional chitosan-based hydrogels through dual cross-linking with borate esters and hydrogen bonds, achieving glucose-sensitive controlled release, providing a new method for the design of polysaccharide-based hydrogels [1]. Zhou et al. designed a flexible, low-cost, biomimetic spiral hollow bacterial cellulose-chitosan fibre, which could be an attractive candidate to replace other petroleum-based sutures [2].

Li et al. developed an emulsion carrying thyme essential oil via inducing cross-linking of chitosan particles through hydrogen bonds and electrostatic interactions. This emulsion has great potential for prolonging the storage life of strawberries [3]. Shi et al. prepared a hydrogel using chitosan and coumarin as raw materials for the controlled release of taxifolin [4]. Zhang et al. prepared chitosan-based Janus nanofiber membranes as wound dressings, which have vast potential applications in skin tissue engineering [5]. Blanzeanu et al. used extrusion moulding to prepare composite blends of chitosan, providing a sustainability–recycling-based approach to converting seafood waste into cutting-edge functional materials [6].

Li et al. summarize the state-of-the-art developments in chitosan-based dressing materials, highlighting the benefits in burn–wound treatment and examining the key challenges and potential future directions for chitosan-based dressing materials. This mini-review provides a new viewpoint on the evolution of wound dressings for burn care [7].

Zhou et al. provide a broad overview of chitosan extraction and modification technologies, with a focus on their applications in environment, energy, and biomedicine. It uses a novel classification framework to provide readers with the most detailed analysis for a systematic understanding of the latest research progress [8].

Together, these eight articles showcase the versatility of chitosan materials and discuss potential preparations and applications, touching upon the variety of ways in which they could be used in many fields, ranging from environmental sustainability to medical therapeutics.

3. Conclusions

The breadth of chitosan-based technologies and their potential to solve practical problems are highlighted in this Special Issue. From sophisticated biomedical systems to ecofriendly preservation solutions, together, these studies emphasize the versatility, biodegradability, and efficacy chitin and chitosan. As research progresses, further optimization of chitin and chitosan properties may result in more relevant and effective applications in industrial and clinical settings.

Funding: This research received no external funding.

Conflicts of Interest: The authors declare no conflict of interest.

References

1. Liao, X.Y.; Feng, X.Y.; Xiao, Z.Y.; Williams, G.R.; Huang, X.Z.; Shi, Y.Y.; Qin, H.; Liu, Y. Multifunctional phenylboric acid modified carboxymethyl chitosan based hydrogel crosslinked by tannic acid. *Int. J. Biol. Macromol.* **2025**, *304*, 140958. [CrossRef] [PubMed]
2. Zhou, Y.T.; Liu, X.T.; Yang, M.; Song, G.H.; Wang, Y.T.; Sun, H.; Yuan, T.; Rao, J.; Lü, B.Z.; Yao, C.L.; et al. Bio-inspired helical-hollow bacterial cellulose fiber for suture materials. *Chem. Eng. J.* **2025**, *505*, 159670. [CrossRef]
3. Li, H.Y.; Liu, M.Z.; Han, S.Y.; Hua, S.H.; Zhang, H.J.; Wang, J.; Xia, N.; Liu, Y.J.; Meng, D.K. Edible chitosan-based Pickering emulsion coatings: Preparation, characteristics, and application in strawberry preservation. *Int. J. Biol. Macromol.* **2024**, *264*, 130672. [CrossRef] [PubMed]
4. Shi, H.L.; Ma, D.X.; Wu, D.; Qiu, X.; Yang, S.; Wang, Y.Y.; Xiao, L.; Ji, X.Y.; Zhang, W.; Han, S.Y.; et al. A pH-responsive, injectable and self-healing chitosan-coumarin hydrogel based on Schiff base and hydrogen bonds. *Int. J. Biol. Macromol.* **2023**, *255*, 128122. [CrossRef] [PubMed]
5. Zhang, W.; Guan, X.; Qiu, X.; Gao, T.H.; Yu, W.Y.; Zhang, M.Y.; Song, L.R.; Liu, D.; Dong, J.D.; Jiang, Z.X.; et al. Bioactive composite Janus nanofibrous membranes loading Ciprofloxacin and Astaxanthin for enhanced healing of full-thickness skin defect wounds. *Appl. Surf. Sci.* **2022**, *610*, 155290. [CrossRef]
6. Blanzeanu, E.; Marin, M.; Verziu, M.N.; Dumitru, A.; Vasile, B.S.; Stavarache, C.; Cozorici, D.E.; Luque, R.; Zaharia, C.; Radu, I.C. Chitosan-polylactic acid composites: From seafood waste to advanced functional materials for 3D printing. *Adv. Compos. Hybrid Mater.* **2025**, *8*, 1. [CrossRef]
7. Li, S.Y.; Pan, W.L.; Zhang, M.; Song, K.L.; Zhou, Z.Q.; Zhao, Q.L.; Li, G.Z.; Zhu, C.Y. Chitosan-Based Dressing Materials for Burn Wound Healing. *Polymers* **2025**, *17*, 1647. [CrossRef] [PubMed]
8. Zhou, Y.Y.; Zhang, Y.; Nie, Y.X.; Sun, D.L.; Wu, D.Y.; Ban, L.; Zhang, H.; Yang, S.; Chen, J.S.; Du, H.S.; et al. Recent advances and perspectives in functional chitosan-based composites for environmental remediation, energy, and biomedical applications. *Prog. Mater. Sci.* **2025**, *152*, 101460. [CrossRef]

polymers

MDPI

Article

Chitosan and Cashew Nut Shell Liquid as Sustainable Additives: Enhancing Starch Digestibility and Reducing Methane Emissions in High-Grain Diets for Feedlot Cattle

Raquel Tenório de Oliveira [1], Rafael Henrique de Tonissi e Buschinelli de Goes [1,*], Jefferson Rodrigues Gandra [2], Fernanda Naiara Fogaça da Cruz [1], Nayara Gonçalves da Silva [1], Lara de Souza Oliveira [1], Jaqueline Luiza Royer [1], Lucas Gabriel Batista Domiciano [1], Tainá Lorraine Pereira Azevedo [3] and Carolina Marques Costa Araújo [1]

1 Faculty of Agricultural Sciences, Federal University of Grande Dourados, Dourados 79804-970, Brazil; raqueltennorio@gmail.com (R.T.d.O.); fogaca.fernandaa@gmail.com (F.N.F.d.C.); nayagsm@hotmail.com (N.G.d.S.); veterinarialaraoliveira@gmail.com (L.d.S.O.); jaqueline.royer123@gmail.com (J.L.R.); domicianolucasb@gmail.com (L.G.B.D.); carolinaaraujo@ufgd.edu.br (C.M.C.A.)

2 Faculty of Veterinary Medicine, Institute of Humid Tropic Studies, Federal University of the South and Southeast of Pará, Xinguara 68555-970, Brazil; jeffersongandra@unifesspa.edu.br

3 Veterinary Medicine Department, State University of Maringá, Umuarama 87507-190, Brazil; tainalpazevedo@gmail.com

* Correspondence: rafaelgoes@ufgd.edu.br

Abstract: Chitosan and technical cashew nutshell liquid (CNSLt) have emerged as promising natural compounds due to their antimicrobial, immunomodulatory, and fermentation-modulating properties. This study aimed to evaluate the inclusion of chitosan and CNSLt as potential substitutes for the ionophore monensin on feed intake, ruminal fermentation, nitrogen balance, and microbial protein synthesis in steers. Five crossbred steers (*Bos taurus*), 18 months old with an average body weight of approximately 350 kg and fitted with permanent ruminal cannulas, were assigned to a 5×5 Latin square design. The experimental diets consisted of: (1) control (CON), (2) monensin (MON; 25 mg/kg of dry matter [DM]), (3) chitosan (CHI; $\geq$850 g/kg deacetylation degree, 375 mg/kg DM), (4) CNSLt (500 mg/kg DM), and (5) CNSLt + CHI (500 + 375 mg/kg DM). Supplementation with CHI or CNSLt + CHI reduced the intake of dry matter, crude protein, and neutral detergent fiber. Additionally, fecal excretion of whole corn kernels increased in these treatments. Ruminal fermentation parameters were affected, with the CNSLt + CHI treatment promoting higher molar proportions of propionate and acetate, along with reduced estimated methane emissions. However, purine derivatives, microbial protein synthesis, and nitrogen balance were not significantly affected by any of the treatments. These results suggest that CNSLt and CHI, particularly when combined, may serve as effective natural alternatives to monensin in high-grain diets for ruminants.

Keywords: methane; natural additives; ruminal fermentation; beef cattle

1. Introduction

A high-grain diet is commonly used in beef cattle to enhance weight gain and productivity. However, the increased availability of rapidly fermentable carbohydrates leads to elevated production of volatile fatty acids (VFAs) and the accumulation of lactic acid, which can result in subacute ruminal acidosis (SARA). The occurrence and persistence of SARA can disrupt the ruminal environment by causing the death of Gram-negative bacteria and promoting the proliferation of acid-tolerant bacteria that thrive at a ruminal pH significantly lower than the recommended level [1].

Alterations in ruminal fermentation can compromise the efficient utilization of diets. In this context, feed additives serve as an alternative to mitigate the risk of acidosis, improve dietary utilization efficiency, and enhance both animal performance and health [2,3]. Among the commonly used additives, ionophores such as monensin are widely applied. However, restrictions on the use of antimicrobials in animal production [4] and food safety regulations imposed by importing countries have prompted the search for alternatives to the use of ionophores [5].

Optimizing ruminal fermentation is crucial for improving feed efficiency and enhancing overall ruminant productivity. Traditional strategies, such as the use of ionophores (i.e., monesin and lasalocin), have been widely employed to regulate microbial populations and fermentation patterns. However, increasing regulatory restrictions and growing consumer concerns regarding antibiotic residues have intensified the search for alternative feed additives. Chitosan, a deacetylated derivative of crustacean chitin, has emerged as a promising natural compound due to its antimicrobial, immunomodulatory, and fermentation-enhancing properties [6]. Studies have shown that chitosan supplementation in ruminant diets can modify the ruminal microbiota, increase propionate production while reducing acetate and methane emissions [7,8]. Additionally, its ability to inhibit fibrinolytic bacteria and protozoa suggests a potential role in enhancing nitrogen utilization and reducing ammonia concentrations in the rumen [9].

Despite these promising results, the mechanisms by which chitosan interacts with ruminal microorganisms and fermentation processes remain not fully understood. Further research is needed to refine dosage strategies and evaluate the long-term effects of chitosan supplementation on animal performance, nutrient digestibility, and environmental sustainability.

Technical cashew nut shell liquid (CNSLt) has emerged as a promising natural additive with multiple benefits for ruminant nutrition and methane mitigation. Rich in phenolic compounds such as anacardic acid, cardanol, and cardol, CNSLt exhibits strong antimicrobial activity, selectively inhibiting methanogenic archaea while promoting propionate-producing microorganisms [10]. In vitro studies have shown that CNSLt supplementation can reduce methane emissions by up to 70.1% and increase propionate production by 44.4%, effectively shifting ruminal fermentation toward a more energy-efficient profile [10].

Additionally, CNSLt influences microbial populations by disrupting the cell membranes of hydrogen-producing bacteria, resulting in a more favorable fermentation pattern characterized by reduced acetate and butyrate production [11]. These findings underscore CNSLt's potential as an environmentally sustainable alternative to conventional feed additives, such as ionophores, for improving ruminant productivity and reducing greenhouse gas emissions.

Based on this evidence, we hypothesized that the combined supplementation of chitosan and technical cashew nutshell liquid (CNSLt) would enhance ruminal fermentation and digestion without negatively affecting feed intake, nitrogen balance, or microbial protein synthesis in high-grain diets for feedlot cattle. Therefore, the objectives of this study were to evaluate the effects of chitosan and CNSLt, individually and in combination, on ruminal fermentation, nutrient digestion, and microbial protein synthesis. Additionally, this study aimed to assess the impact of these additives on feed intake and nitrogen balance, supporting their potential as sustainable alternatives to conventional feed additives in high-grain diets.

2. Materials and Methods

2.1. Location, Animals, and Treatments

The experiment was conducted at the Ruminant Nutrition Sector, Animal Nutrition Laboratory, and By-Product Evaluation Laboratory (LAPAC/FINEP) of the Federal University of Grande Dourados (UFGD), Brazil.

Five crossbred steers (Holstein × Zebu), aged 18 months, castrated, with an average body weight of approximately 350 kg, and fitted with permanent ruminal cannulas, were used. The animals were housed in individual covered pens (24 m^2; 4 × 6 m) with concrete flooring, each equipped with individual feed troughs and automatic waterers with a supply of 60 L/animal/day. A 5 × 5 Latin square design was employed, with animals randomly assigned to treatments across five experimental periods. Each period lasted 19 days, including 10 days of dietary adaptation and 9 days of data collection.

The diet consisted of 85% corn grain and 15% protein pellet, formulated to meet the nutritional requirements of growing steers. Treatments were based on the inclusion of chitosan (CHI) [12] and technical cashew nut shell liquid (CNSLt) [13], either alone or in combination. Chitosan (≥850 g/kg deacetylation degree, 0.32 g/mL density, pH 7.90, viscosity) was supplied by Polymar Indústria e Comércio de Importação e Exportação LTDA (Fortaleza, CE, Brazil), and CNSLt was obtained from Usibras Company (Aquiraz, CE, Brazil). The CNSLt used contained 10.03 mg/g of anacardic acid, 540.77 mg/g of cardanol, 102.34 mg/g of cardol, and 19.17 mg/g of 2-methylcardol, and its composition was confirmed by chemical analysis [13]. The experimental treatments were as follows: CON: control diet without additives; MON: monensin at 25 mg/kg of dry matter (DM); CHI: chitosan at 375 mg/kg DM; CNSLt: CNSLt at 500 mg/kg DM; and CNSLt + CHI: combination of CNSLt (500 mg/kg DM) and CHI (375 mg/kg DM).

Animals were fed one time daily (08:00), based on the previous day's dry matter intake. Feed offered and refusals (orts) were recorded daily, maintaining a 5–10% surplus to avoid feed restriction. The two dietary components (whole corn and the pellet) were manually mixed in the feed trough and offered as a total mixed ration (TMR) (Table 1).

Table 1. Ingredient proportions and chemical composition of the supplements utilized.

Ingredients		**(g/kg de MS)**	
Whole corn		850	
Protein, vitamin, and mineral pellet [1]		150	
		Chemical composition	
g/kg DM	Corn	Pellet	Diet
Dry matter	850	912.1	845.8
Organic matter	957	822.5	933.2
Crude protein	95	394.7	137
Neutral detergent fiber	96	372.9	241
Acid detergent fiber	18	236.9	63.5
Starch	750	38.9	651
Total digestible nutrients	797.86	682.36	737.38

[1] Ca, 43 g/kg; P, 10 g/kg; S, 4 g/kg; Mg, 0.7 g/kg; K, 2.7 g/kg; Na, 9.7 g/kg; Co, 5 mg/kg; Cu, 175 mg/kg; Cr, 1.4 mg/kg; F, 130 mg/kg; I, 5 mg/kg; Mn, 182 mg/kg; Mo, 0.35 mg/kg; Zn, 421 mg/kg; Vitamin A, 21.000 U.I; Vitamin D, 3.000 U.I; Vitamin E, 140 U.I.

2.2. Nutrient Intake and Apparent Total Digestibility

Dry matter intake (DMI) was determined by calculating the difference between the amount of feed offered and feed refused (orts), and it was also estimated based on total fecal dry matter (DM) excretion. To estimate fecal DM output, titanium dioxide (TiO_2) was used as an external marker. TiO_2 was administered daily in paper cartridges at a dosage of 5 g/day and introduced directly into the rumen via the cannula once daily at 08:00 for 10 consecutive days. The initial five days were designated for external marker adaptation, while fecal sampling occurred over the following five days [14].

Beginning on the seventh day of each experimental period, fecal samples (~200 g) were collected directly from the rectal ampulla at multiple time points (08:00, 10:00, 12:00, 14:00, and 16:00). Samples were stored in labeled plastic trays and pre-dried in a forced-air oven at 55 °C. At the end of each period, samples from each animal were pooled to form a composite sample per period. Fecal TiO_2 concentrations were determined by UV/Vis spectrophotometry (380 nm), as described by [15].

Fecal excretion was estimated using the following equation:

$$FE = \frac{FI}{TFC}$$

where FE = daily fecal excretion (g/day), FI = titanium dioxide intake (g/day), and TFC = titanium dioxide concentration in feces (g/g DM).

Apparent total digestibility coefficients were determined for dry matter (DM), crude protein (CP), and organic matter (OM). Analyses were conducted following standard AOAC procedures: DM (method 930.15), CP (N $\times$ 6.25; method 984.13), and ash (ASH; method 942.05), with OM calculated as OM = 100−ASH [16]. Fiber fractions, including neutral detergent fiber (NDF) and acid detergent fiber (ADF), were analyzed according to Van Soest et al. [17]. Starch content was measured using the enzymatic colorimetric method described by [18].

To quantify corn grain excretion in feces, samples were collected directly from the rectal ampulla of each animal between 09:00 and 11:00 on the 16th day of each experimental period. A 300 g fecal sample was weighed using an analytical balance and washed under running water through a 2.00 mm sieve (Granutest, Tyler 9, ABNT 10, Prolab, São Paulo, Brazil). The recovered corn grain particles were manually collected, weighed, and dried in a forced-air oven at 55 °C for 72 h to determine their dry matter content [19].

2.3. Ruminal Fermentation

On the 19th day of each experimental period, ruminal fluid samples were manually collected to determine ruminal pH, ammonia nitrogen ($N\text{-}NH_3$) concentration, short-chain fatty acid (SCFA), and branch-chain fatty acid (BCFA) profiles. Sampling occurred immediately before feeding and at 2, 4, 6, and 8 h post-feeding, with samples obtained from the liquid–solid interface of the rumen. The collected samples were filtered through triple-layered gauze. Ruminal pH was measured using a portable digital pH meter (Meta Química, Meta 210P, São Paulo, Brazil).

For SCFA analysis, 20 mL of ruminal fluid was centrifuged at 3500 rpm for 5 min. From the supernatant, 1800 μL was mixed with 100 μL of 20% ortho-phosphoric acid and frozen for later analysis. Additionally, 1600 μL aliquots were mixed with 400 μL of formic acid (98–100%) and centrifuged at 7000× g for 15 min at 4 °C.

SCFA concentrations were determined using a gas chromatograph (GC-2010 Plus, Shimadzu, Barueri, Brazil) equipped with an automatic injector (AOC-20i), a Stabilwax-DA™ capillary column (30 m, 0.25 mm ID, 0.25 μm df; Restek®, Bellefonte, PA, USA), and a flame ionization detector. Samples were acidified with 1 M ortho-phosphoric acid (Merck®, Rahway, NJ, USA, Ref. 100573) and fortified with a mixture of free volatile acids (Supelco®, St. Louis, MO, USA, Ref. 46975). A 1 μL aliquot was injected with a split ratio of 40:1, using helium as the carrier gas at a linear velocity of 42 $cm \cdot s^{-1}$. The total chromatographic run time was 11.5 min. The injector and detector temperatures were set at 250 °C and 300 °C, respectively. The column temperature program began at 40 °C, ramping to 120 °C at 40 °C/min, to 180 °C at 10 °C/min, and finally to 240 °C at 120 °C/min, where it was held for 3 min. Quantification was performed using dilutions of the WSFA-2 standard (Supelco®, Ref. 47056) and glacial acetic acid (Sigma-Aldrich®, St. Louis, MO, USA, Ref. 33209), with peak identification and integration conducted using GCsolution v.2.42.00 software (Shimadzu®).

For $N\text{-}NH_3$ analysis, 40 mL of ruminal fluid was preserved with 1 mL of 1:1 HCl and frozen at −18 °C. Ammonia nitrogen concentration was determined via distillation using 2N KOH as the distillation base, following prior centrifugation at 1000× g for 15 min, without acid digestion [20].

Methane production (mM/L) was estimated using the following equation [21]:

$$CH_4 = 0.45(C_2) - 0.275(C_3) + 0.40(C_4)$$

where C_2, C_3, and C_4 represent acetate, propionate, and butyrate concentrations (mM), respectively.

2.4. Microbial Protein Synthesis

Urine samples were collected on the 15th, 16th, 17th, and 18th day of each experimental period using the spot sampling method during spontaneous urination, 4 h after supplement feeding [22]. To determine concentrations of creatinine, urea, uric acid, and allantoin, a 10 mL urine aliquot was diluted in 40 mL of 0.036 N sulfuric acid. An additional 40 mL aliquot was preserved in 1 mL of concentrated sulfuric acid (36 N) for total urinary nitrogen (N) analysis. All samples were properly labeled and stored at −18 °C.

Allantoin concentration was determined using a colorimetric method [23,24]. Commercial assay kits (Labtest®, Lagoa Santa, Brazil; Gold Analisa® Diagnóstica Ltd.a, Belo Horizonte, Brazil) were used to quantify creatinine, urea, and uric acid concentrations.

The total excretion of purine derivatives (PD, mmol/day) was calculated by summing urinary allantoin and uric acid. The absorbed microbial purines (Pabs, mmol/day) were estimated using the following equation [25]:

$$PD = 0.85 * Pabs + 0.385 * BW^{0.75}$$

where BW is the body weight of the animal in kilograms.

Daily excretion of urea nitrogen (N-urea) and creatinine nitrogen (N-creatinine) was calculated by multiplying the concentrations of urea and creatinine in the spot urine sample by the estimated 24 h urinary volume, using correction factors of 0.466 and 0.3715, respectively, which correspond to the nitrogen content of urea and creatinine.

The estimated daily urinary volume (UV, L/day) was calculated using the following equation [26]:

$$UV = \frac{27.36 * BW}{[Creatinine]}$$

where BW is the animal's body weight (kg), and [Creatinine] is the creatinine concentration (mg/L) in the spot urine sample. The constant 27.36 represents the average daily creatinine excretion (mg/kg BW/day) for crossbred and Zebu steers. Microbial protein synthesis = (Urinary excretion of purine derivatives) × (a constant value) [26].

Nitrogen balance (NB) was calculated as the difference between total nitrogen intake and total nitrogen excretion in urine and feces. Nitrogen concentrations in urine and feces were determined using the micro-Kjeldahl method. Retained nitrogen (NRet) was estimated by subtracting Absorbed N–Urine N.

2.5. Urea and Creatinine Metabolism

On the 17th day of each experimental period, blood samples were collected from the caudal vein four hours after feeding. Heparin was used as an anticoagulant. Immediately after collection, the samples were centrifuged at 5000 rpm for 15 min to separate the plasma. The supernatant was then labeled and stored at −18 °C until analysis. Plasma urea and creatinine concentrations were determined using a commercial enzymatic assay kit (Gold Analisa® Diagnóstica Ltda, Belo Horizonte, Brazil).

2.6. Statistical Analysis

Data were analyzed using the MIXED procedure of SAS (Statistical Analysis System, version 9.4, SAS Institute Inc., Cary, NC, USA) in a 5 × 5 Latin square design, considering fixed effects of treatment, period, and random effects of animal and residual error, with the following model:

$$Y_{ijl} = \mu + A_i + P_j + D_l + err_{ijl}$$

where Y_{ijl} = dependent variable; μ = overall mean; A_i = animal effect (i = 1 to 5); P_j = period effect (j = 1 to 5); D_l = diet effect; and e_{ijl} = experimental error.

Repeated measures over time (for variables such as pH, N-NH_3, and SCFA) were analyzed using the REPEATED statement, and the best covariance structure was selected based on the lowest Akaike Information Criterion (AIC) value, according to the following model:

$$Y_{ijk} = \mu + A_i + P_j + D_k + T_y + T_y(D_k) + err_{ijk}$$

where Y_{ijk} = dependent variable; μ = overall mean; A_i = animal effect (i = 1 to 5); P_j = period effect (j = 1 to 5); D_k = treatment effect (k = 1 to 5); T_y = time effect (y = 1 to 5); T_y (D_k) = interaction between diet and time; and err_{ijk} = experimental error.

Treatment means were compared using Tukey's test at a significance level of $p < 0.05$. Data normality and homoscedasticity were verified using the Shapiro–Wilk and Levene's tests, respectively. Results are presented as least squares means ± standard error of the mean (SEM).

3. Results

Steers supplemented with CHI + CNSLt exhibited lower ($p \leq 0.032$) intakes of dry matter (DM), corn kernels, pellets, organic matter (OM), crude protein (CP), and neutral detergent fiber (NDF) compared to those fed the CON, CNSLt, and MON diets (Table 2). However, no significant differences were observed in comparison to steers supplemented with CHI alone. Additionally, CHI-supplemented steers did not differ from those on the other experimental diets.

Steers supplemented with CHI + CNSLt exhibited greater ($p = 0.008$) starch digestibility compared to those receiving the CON, CNSLt, and MON diets. However, no significant differences were observed in comparison to steers supplemented with CHI alone.

Additionally, CHI + CNSLt supplementation resulted in lower ($p = 0.012$) corn kernel excretion in feces compared to steers receiving the CON and CNSLt treatments, with no significant differences observed relative to CHI or MON.

Furthermore, CHI + CNSLt-supplemented steers showed lower ($p = 0.014$) ruminal ammonia nitrogen concentrations compared to all other treatment groups. Regarding short-chain fatty acid concentrations, steers supplemented with CHI + CNSLt exhibited lower ($p \leq 0.045$) acetate and butyrate concentrations, higher ($p = 0.038$) propionate concentrations, and consequently lower ($p = 0.018$) estimated methane concentrations compared to the other treatment groups (Table 3).

Table 2. Intake and digestibility according to experimental diets.

Item	Experimental Diets [1]					SEM [2]	p-Value
	CON	CHI	CSNLt	CHI + CSNLt	MON		
			Intake (g/d)				
Dry matter	7.46 [a]	7.23 [ab]	8.09 [a]	6.60 [b]	7.41 [a]	0.387	0.032
Dry matter, %BW	2.24	1.89	2.10	2.26	2.20	0.122	0.554
Corn kernel	6.34 [ab]	6.15 [b]	6.87 [a]	5.61 [b]	6.29 [ab]	0.329	0.017
Pellet	1.12 [ab]	1.08 [b]	1.21 [a]	0.990 [b]	1.11 [ab]	0.058	0.026
Organic matter	7.16 [a]	6.95 [ab]	7.77 [a]	6.34 [b]	7.11 [a]	0.376	0.018
Crude protein	0.984 [a]	0.954 [ab]	1.06 [a]	0.872 [b]	0.984 [a]	0.051	0.026
NDF	2.45 [a]	2.32 [ab]	2.83 [a]	1.94 [b]	2.42 [a]	0.231	0.017
Starch	3.86 [a]	3.71 [ab]	3.27 [a]	3.30 [b]	3.82 [a]	0.387	0.032
			Digestibility (g/kg)				
Dry matter	534	454	504	568	504	4.194	0.124
Organic matter	568	488	530	596	535	4.398	0.147
Crude protein	700	616	686	718	702	5.470	0.149
NDF	486	404	456	502	445	7.347	0.161
Starch	792 [b]	826 [a]	806 [b]	854 [a]	815 [b]	3.143	0.008
			Excretion of intact corn kernel (g/kg)				
Corn kernel	306.5 [a]	268.2 [ab]	301.6 [a]	241.6 [b]	276.7 [ab]	2.985	0.012

[1] CON (no additive); CHI (supplemented with 375 mg/kg DM); CNSLt (supplemented with 500 mg/kg DM of technical cashew nutshell liquid); CHI + CNSLt (supplemented with 375 mg/kg DM of chitosan + 500 mg/kg DM of technical cashew nutshell liquid) and MON (supplemented with 25 mg/kg DM of sodic monensin). [2] SEM (standard error of mean). [a,b] Means followed by different letters on the same line differ by 5% in the TUKEY test adjusted by SAS PROC MIXED.

Table 3. Ruminal fermentation according to experimental diets.

Item	Experimental Diets [1]					SEM [2]	p-Value
	CON	CHI	CSNLt	CHI + CSNLt	MON		
pH	6.21	6.22	6.20	6.28	6.28	0.006	0.654
$N\text{-}NH_3$, mg/dL	20.96 [a]	25.82 [a]	23.83 [a]	15.14 [b]	22.61 [a]	0.026	0.014
			mmol/L				
Acetate	44.56 [b]	57.94 [a]	54.37 [a]	37.75 [c]	47.53 [b]	0.120	0.045
Propionate	30.16 [a]	26.14 [ab]	24.57 [b]	33.51 [a]	26.23 [ab]	0.131	0.038
Butyrate	13.72 [b]	19.04 [a]	20.25 [a]	10.65 [c]	11.71 [bc]	0.087	0.012
Isobutyrate	1.18	1.23	1.40	1.03	1.21	0.018	0.236
Isovalerate	4.05	4.37	3.66	3.07	4.61	0.040	0.414
Valerate	1.97	1.65	2.07	3.10	2.48	0.042	0.447
Brach chain fatty acids	7.20	7.24	7.13	7.19	8.30	0.046	0.484
Total	95.64	110.36	106.32	89.11	93.76	0.173	0.784
Acetate/propionate	1.93	2.24	2.47	1.79	2.06	0.028	0.654
Methane	17.25 [ab]	25.81 [a]	26.50 [a]	12.03 [b]	18.86 [ab]	0.097	0.018

[1] CON (no additive); CHI (supplemented of 375 mg/kg DM); CNSLt (supplemented of 500 mg/kg DM of technical cashew nutshell liquid); CHI + CNSLt (supplemented of 375 mg/kg DM of chitosan + 500 mg/kg DM of technical cashew nutshell liquid) and MON (supplemented of 25 mg/kg DM of sodic monensin). [2] SEM (standard error of mean). [a–c] Means followed by different letters on the same line differ by 5% in the TUKEY test adjusted by SAS PROC MIXED.

Additionally, steers receiving CHI or CNSLt individually showed higher ($p \leq 0.045$) concentrations of acetate, butyrate, and methane than those in the other groups. Steers supplemented with MON exhibited intermediate ($p \leq 0.045$) concentrations of acetate, propionate, butyrate, and methane compared to those receiving CHI, CNSLt, or CHI + CNSLt.

Steers supplemented with CHI + CNSLt also demonstrated lower ($p = 0.027$) nitrogen intake than those receiving the CON, CNSLt, and MON diets (Table 4); however, no significant differences were observed when compared to steers supplemented with CHI alone.

Table 4. Nitrogen balance and microbial protein synthesis according to experimental diets.

Item	Experimental Diets [1]					SEM [2]	p-Value
	CON	CHI	CSNLt	CHI + CSNLt	MON		
			g/d				
N-intake	157.44 [a]	152.64 [ab]	169.60 [a]	139.42 [b]	156.26 [a]	0.227	0.027
N-feces	16.13	21.68	25.63	14.23	19.53	0.149	0.224
N-urine	21.40	27.37	12.49	13.66	6.08	0.132	0.324
N-absorbed	141.31	130.96	143.97	125.19	136.73	0.242	0.442
N-retained	119.91 [b]	103.59 [c]	131.48 [a]	111.53 [b]	130.65 [a]	0.246	0.007
			mmol/d				
Allantoin	221.42	162.40	221.64	201.87	202.18	0.423	0.841
Uric acid	47.43	14.13	29.26	35.23	55.46	0.207	0.751
Total purines	268.81	176.54	250.84	237.09	257.64	0.443	0.801
Purines absorbed	303.45	193.47	280.71	267.42	198.37	0.482	0.872
			g/d				
Microbial nitrogen	220.62	140.66	204.09	194.43	183.77	0.411	0.847
Microbial protein	1378.90	879.14	1275.54	1215.18	1198.59	1.028	0.847

[1] CON (no additive); CHI (supplemented of 375 mg/kg DM); CNSLt (supplemented of 500 mg/kg DM of technical cashew nutshell liquid); CHI + CNSLt (supplemented of 375 mg/kg DM of chitosan + 500 mg/kg DM of technical cashew nutshell liquid) and MON (supplemented of 25 mg/kg DM of sodic monensin). [2] SEM (standard error of mean). [a–c] Means followed by different letters on the same line differ by 5% in the TUKEY test adjusted by SAS PROC MIXED.

Additionally, steers supplemented with CHI + CNSLt exhibited intermediate nitrogen retention ($p = 0.007$) compared to the other groups. Steers supplemented with CNSLt alone showed the highest nitrogen retention, whereas those receiving CHI alone presented the lowest values.

No significant differences were observed among treatments regarding microbial protein synthesis (Table 4), as well as urea and creatinine metabolism parameters (Table 5).

Table 5. Urea and creatinine metabolism according to the experimental diets.

Item	Experimental Diets [1]					SEM [2]	p-Value
	CON	CHI	CSNLt	CHI + CSNLt	MON		
			Urine (mg/dL)				
Urea	801.28	813.02	834.42	802.35	793.49	28.56	0.357
Creatinine	1.42	1.86	1.78	1.77	2.43	0.12	0.328
N-urea	352.69	354.35	361.64	355.97	338.55	12.98	0.325
N-Creatinine	0.529	0.746	0.802	0.722	0.688	0.015	0.357
			Blood (mg/dL)				
Urea	20.75	20.79	22.59	23.22	24.08	1.88	0.335
Creatinine	2.76	2.57	2.00	2.31	2.92	0.74	0.247
N-urea	9.60	9.35	10.92	10.88	10.15	1.05	0.635
N-Creatinine	1.19	1.12	1.18	1.13	1.16	0.32	0.345
			Excretion (mg/kg BW)				
Urea	887.58	820.64	783.50	889.66	669.22	15.89	0.868
Creatinine	27.47	27.75	27.41	27.53	27.51	3.21	0.865
			Clearance (mg/kg BW)				
Urea	45.45	44.85	42.00	43.96	44.00	2.65	0.881
Creatinine	5.95	6.07	6.06	5.98	6.25	1.23	0.865
			Fractional excretion (%)				
Urea	69.95	67.78	66.23	68.23	67.18	4.88	0.885

[1] CON (no additive); CHI (supplemented of 375 mg/kg DM); CNSLt (supplemented of 500 mg/kg DM of technical cashew nutshell liquid); CHI + CNSLt (supplemented of 375 mg/kg DM of chitosan + 500 mg/kg DM of technical cashew nutshell liquid) and MON (supplemented of 25 mg/kg DM of sodic monensin). [2] SEM (standard error of mean).

4. Discussion

Our hypothesis proposed that combined supplementation with chitosan and cashew nutshell liquid (CNSL) would enhance ruminal fermentation and digestion without adversely affecting feed intake, nitrogen balance, or microbial protein synthesis in high-grain diets for feedlot cattle. The results provide partial support for this hypothesis. Supplementation with CHI + CSNLt improved starch digestibility and altered ruminal fermentation patterns, evidenced by increased propionate concentrations and reduced levels of acetate, butyrate, and methane. These changes suggest a shift toward more efficient energy utilization within the rumen.

Additionally, the reduced fecal excretion of corn kernels suggests enhanced starch utilization. However, contrary to expectations, nitrogen retention in the CHI + CSNLt group was intermediate, with CSNLt alone resulting in the highest retention and CHI alone the lowest. Furthermore, microbial protein synthesis and urea and creatinine metabolism remained unaffected, suggesting that CHI + CSNLt supplementation did not significantly enhance nitrogen utilization as anticipated. While the combined additives improved digestion and fermentation efficiency, their effects on nitrogen metabolism and microbial protein synthesis were less pronounced. Therefore, the hypothesis is supported primarily regarding fermentation and digestibility improvements, warranting further investigation into their influence on nitrogen metabolism and protein dynamics in high-concentrate diets.

The observed reduction in dry matter, organic matter, and fiber intake in CHI + CSNLt-supplemented steers may be attributed to metabolic regulation of intake, particularly influenced by increased ruminal propionate concentrations [27–29]. Propionate, a key gluconeogenic volatile fatty acid (VFA), contributes to satiety signaling via hepatic oxidation, potentially resulting in earlier satiety and reduced intake [30].

The improved total starch digestibility and decreased fecal corn kernel excretion suggest enhanced ruminal starch degradation and post-ruminal absorption, possibly linked to the antimicrobial effects of chitosan and CNSL [12]. These additives may modulate ruminal microbial populations by promoting amylolytic bacteria while suppressing excessive proteolysis and methanogenesis. The shift toward higher propionate and lower acetate and butyrate levels supports this mechanism.

Despite reduced intake, the greater energy yield from increased propionate availability may have compensated for lower energy intake, maintaining the energy supply. However, the absence of significant differences between CHI + CSNLt and CHI alone suggests that chitosan may have had a predominant role in these metabolic effects, while CNSL may have offered an additive, but not synergistic, benefit for fermentation and digestive efficiency.

The physical form of the diet may have also supported ruminal pH maintenance. Whole corn kernels digest more slowly than processed corn, and pellets provide a source of NDF that supports ruminal motility, saliva production, and pH buffering [12]. The high propionate concentration typical of corn-based diets [31] was likely further increased by chitosan's ability to shift fermentation toward propionate production through inhibition of Gram-positive bacteria [32,33]. The combination of CHI and CNSLt may have intensified this effect, as anacardic acid in CNSLt has been reported to enhance propionic acid production [10].

The reduction in ruminal ammonia nitrogen concentrations in steers supplemented with CHI + CSNLt may reflect the modulation of proteolytic microbial activity. Both chitosan and CNSLt selectively inhibit Gram-positive bacteria, including hyper-ammonia-producing species, thereby potentially reducing deamination rates and improving nitrogen use efficiency [34].

The changes in SCFA profiles—reduced acetate and butyrate alongside elevated propionate—further support the occurrence of microbial modulation. Enhanced starch digestibility and lower fecal starch loss in the CHI + CSNLt group are consistent with this shift toward greater fermentation efficiency and energy utilization [35]. Additionally, the decrease in methane concentration may be due to the propionate-promoting effects of the additives, as propionate serves as a competitive hydrogen sink, limiting substrate availability for methanogenesis [10]. The intermediate fermentation profile in MON-supplemented steers suggests only partial modulation, while the elevated acetate, butyrate, and methane levels in CHI- and CSNLt-supplemented groups suggest that the additives alone were less effective in promoting a glucogenic profile compared to their combined use.

The CHI + CSNLt combination demonstrated effects on ruminal fermentation similar to those of ionophores [35,36]. The observed reduction in acetate likely reflects the inhibition of Gram-positive acetate-producing bacteria, promoting propionate synthesis [32,33]. The VFA profile in the rumen is influenced by diet composition and microbial populations, and high-starch diets, such as those used in this study, favor succinate- and propionate-producing bacteria [37].

Reduced methane production with CHI + CSNLt supplementation may involve several mechanisms. Chitosan has been shown to reduce H_2 ion concentrations and inhibit methanogenic bacteria [32], while anacardic acids in CNSL exert strong anti-methanogenic effects [38]. Additionally, by enhancing propionate production, CNSL indirectly reduces hydrogen availability for methanogenesis, further decreasing CH_4 emissions [13,34].

The reduction in methane production observed with CHI + CSNLt supplementation may be associated with the combined mechanisms of action of these additives. Chitosan creates a ruminal environment with lower H_2 ion concentrations and exerts antimicrobial activity against methanogenic bacteria [32]. Meanwhile, CNSLt contains anacardic acids, which have a significant inhibitory effect on methanogenic bacteria [38]. Furthermore, CNSLt enhances propionate production, and this VFA competes directly with methane for the available H_2, thereby reducing hydrogen availability for methanogenesis [13,34].

The lower nitrogen intake in the CHI + CSNLt group was consistent with the reduced dry matter and crude protein intake, suggesting feed intake regulation by these additives. Despite this, the intermediate nitrogen retention observed may suggest a tendency toward improved nitrogen utilization efficiency [13,35]. This is supported by the reduced ruminal ammonia concentrations in this group, which suggests lower proteolysis and deamination rates, likely due to suppression of hyper-ammonia-producing microbes. Enhanced starch digestibility may have promoted a more synchronized energy-to-nitrogen ratio, favoring microbial protein synthesis [9].

The greater nitrogen retention observed in steers supplemented with CSNLt alone may reflect moderate modulation of fermentation, allowing for more effective microbial nitrogen capture. In contrast, lower retention in CHI-supplemented animals may suggest that its antimicrobial action disrupted microbial protein synthesis or altered nitrogen recycling dynamics [9].

In summary, CHI + CSNLt supplementation significantly modulated ruminal fermentation by promoting a glucogenic profile with increased propionate production and reduced acetate, butyrate, and methane emissions. These changes, along with improved starch digestibility and lower fecal starch loss, highlight the potential of CHI + CSNLt as an effective strategy to optimize fermentation and feed efficiency in high-starch, low-fiber diets. Although the effects on nitrogen metabolism were less pronounced, the combination showed promise in improving nitrogen utilization efficiency.

Future studies should investigate the long-term effects of CHI + CSNLt supplementation on performance, microbial community dynamics, and metabolic health. Further exploration into dose optimization, interactions with various starch sources, and impacts on post-ruminal digestibility and systemic metabolism may enhance its practical application in intensive feedlot systems.

5. Conclusions

The combined supplementation of chitosan and cashew nut shell liquid (CHI + CSNLt) proved to be an effective strategy for modulating ruminal fermentation in high-grain diets. It enhanced starch digestibility, reduced methane emissions, and optimized nitrogen utilization. These findings indicate that CHI + CSNLt may serve as a viable alternative to monensin sodium, particularly in diets characterized by high levels of fermentable starch and low physically effective fiber, where metabolic disorders such as subacute ruminal acidosis are of concern.

By promoting a more glucogenic fermentation profile while simultaneously mitigating nitrogen losses and enteric methane production, CHI + CSNLt emerges as a promising and sustainable feed additive for intensive feedlot systems. Nonetheless, further research is needed to determine optimal dosing strategies, assess long-term effects on the ruminal microbiome, and evaluate the impacts on overall animal performance and health. Despite these considerations, under extreme dietary conditions, the CHI + CSNLt combination presents a compelling alternative to ionophores, offering both performance advantages and potential improvements in the environmental sustainability of ruminant production.

Author Contributions: R.T.d.O.—investigation, methodology, and formal analysis; R.H.d.T.e.B.d.G.—funding acquisition, supervision, investigation, project administration, and writing—review; J.R.G.—data analysis and writing—review; F.N.F.d.C.—writing—original draft, writing—review and editing; N.G.d.S.—investigation and methodology; J.L.R.—data curation and formal analyses; L.d.S.O.—formal analyses; L.G.B.D.—investigation and data curation; T.L.P.A.—formal analyses; C.M.C.A.—validation, data analysis, and writing—review. All authors have read and agreed to the published version of the manuscript.

Funding: The project was financially supported by Conselho Nacional de Pesquisa e Desenvolvimento Científico e Tecnologico, Process: (310837/2020-0/309888/2023-9), Coordenação de Aperfeiçoamento de Pessoal de Nível Superior, Brasília, Brazil, the Foundation for Support and Development of Education, Science and Technology of the State of Mato Grosso do Sul (FUNDECT; Process: 71/700.134/2018; 71/032.380/2022; 83/035.027/2024).

Institutional Review Board Statement: The methods and protocols used during the development of this study were approved by the Animal Use Ethics Committee of the Federal University of Grande Dourados (CEUA/UFGD) under protocol number 023/2015-3.

Informed Consent Statement: Not applicable.

Data Availability Statement: The original contributions presented in this study are included in the article. Further inquiries can be directed to the corresponding authors.

Acknowledgments: The authors thank Claudia Andrea Lima Cardoso of the State University of Mato Grosso do Sul (UEMS) for her support in the chromatographic analysis of the liquid from the technical cashew nut shell liquid (CNSLt). In addition, the authors express appreciation to Antonio Ferriani Branco for providing the CNLSt used. To the Federal University of Grande Dourados (UFGD) for granting space for the development of research. To the members of the NERU/UFGD research group for their commitment to carrying out the proposed work.

Conflicts of Interest: The authors declare no conflict of interest.

References

1. Ogata, T.; Makino, H.; Ishizuka, N.; Iwamoto, E.; Masaki, T.; Ikuta, K.; Kim, Y.; Sato, S. Long-term high-grain diet altered ruminal pH, fermentation, and rumen bacterial community composition and functions, leading to increased lactic acid production in Japanese Black cattle during fattening. *PLoS ONE* **2019**, *14*, e0225448. [CrossRef] [PubMed]
2. Golder, H.M.; Celi, P.; Rabiee, A.R.; Lean, I.J. Effects of feed additives on rumen and blood profiles during a starch and fructose challenge. *J. Dairy Sci.* **2014**, *97*, 985–1004. [CrossRef]
3. Honan, M.; Feng, X.; Tricarico, J.M.; Kebreab, E. Feed additives as a strategic approach to reduce enteric methane production in cattle: Modes of action, efficacy and safety. *Anim. Prod. Sci.* **2021**, *62*, 1303–1317. [CrossRef]
4. Percio, C.; Barreta, D.A.; Silva, E.R.; Zotti, C.A. Bovinocultura de corte brasileira sem o uso de antibióticos: Consequências e alternativas. *Horiz. Das Ciênc. Soc. Rurais* **2019**, *2*, 306–321.
5. Perna Junior, F.; Cassiano, E.C.O.; Martins, M.F.; Romero, L.A.; Zapata, D.C.V.; Pinedo, L.A.; Marino, C.T.; Rodrigues, P.H.M. Effect of tannins-rich extract from *Acacia mearnsii* or monensin as feed additives on ruminal fermentation efficiency in cattle. *Livest. Sci.* **2017**, *203*, 21–29. [CrossRef]
6. Shah, S.; Joshi, R.; Rai, N.; Adhikari, R.; Pandit, R. Microstructural analysis of biowaste-derived hydroxyapatite-chitosan nanocomposites. *Micro Nano Lett.* **2022**, *17*, 369–376. [CrossRef]
7. Goiri, I.; Oregui, L.M.; Garcia-Rodriguez, A. Use of chitosans to modulate ruminal fermentation of a 50:50 forage-to-concentrate diet in sheep. *J. Anim. Sci.* **2010**, *88*, 749–755. [CrossRef]
8. Belanche, A.; Pinloche, E.; Preskett, D.; Newbold, C.J. Effects and mode of action of chitosan and ivy fruit saponins on the microbiome, fermentation, and methanogenesis in the rumen simulation technique. *FEMS Microbiol. Ecol.* **2016**, *92*, fiv160. [CrossRef]
9. Dias, A.O.C.; Goes, R.H.T.B.; Gandra, J.R.; Takiya, C.S.; Branco, A.F.; Jacauna, A.G.; Oliveira, R.T.; Souza, C.J.S.; Vaz, M.S.M. Increasing doses of chitosan to grazing beef steers: Nutrient intake and digestibility, ruminal fermentation, and nitrogen utilization. *Anim. Feed. Sci. Technol.* **2017**, *151*, 215–227. [CrossRef]
10. Watanabe, Y.; Suzuki, R.; Koike, S.; Nagashima, K.; Mochizuki, M.; Forster, R.J.; Kobayashi, Y. In vitro, the evaluation of cashew nut shell liquid as a methane-inhibiting and propionate-enhancing agent for ruminants. *J. Dairy Sci.* **2010**, *93*, 5258–5267. [CrossRef]
11. Oh, S.S.; Suzuki, Y.; Hayashi, S.; Suzuki, Y.; Koike, S.; Kobayashi, Y. Potency of cashew nut shell liquid in rumen modulation under different dietary conditions and indication of its surfactant action against rumen bacteria. *J. Anim. Sci. Technol.* **2017**, *59*, 1–7. [CrossRef] [PubMed]
12. Silva, T.I.S.; Goes, R.H.T.B.; Cônsolo, N.R.B.; Gandra, J.R.; Osmari, M.P.; Silva, C.C.; Oliveira, L.E.F.; Souza, T.M.; Barbosa, L.C.G.S.; Anschau, D.G.; et al. Effects of replacing dietary antibiotic supplementation with chitosan levels on rumen metabolism and nitrogen use in finishing steers fed forage-free diets. *Trop. Anim. Sci. J.* **2024**, *47*, 225–234. [CrossRef]
13. de Queiroz Vieira, E.R.; Goes, R.H.T.B.; Diaz, T.G.; Osmari, M.P.; Seno, L.O.; Itavo, L.C.V.; Gandra, J.R.; Anschau, D.G.; Oliveira, R.T.; Silva, N.G.; et al. Chitosan combined with technical cashew nut shell liquid improves in vitro ruminal parameters and gas production kinetics. *Rev. Bras. Zootec.* **2022**, *51*, e20200186.
14. Ferreira, M.A.; Valadares Filho, S.C.; Marcondes, M.I.; Paixão, M.L.; Paulino, M.F.; Valadares, R.F.D. Evaluation of markers in ruminant trials: Digestibility. *Rev. Bras. Zootec.* **2009**, *38*, 1568–1573. [CrossRef]
15. Myers, W.D.; Ludden, P.A.; Nayigihugu, V.; Hess, B.W. Technical note: A procedure for the preparation and quantitative analysis of samples for titanium dioxide. *J. Anim. Sci.* **2004**, *82*, 179–183. [CrossRef]
16. AOAC, Association of Official Analytical Chemists. *Official Methods of Analysis*, 15th ed.; AOAC International: Arlington, VA, USA, 1990.
17. Van Soest, P.J.; Roertson, J.B.; Lewis, B.A. Methods for dietary fiber, neutral detergent fiber, and no starch polysaccharides in relation to animal nutrition. *J. Dairy Sci.* **1991**, *74*, 3583–3597. [CrossRef]
18. Hendrix, D.L. Rapid extraction and analysis of nonstructural carbohydrates in plant tissues. *Crop Sci.* **1993**, *33*, 1306–1311. [CrossRef]
19. Oliveira, R.T.; Goes, R.H.T.B.; Sabedot, M.R.; Osmari, M.P.; Silva, N.G.; Anschau, D.G.; Oliveira, S.S.; Picanço, Y.S. Intake, digestibility and recovery of grains from feces of steers fed different oilseeds. *Rev. Bras. Saúde Produção Anim.* **2020**, *21*, e2121272020. [CrossRef]
20. Detmann, E.; Costa ESilva, L.F.; Rocha, G.C.; Palma, M.N.N.; Rodrigues, J.P.P. *Métodos Para Análise de Alimentos-INCT*, 2nd ed.; Suprema Gráfica: Viçosa, Brazil, 2021.
21. Moss, A.; Jouany, J.P.; Newbols, J. Methane production by ruminants: Its contribution to global warming. *Ann. Zootech. INRA/EDP Sci.* **2000**, *49*, 231–253. [CrossRef]
22. Chizzotti, M.L.; Valadares Filho, S.C.; Valadares, R.F.D.; Chizzotti, F.H.M.; Campos, J.M.S.; Marcondes, M.I.; Fonseca, M.A. Intake, digestibility and urinary excretion of urea and purine derivatives in heifers with different body weights. *Rev. Bras. Zootec.* **2006**, *35*, 1813–1821. [CrossRef]

23. Fujihara, T.; Ørskov, E.R.; Reeds, P.J.; Kyle, D.J. The effect of protein infusion on urinary excretion of purine derivatives in ruminants nourished by intragastric nutrition. *J. Agric. Sci.* **1987**, *109*, 7–12. [CrossRef]
24. Chen, X.B.; Gomes, M.J. *Estimation of Microbial Protein Supply to Sheep and Cattle Based on Urinary Excretion of Purine Derivatives—An Overview of the Technical Details*; International Feed Research Unit, Rowett Research Institute: Aberdeen, UK, 1992; p. 22.
25. Verbic, J.; Chen, X.B.; Macleod, N.A.; Ørskov, E.R. Excretion of purine derivatives by ruminants. Effect of microbial nucleic acid infusion on purine derivative excretion by steers. *J. Agric. Sci.* **1990**, *114*, 243–248. [CrossRef]
26. Rennó, L.N.; Valadares, R.F.D.; Valadares Filho, S.C.; Leão, M.I.; Coelho da Silva, J.F.; Cecon, P.R.; Gonçalves, L.C.; Dias, H.L.C.; Linhares, R.S. Concentração plasmática de ureia e excreções de ureia e creatinina em novilhos. *Rev. Bras. Zootec.* **2000**, *29*, 1235–1243. [CrossRef]
27. Osmari, M.P.; Branco, A.F.; Diaz, T.G.; Matos, L.F.; Goes, R.H.T.B.; Teodoro, A.L. Technical cashew nut shell liquid associated with non-protein nitrogen sources in high-grain diets for ruminants: Intake and digestibility of nutrients, ruminal fermentation, and microbial protein synthesis. *Semin. Ciênc. Agrárias* **2019**, *40*, 259–270. [CrossRef]
28. Coutinho, D.A.; Branco, A.F.; Dantos, G.T.; Osmari, M.P.; Teodoro, A.L.; Diaz, T.G. Intake, digestibility of nutrients, milk production and composition in dairy cows fed on diets containing cashew nut shell liquid. *Acta Scientiarum Anim. Sci.* **2014**, *36*, 311–316. [CrossRef]
29. Zanferari, F.; Vendramini, T.H.A.; Rentas, M.F.; Gardinal, R.; Calomeni, G.D.; Mesquita, L.G.; Takiya, C.S.; Rennó, F.P. Effects of chitosan and whole raw soybeans on ruminal fermentation and bacterial populations, and milk fatty acid profile in dairy cows. *J. Dairy Sci.* **2018**, *101*, 10939–10952. [CrossRef] [PubMed]
30. Allen, M.S. Effects of diet on short-term regulation of feed intake by lactating dairy cattle. *J. Dairy Sci.* **2000**, *83*, 1598–1624. [CrossRef] [PubMed]
31. Vyas, D.; McGeough, E.J.; Mohammed, R.; McGinn, S.S.; McCallister, T.A.; Beauchemin, K.A. Effects of Propionibacterium strains on ruminal fermentation, nutrient digestibility and methane emissions in beef cattle fed a corn grain finishing diet. *Animal* **2014**, *8*, 1807–1815. [CrossRef]
32. Jiménez-Ocampo, R.; Valencia-Salazar, S.; Pinzón-Diaz, C.E.; Herrera-Torres, E.; Aguilar-Pérez, C.F.; Arango, J.; Ku-Wera, J.C. The role of chitosan as a possible agent for enteric methane mitigation in ruminants. *Animals* **2019**, *9*, 942. [CrossRef]
33. Santos, M.V.; Goes, R.H.T.B.; Takiya, C.S.; Cabral, L.S.; Mombach, M.A.; Oliveira, R.T.; Silva, N.G.; Anschau, D.G.; Freitas Júnior, J.E.; Araújo, M.L.G.M.L.; et al. Effect of increasing doses of chitosan to grazing beef steers on the relative population and transcript abundance of Archaea and cellulolytic and amylolytic bacteria. *Anim. Biotechnol.* **2023**, *34*, 246–252. [CrossRef]
34. Gandra, J.R.; Del Valle, T.A.; Takiya, C.S.; Freitas Junior, J.E.; Oliveira, E.R.; Gandra, E.R.S.; Pedrini, C.A.; Mendes, P.V.C. Effects of ricinoleic acid from castor oil and cashew nutshell liquid on nutrient digestibility and ruminal fermentation in dairy heifers. *Rev. Bras. Saúde Produção Anim.* **2022**, *23*, 23–33. [CrossRef]
35. Pereira, D.C.; Goes, R.H.T.B.; Martinez, A.C.; Gandra, J.R.; Presendo, E.; Santos, M.V.; Oliveira, R.T.; Silva, N.G.; Ribeiro, M.G.; Alvez, J.L.R. In vitro evaluation of the association of chitosan and cashew nut shell liquid as additives for ruminants. *Rev. Bras. Saúde Produção Anim.* **2019**, *20*, e05102019. [CrossRef]
36. Osmari, M.P.; Branco, A.F.; Goes, R.H.T.B.; Diaz, T.G.; Matos, L.F. Increasing dietary doses of cashew nut shell liquid on rumen and intestinal digestibility of nutrients in steers fed a high-grain diet. *Arch. Zootec.* **2017**, *66*, 375–381.
37. Wang, L.; Zhang, G.; Li, Y.; Zhang, Y. Effects of high forage/concentrate diet on volatile fatty acid production and the microorganisms involved in VFA production in cow rumen. *Animals* **2020**, *10*, 223. [CrossRef]
38. Branco, A.F.; Gianllongo, F.; Frederick, T.; Weeks, H.; Oh, J.; Hristov, A.N. Effect of cashew nut shell technical liquid on ruminal methane emission and lactation performance of dairy cows. *J. Dairy Sci.* **2015**, *98*, 4030–4040. [CrossRef]

polymers

MDPI

Article

Exploring the Characterization, Physicochemical Properties, and Antioxidant Activities of Chitosan-Encapsulated Green Tea Extract Microsphere Resin

Lina Yu [1], Siyu Feng [2], Yu Song [1], Jie Bi [1], Yuan Gao [1], Luhui Wang [1], Chen Jiang [1] and Mingqing Wang [1,*]

[1] Shandong Peanut Research Institute, Qingdao 266100, China; lhtyln0626@163.com (L.Y.); songyuhss@163.com (Y.S.); bj.baby@163.com (J.B.); gaoyuan_hah@163.com (Y.G.); wlhfood@sina.com (L.W.); jiangchen_qd@126.com (C.J.)

[2] College of Biotechnology, Jiangsu University of Science and Technology, Zhenjiang 212000, China; 13614505512@163.com

* Correspondence: annabelle2010@163.com

Abstract: Chitosan, a naturally occurring alkaline polysaccharide with excellent biocompatibility, non-toxicity, and renewability, has the ability to undergo cross-linking reactions with polyphenolic compounds. In this study, chitosan-encapsulated green tea extract microsphere resin (CS-GTEMR) was successfully prepared using chitosan and green tea extract via reversed-phase suspension cross-linking polymerization. The structural characterization of CS-GTEMR was conducted using Fourier Transform Infrared Spectroscopy (FTIR) and Differential Scanning Calorimetry (DSC). Additionally, its physical properties, swelling behavior, polyphenol content, and antioxidant activities were investigated. The results indicate that CS-GTEMR consists of reddish-brown microspheres with a smooth surface and dense pores. The study found that the total content of polyphenolic compounds encapsulated in CS-GTEMR was 50.485 ± 0.840 μg/g. The characteristic absorption peak of phenolic hydroxyl groups appeared in the FTIR spectrum, suggesting that the polyphenolic compounds had been successfully encapsulated within the CS-GTEMR. The equilibrium swelling ratio of CS-GTEMR was determined to be 229.7%, indicating their suitability for use in solutions with a pH range of 1–13. In simulated gastric and intestinal fluids, the release rates of polyphenolic compounds from CS-GTEMR were 24.934% and 3.375%, respectively, indicating that CS-GTEMR can exert a sustained-release effect on polyphenolic compounds. CS-GTEMR demonstrated antioxidant activities such as scavenging DPPH radicals, superoxide anion radicals, hydroxyl radicals, and hydrogen peroxide, as well as exhibiting iron-reducing and molybdenum-reducing powers. With its high mechanical strength, acid resistance, and organic solvent resistance, CS-GTEMR can protect polyphenolic compounds from damage. Therefore, CS-GTEMR can be utilized as a natural antioxidant or preventive agent in food, expanding the application scope of green tea extracts.

Keywords: chitosan-encapsulated green tea extract microsphere resin; preparation; characterization; physical properties; swelling properties; polyphenolic content; antioxidant activities

1. Introduction

The important role of polyphenolic compounds in human health is of current interest. The main sources of polyphenolic compounds in diet are fruits and beverages, especially in tea and coffee [1–3]. Li et al. quantified 19 polyphenolic compounds from green tea extract by ultra-high performance liquid chromatography coupled with quadrupole-time of flight mass spectrometry (UPLC-Q-TOF MS) [4]. This suggests that tea products are a good source of polyphenolic compounds. Polyphenolic compounds generally have significant antioxidant and pharmacological capabilities, including scavenging free radicals, antibacterial, anti-inflammatory, and cholesterol-lowering activities. When polyphenolic compounds enter the plasma, they contribute to increasing the antioxidant capacity of the plasma and reduce the risk of cardiovascular disease by preventing the oxidation of plasma low-density lipoprotein. Rajapaksha et al. obtained eight polyphenolic compounds from *Schinus terebinthifolia* fruit, which exhibited significant antioxidant, cytotoxic, anti-inflammatory, analgesic, and antimicrobial properties, and also proliferation inhibition in MCF-7 human cancer cell lines [5]. However, the concentration and antioxidant activities of polyphenolic compounds are reduced in alkaline pH environments such as intestinal fluid in the human body, which greatly limits their application.

In order to solve the problem that polyphenolic compounds are easily lost in the human body, they can be encapsulated in polymers to reduce the influence of the pH value of the environmental solution on them. In this way, the release time of polyphenolic compounds can be controlled or prolonged, and their concentration levels in plasma can be maintained, so as to give full play to their physiological activities. Chitosan is a natural, non-toxic, non-immunogenic, alkaline polysaccharide that is biocompatible, renewable, and biodegradable. Due to the biocompatibility and adhesion properties of chitosan, bioactive substances can be encapsulated in chitosan, which contributes to the transport and sustained release of bioactive substances. Liu et al. incorporated inulin and chitosan into alginate-based microspheres for targeted delivery and release of quercetin into the colon [6]. Quercetin encapsulated in microspheres retained 80.3% after in vitro gastrointestinal digestion. Colonic fermentation experiments showed that quercetin release was delayed, but fermentation occurred within 3 h and was completely metabolized by microorganisms within 24 h. Yu et al. synthesized astaxanthin-loaded nanoparticles by the amidation reaction of carboxymethyl chitosan and methionine [7]. The loading capacity of astaxanthin reached 39.68 μg/mL, which realized the controllable release of astaxanthin in the simulated intestinal high-concentration ROS environment. In vitro cell experiments showed that the nanoparticles could effectively alleviate the oxidative damage caused by H_2O_2 on the intestinal epithelial cell line No.6 (IEC-6 cells). It is suggested that if polyphenolic compounds are encapsulated in chitosan microspheres, they will also have a controllable release effect and can give full play to the physiological activity of polyphenolic compounds.

Green tea extract, a product derived from the tea processing industry, primarily consists of polyphenolic compounds and also contains bioactive components such as tea polysaccharides and amino acids. It exhibits significant potential in promoting human health, with demonstrated effects including the inhibition of atherosclerosis, reduction in blood pressure, modulation of insulin secretion levels, enhancement of memory function, suppression of prostate cancer cell proliferation, and improvement of intestinal health. Hossain et al. conducted a multidimensional investigation encompassing cellular, animal, and molecular mechanistic studies to elucidate the effects of green tea extract on atherosclerosis [8]. Their findings demonstrated that green tea extract could mitigate oxidative stress and enhance cell survival rates while significantly improving the blood lipid profile in high-fat diet-fed mice. Molecular docking analysis revealed that catechin compounds exhibited strong binding affinities with key targets, including LOX-1, HMG-CoA reductase, caspase-3, and Nrf2. Meanwhile, de la Fuente-Muñoz et al. demonstrated that the combined extract of black tea and green tea effectively alleviated angiotensin II-induced cardiovascular dysfunction in mice through anti-inflammatory, antioxidant, and anti-apoptotic effects, while exhibiting antihypertensive properties [9]. These findings indicate that green tea extract may serve as a potential natural agent for preventing atherosclerosis and reducing cardiovascular risk. Sulaimani et al. revealed that green tea extract exerted no significant impact on postprandial blood glucose in healthy adults but delayed postprandial insulin secretion in the morning rather than at night [10]. This phenomenon untangles the time-dependent regulatory effect of green tea extract on insulin sensitivity. Joo et al. administered green tea extract to patients with subjective memory complaints for 12 weeks [11]. The results demonstrated enhanced memory function and strengthened functional connectivity within the default mode network, providing critical evidence for the cognitive-improving effects of green tea extract. In a separate study, Moalemi et al. treated PC3 prostate cancer cells with green tea extract for 48 h, revealing suppressed expression of cyclin B1, p-AR, CDK1, p-AKT, PSA, c-Myc, and p-CDK1 [12]. These findings offer a theoretical foundation for the design of antitumor treatment plans. In addition, green tea extract demonstrates remarkable efficacy in maintaining intestinal health. For instance, it significantly elevates the levels of antioxidant factors in the intestinal tract of mice while reducing the content of pro-inflammatory cytokines [13]. This intervention alleviates antibiotic-induced weight loss and intestinal pathological damage, promotes the colonization of *Bifidobacterium* and *Lactobacillus* genera in the murine gut microbiota, and concurrently inhibits the proliferation of *Escherichia coli* and *Enterococcus* species. The study investigating the effects of green tea extract on growth performance, *Clostridium perfringens* colonization, and inflammatory responses in broilers with experimental subclinical necrotic enteritis demonstrated that dietary supplementation with green tea extract alleviated characteristic growth suppression in chicks during subclinical necrotic enteritis, reduced the severity of *C. perfringens* infection in the intestine, and modulated inflammatory responses [14]. The aforementioned research findings comprehensively demonstrate the potent protective and health-enhancing capabilities of green tea extract in biological systems. It can be anticipated that green tea extract holds promising prospects as a potential natural therapeutic agent.

However, one critical issue warrants attention: among these diverse biological effects, green tea extract needs to exert its actions at distinct target sites, including but not limited to arterial tissues and the intestinal tract. If green tea extract can reach the target sites in the organism at sufficiently high concentrations, it can effectively exert its preventive and therapeutic effects. However, when administered in its current powdered or solution form, green tea extract is evidently unable to achieve the objective of reaching the target sites at high concentrations and functioning efficiently within the complex physiological environment of the organism. The development of an appropriate delivery system for green tea extract is crucial to achieve targeted release at specific sites with high bioavailability in vivo. Chitosan has emerged as the material of choice for in vivo green tea extract delivery due to its remarkable functional properties, excellent biocompatibility, and absence of toxic side effects. Bavi et al. fabricated chitosan–gelatin–green tea extract composite particles using an electrospray system [15]. The results demonstrated that these composite particles achieved sustained release of green tea extract for up to 9 days under pH 7.4 conditions, significantly enhanced cell viability, and effectively inhibited apoptosis. This system provides an excellent natural biomaterial for Parkinson's disease treatment. Piran et al. encapsulated green tea extract in chitosan–citric acid nanoparticles, and their results showed that the antioxidant activity of tea polyphenols was enhanced [16]. This confirmed the potential of nanoparticles to deliver green tea polyphenols in food. Chuysinuan et al. investigated the controlled release behavior and antioxidant activity of green tea extract embedded in a composite of cyclodextrin and chitosan [17]. Through the evaluation of swelling properties, degradation curves, and the scavenging activities of DPPH and ABTS free radicals, it was proved that this method had a protective effect on green tea extract.

The above research results show that chitosan hydrogel has a protective effect on green tea extract and can exert its antioxidant activity. However, chitosan can only be dissolved under acidic pH conditions, which makes the preparation conditions of chitosan hydrogel extremely harsh. Thereby, the product of chitosan green tea extract is prone to degradation in physiological solutions. Kudłacik-Kramarczyk et al. evaluated the interaction between chitosan-encapsulated yellow tea extract hydrogel and simulated fluids by the change rule of pH value [18]. They found that incubation in simulated physiological solution led to a decrease in the contact angle of the hydrogel, which could even decrease by 60%. As can be seen, the direct encapsulation of green tea extract in chitosan hydrogel will affect its utilization rate in the organism. In order to overcome this disadvantage of chitosan hydrogel, chitosan should be modified to ensure that the prepared chitosan-encapsulated tea extract product has a stable structure and properties. Reversed-phase suspension cross-linking polymerization is an effective method for chitosan modification. When the cross-linking agent is added to the emulsion formed by chitosan acetate solution and liquid paraffin, the amino and hydroxyl groups in chitosan are cross-linked with the aldehyde group of the cross-linking agent, thus forming microspheres with dense surface and porosity. The cross-linking reaction can change the crystal structure of chitosan, making its products resistant to acids, alkalis, and chemicals. It is foreseeable that if the green tea extract is encapsulated in chitosan microspheres, the chitosan microspheres can resist the physiological fluid, allowing the green tea extract to exert its beneficial effects. In this paper, chitosan microspheres loaded with green tea extract were prepared by a reverse-phase suspension cross-linking polymerization method. The morphology was observed by a high-resolution three-dimensional microscope, and the size distribution was analyzed by a particle size analyzer. The structural characteristics of the polyphenolic compounds in the microspheres combined with the chitosan matrix were analyzed by FTIR and DSC. The feasibility of the reverse-phase suspension cross-linking polymerization method in the application of chitosan microspheres encapsulation of green tea extract was explored by evaluating the swelling properties of microspheres, the release properties, and the antioxidant activities of polyphenolic compounds. Compared with existing studies on chitosan-encapsulated polyphenolic compounds microspheres, the novelty of this research involves designing a reverse-phase suspension two-step cross-linking polymerization method for preparing chitosan-encapsulated green tea extract microspheres. The advantage of this approach is that a hydrogel is first formed through pre-cross-linking, followed by a secondary cross-linking reaction to generate stable and robust chitosan-encapsulated green tea extract microspheres. They exhibit the microstructural characteristics of the microspheres, demonstrating resistance to acids, alkalis, and chemicals, thereby facilitating more controlled release of the encapsulated compounds. Given the superior antioxidant activity of green tea extract and its significant potential in protecting and enhancing organismal health, this study aimed to develop a chitosan-encapsulated green tea extract microsphere resin (CS-GTEMR) capable of protecting and controllably releasing polyphenolic compounds. The primary objective of this research was to establish CS-GTEMR as a promising natural therapeutic agent for the prevention and treatment of certain diseases, as well as a nutraceutical supplement to safeguard human health. Furthermore, CS-GTEMR could be developed into a functional food ingredient for meat products, baked goods, and beverages, where it would serve as a natural antioxidant to extend shelf life and thereby broaden the application scope of green tea extract.

2. Materials and Methods

2.1. Chemical Reagents

Chitosan with a viscosity average molecular weight of 5.3×10^5 and deacetylation degree (DDA) 85% was provided by Lizhong Chitosan Co., Ltd., Qingdao, China, and used without any purification. Green tea extract (GTE) was obtained from Hainan Qunli Pharmaceutical Co., Ltd., Sanya, China. Acetic acid, liquid paraffin, ethyl acetate, Span-80, formaldehyde solution, 50% glutaraldehyde solution, petroleum ether (boiling range 60–90 °C), acetone, and anhydrous ethanol were analytical reagents produced by Sinopharm Chemical Reagent Co., Ltd., Shanghai, China. All other reagents were of analytical grade and used as received.

2.2. Preparation of CS-GTEMR and Chitosan Microsphere Resin (CS-MR)

The preparation of CS-GTEMR by using reversed-phase suspension cross-linking polymerization technology was carried out according to the method of Yu et al. [19]. The process involves 8 steps, which are dissolution, dispersion, emulsification, pre-cross-linking, cross-linking, filtration, washing, and drying. First, green tea extract (0.5% w/v) was dissolved in distilled water, and then acetic acid (2% v/v) was added to the green tea extract solution with stirring. Chitosan (5% w/v) was added to the mixture with a glass rod and stirred to form a viscous red-brown chitosan green extract mixture with many bubbles. After standing overnight, the mixed liquid is degassed in a vacuum. Secondly, the above-mentioned mixed liquid was added to liquid paraffin (50% v/v). The mixed liquid was stirred in a water bath at 40 °C at 300 r/min for 20 min to form a suspension with many small liquid droplets. Thirdly, Span-80 (0.6% w/v) was added to the suspension, followed by the addition of ethyl acetate (10% v/v). In a water bath at 40 °C, the mixed liquid was stirred at 300 r/min for 20 min to form an emulsion. Fourthly, formaldehyde solution (10% v/v) was added to the emulsion, in a water bath at 50 °C, and the emulsion was stirred at 300 r/min for 20 min. Fifth, after adding the glutaraldehyde solution (5% v/v) to the above solution, the stirring speed was immediately reduced to 175 r/min, and the temperature was increased to 60 °C. The pH value of the reaction mixture was adjusted to 7.5 with 2 mol/L NaOH solution, and then stirred for 3 h. Sixth, when the reaction is over, a proper amount of petroleum ether is poured into the reaction mixture and mixed evenly. After CS-GTEMR settling, discard the upper layer of liquid. Then, pour in an appropriate amount of petroleum ether, mix evenly, and remove the petroleum ether with a vacuum filter to obtain CS-GTEMR. Seventh, wash the CS-GTEMR with acetone, anhydrous ethanol, and distilled water in turn. Then, CS-GTEMR was soaked in distilled water, and after CS-GTEMR settled to the bottom of the container, the upper layer of distilled water and CS-GTEMR suspended on the surface was decanted. The process was repeated five times. Finally, the CS-GTEMR was dried under vacuum at 50 °C.

CS-MR was also prepared by reversed-phase suspension cross-linking polymerization technology. When preparing CS-MR, chitosan (5% w/v) was added directly to the acetic acid solution (2% v/v) instead of the green tea extract acetic acid solution. The other steps were the same as the preparation method of CS-GTEMR.

2.3. Determination of Physical Properties

Physical properties of CS-GTEMR and CS-MR were determined according to the method of Yu et al. [20].

2.3.1. Determination of Water Absorption Capacity

The 0.1 g samples were fully swelled in distilled water and filtered. Then, the surface water of the samples was sucked dry and weighed (W_1, g). Finally, the samples were dried at 105 °C for 4 h and weighed (W_2, g). The water absorption capacity (M, %) was calculated using Formula (1).

$$M(\%) = \frac{W_1 - W_2}{W_1} \times 100 \tag{1}$$

2.3.2. Determination of Pile-Up Density

Pile-up density is the weight of the samples per unit volume, in which the volume includes the skeleton, pore, and interstitial volume of the samples. Take a sample of volume 2 mL (V_P, mL) and weigh its mass (W, g). The pile-up density (ρ_p, g/mL) was calculated using Formula (2).

$$\rho_P(g/mL) = \frac{W}{V_P} \tag{2}$$

2.3.3. Determination of Skeleton Density

Add 5 mL of n-heptane to a 10 mL graduated cylinder and measure its mass (W_1, g). Then, pour out the n-heptane. Add 0.1 g (W, g) of the sample into the cylinder, followed by 2 mL of n-heptane, and let it sit for 2 h. Afterward, top up the cylinder with n-heptane to the 5 mL mark and measure its mass (W_2, g) again. The skeletal volume (V_T, cm^3) and skeletal density (ρ_T, g/cm^3) can be calculated using Formulas (3) and (4), respectively.

$$V_T = \frac{W_1 - W_2 + W}{dt} \tag{3}$$

where dt is the n-heptane density, 0.6830 g/cm^3.

$$\rho_T(g/cm^3) = \frac{W}{V_T} \tag{4}$$

2.3.4. Determination of Pore Degree

The pore degree (P) can be calculated using Formula (5), where ρ_T and M are skeletal density and water absorption capacity, respectively.

$$P = \frac{\rho_T M}{\rho_T M + 1 - M} \tag{5}$$

2.3.5. Determination of Free Aldehyde Group Content

After fully swelling 0.1 g (W, g) of the sample in water, the water was removed by suction filtration. Then, 10.0 mL of hydroxylamine reagent was added, and the mixture was oscillated in a 25 °C water bath for 1 h. Following this, two drops of 0.05% bromophenol blue indicator were added, and the solution was titrated to the endpoint with a standard hydrochloric acid solution (N, 0.02 mol/L). The volume of standard hydrochloric acid consumed by the blank control was denoted as V_0 (mL), while the volume consumed by the sample was denoted as V_1 (mL). The free aldehyde group content was calculated using Formula (6).

$$\text{Free aldehyde group content (mmol/g)} = \frac{N \times (V_0 - V_1)}{W} \tag{6}$$

2.3.6. Determination of Weak Basic Exchange Capacity

After fully swelling 0.1 g (W, g) of the sample in water, it was filtered to remove moisture, and then oscillated with 20.0 mL of standard hydrochloric acid solution (N_1, 0.05 mol/L) in a 25 °C water bath for 1 h. Following this, 15.0 mL of the supernatant was taken, and two drops of 0.2% phenolphthalein indicator were added. The solution was then titrated to the endpoint using a standard sodium hydroxide solution (N_2, 0.05 mol/L).

The volume of standard hydrochloric acid solution consumed was denoted as V_1 (mL), and the volume of standard sodium hydroxide solution consumed was denoted as V_2 (mL). The weak basic exchange capacity was calculated using Formula (7) below.

$$\text{Weak basic exchange capacity (mmol/g)} = \frac{N_1 \times V_1 - N_2 \times V_2 \times \frac{20}{15}}{W} \quad (7)$$

2.4. Structure and Thermal Stability Characterization

The morphological characterization of the sample was carried out by a super depth of field three-dimensional microscopy system (VHX-950F, KEYENCE, Osaka, Japan). Average particle diameter and particle size distribution of the samples were determined by a laser diffraction particle size analyzer (LS200, Beckman Coulter, Inc., Fullerton, CA, USA). The FTIR spectra of samples were determined by a Fourier transform infrared spectrometer (Nicolet NEXUS 470, Thermo Nicolet Corporation, Madison, WI, USA). DSC curves of samples were determined by differential scanning calorimeter (200PC, NETZSCH-Gerätebau GmbH, Selb, Germany), and their thermal stability was analyzed.

2.5. Determination of the Polyphenolic Compounds Content

Polyphenolic compounds content in CS-GTEMR was determined according to the method of Pérez et al. [21]. The 0.1 g sample was fully swollen in distilled water and then filtered to remove excess water. A total of 0.5 mL of Folin–Ciocalteu reagent was added to the sample, mixed well, and allowed to stand in the dark for 5 min. Subsequently, 2.5 mL of a Na_2CO_3 solution (20% w/v) was added to the mixture, which was then mixed uniformly and left to stand in the dark for 1 h. The absorbance was determined at 725 nm, and the polyphenolic compounds content of the sample was calculated based on a standard curve (pyrogallol). The reported polyphenolic compounds content represents the Mean ± SD of three independent experiments.

2.6. Determination of Antioxidant Activities

2.6.1. Determination of DPPH Free Radical Scavenging Activity

DPPH free radical scavenging activity was determined according to the method of Yu et al. [22]. Three groups of different reaction solutions were prepared, involving solution 1 (2 mL sample solution and 2 mL DPPH solution (0.2 mmol/L)), solution 2 (2 mL sample solution and 2 mL anhydrous ethanol), and solution 3 (2 mL DPPH solution (0.2 mmol/L) and 2 mL distilled water). All three mixture solutions were kept in the dark at room temperature for 20 min. Then, the absorbance values of A_i (solution 1), A_j (solution 2), and A_0 (solution 3) were determined at 517 nm. The DPPH free radical scavenging rate (8) was calculated as follows:

$$\text{DPPH free radical scavenging rate (\%)} = (1 - \frac{A_i - A_j}{A_0}) \times 100 \quad (8)$$

2.6.2. Determination of Superoxide Anion Free Radical Scavenging Activity

Superoxide anion free radical scavenging activity was determined according to the method of Zhang et al. [23]. To a 2.0 mL Tris-HCl buffer solution (pH 8.2, 0.1 mmol/L), add 2.0 mL of distilled water, mix well, and let it stand in a water bath at 25 °C for 20 min. Then, add 0.1 mL of preheated 3 mmol/L pyrogallol at 25 °C, mix well, and immediately measure the absorbance at 325 nm every 30 s for a total of 11 measurements. Calculate the rate of change, ΔA_0, of pyrogallol absorbance over time. A total of 2.0 mL of sample solution was mixed with 2.0 mL of distilled water and allowed to stand at 25 °C for 20 min after thorough mixing. Then, 0.1 mL of 3 mmol/L pyrogallol, preheated in a 25 °C water bath, was added. The mixture was immediately measured for absorbance at 325 nm every

30 s, with a total of 11 measurements taken. The rate of change in absorbance (ΔA_S) of the sample solution over time was calculated. The superoxide anion radical scavenging rate E (%) was determined using Formula (9).

$$E(\%) = \frac{\Delta A_0 - \Delta A_S}{\Delta A_0} \times 100 \tag{9}$$

2.6.3. Determination of Hydroxyl Radical Scavenging Activity

Hydroxyl free radical scavenging activity was determined according to the method of Falcone et al. [24]. (1) Mix 2 mL of phosphate buffer (pH = 7.4, 0.1 mol/L), 1 mL of distilled water, and 1 mL of $FeSO_4$ solution (0.75 mmol/L) uniformly. Then, add 1 mL of H_2O_2 (0.01%) and mix for 1 min before adding 1 mL of 1,10-phenanthroline anhydrous ethanol solution (0.75 mmol/L). Allow the mixture to stand in a water bath at 37 °C for 60 min. Determine the absorbance value A_P at 536 nm;

(2) Replace the H_2O_2 in step (1) with 1 mL of distilled water, keeping all other conditions the same as in (1), and measure the absorbance value A_B;

(3) Substitute the distilled water in step (1) with 1 mL of sample solution, and measure the absorbance value A_S. Calculate the hydroxyl radical scavenging rate D (%) according to Formula (10).

$$D(\%) = \frac{A_S - A_P}{A_B - A_P} \times 100 \tag{10}$$

2.6.4. H_2O_2 Scavenging Activity

H_2O_2 scavenging activity was determined according to the method of Porcher et al. [25]. After fully swelling CS-GTEMR (0.1 g) in distilled water, remove the water and add 4 mL of H_2O_2 solution (2 mmol/L). Let the mixture stand for 10 min. Adjust the zero point with phosphate buffered solution (pH = 7.4) and determine the absorbance values of the H_2O_2 solution and the sample at 203 nm.

2.6.5. Determination of Iron Reducing Power

Iron reducing power was determined according to the method of Yu et al. [22]. A total of 2.5 mL phosphate buffer solution (pH 6.6, 0.1 mol/L) and 2.5 mL $K_3Fe(CN)_6$ (1%) were added to 1.0 mL sample solution, mixed evenly, and stood in a water bath at 50 °C for 20 min. Next, 2.5 mL TCA (10%) was added to the above mixture solution and centrifuged for 5 min at 2800× g. Then, 2.5 mL distilled water and 0.5 mL $FeCl_3$ (0.1%) were added to 2.5 mL supernatant, mixed evenly, and the mixture solution stood at room temperature for 10 min. The effect of iron reducing power was determined by the absorbance value at 700 nm.

2.6.6. Determination of Molybdenum Reducing Power Activity

Molybdenum reducing power was determined according to the method of Yu et al. [22]. After fully swelling CS-GTEMR (0.1 g) in distilled water and removing the excess water, 4 mL of phosphomolybdenum blue reagent solution was added. As a control, 0.1 mL of ascorbic acid solution (100 μg/mL) was added to 4 mL of phosphomolybdenum blue reagent solution. Both the sample and the control were incubated in a water bath at 95 °C for 90 min, then cooled to room temperature, and their absorbance values were determined at 695 nm. A_S is the sample's absorbance, and A_C is the control's absorbance. The molybdenum reducing power activity was calculated according to Formula (11).

$$\text{Molybdenum reducing power activity } (\%) = \frac{A_C - A_S}{A_C} \times 100 \tag{11}$$

2.7. Statistical Analysis

All experiments were performed in triplicate, and data were expressed as means and standard deviations. The SPSS Statistics 17.0 software (IBM Inc., Armonk, NY, USA) was used to analyze the variance of the results with the method of least significant difference (LSD).

3. Results and Discussion

3.1. Characterization of CS-GTEMR

3.1.1. Basic Physical Properties of CS-GTEMR

The three-dimensional super depth of field microscopy image of CS-GTEMR is presented in Figure 1. CS-GTEMR appears as a reddish-brown, spherical body with a smooth surface and a dense, porous structure. This configuration enhances their specific surface area, favoring the exertion of active effects. During the preparation of CS-GTEMR via the reversed-phase suspension cross-linking polymerization technology, chitosan and green tea extract acetic acid solution are introduced into liquid paraffin containing emulsifier and pore-forming agent, resulting in the formation of an oil-in-water emulsion. In the traditional preparation method of chitosan microsphere gel, only one cross-linking reaction is performed using cross-linking agents such as glutaraldehyde, epichlorohydrin, or genipin [26,27]. However, in this study, formaldehyde was used as a pre-cross-linking agent to form a liquid gel of chitosan. Subsequently, glutaraldehyde was employed as the cross-linking agent, and after a secondary cross-linking reaction, the chitosan liquid gel solidified into a hard and porous microsphere resin. Furthermore, during the secondary cross-linking process, the polyphenolic compounds and polysaccharides present in green tea extract can effectively interact with chitosan, resulting in the formation of chitosan-encapsulated green tea extract microsphere resin (CS-GTEMR). The determination of particle size distribution revealed that the average particle size of CS-GTEMR is approximately 316.106 μm, with 100% of the particles being smaller than 549.541 μm. Additionally, 90%, 50%, and 10% of the particles are smaller than 460.905 μm, 309.487 μm, and 237.500 μm, respectively. Based on the particle size distribution diagram in Figure 2, it can be observed that the particle size of CS-GTEMR follows a normal distribution with a relatively uniform distribution. There exist interactions between the chitosan matrix and components such as tea polyphenols from green tea extracts, resulting in smaller particle sizes. These hydrophilic interactions are essential forces for the formation of spherical particles. Additionally, some covalent bonds can also be formed between them; for instance, the amino groups in chitosan can react with the quinone rings in tea polyphenols. These physicochemical interactions have been confirmed in proteins and polyphenols [28].

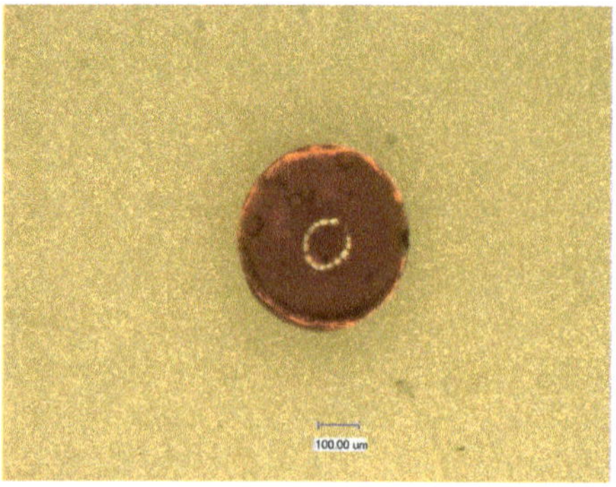

Figure 1. Microphotograph of CS-GTEMR (×100).

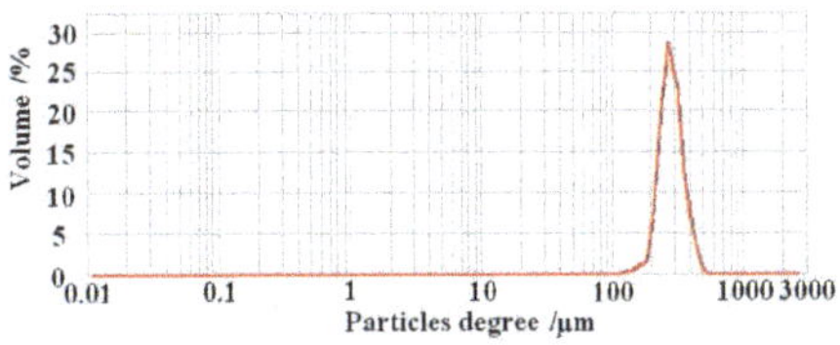

Figure 2. Particle size distribution of CS-GTEMR.

Table 1 presents the physical properties of CS-GTEMR. Yu et al. reported the physical properties of chitosan microsphere resin (CS-MR) prepared by reversed-phase suspension cross-linking polymerization technology [20]. The water adsorption capacity of CS-MR is 51.982 ± 1.944%, indicating that CS-GTEMR exhibits a higher water adsorption capacity. This enhanced capacity is attributed to the presence of green tea extract components in CS-GTEMR. Among these components, groups such as hydroxyl groups from polyphenolic compounds or polysaccharides, amino groups, and carboxyl groups from proteins or amino acids readily bond with water, thereby increasing the water adsorption capacity [29]. The content of free aldehyde groups in CS-GTEMR is lower than that in CS-MR (0.315 ± 0.009 mmol/g). During the cross-linking reaction, the proteins and amino acids, which are the main components of green tea extract, contain free amino groups. These amino groups have the potential to react with either both aldehyde groups of glutaraldehyde or with just one aldehyde group (where the other aldehyde group has already cross-linked with the amino group of chitosan), resulting in the formation of carbon–nitrogen double bonds (Schiff base) [30]. This process contributes to a reduction in free aldehyde groups in CS-GTEMR. The weak basic exchange capacity of CS-GTEMR is greater than that of CS-MR (1.311 ± 0.084 mmol/g). In addition to the free amino groups present in chitosan, the proteins and amino acids from green tea extract in CS-GTEMR also contain amino and carboxyl groups, all of which contribute to its weak basic exchange capacity. Hence, CS-GTEMR exhibits a higher weak basic exchange capacity. Based on the above analysis, CS-GTEMR can be described as a reddish-brown, spherical, porous microsphere resin with water absorption and weak basic exchange capabilities.

Table 1. Physical properties of CS-GTEMR and CS-MR.

Physical Properties	CS-GTEMR	CS-MR [20]
Water absorption capacity/M (%)	64.296 ± 1.588	51.982 ± 1.944
Pile-up density/ρ_P (g/mL)	0.843 ± 0.087	0.862 ± 0.007
Skeletal density/ρ_T (g/cm^3)	1.248 ± 0.480	1.212 ± 0.453
Pore degree/P	0.679 ± 0.075	0.554 ± 0.097
Free aldehyde group/(mmol/g)	0.267 ± 0.012	0.315 ± 0.009
Weak basic exchange capacity/(mmol/g)	1.409 ± 0.084	1.311 ± 0.084

3.1.2. FTIR Spectroscopy Analysis

Figure 3 presents the FTIR spectra of chitosan powder, CS-MR, and CS-GTEMR, with the wave numbers of their main peaks listed in Table 2. The broad peak, formed by the overlapping of the stretching vibration absorption peaks of ν(O-H) and ν(N-H) in chitosan powder, CS-MR, and CS-GTEMR, shifts from 3444 cm^{-1} to approximately 3420 cm^{-1} towards the shorter wave number. This shift is attributed to the presence of the Schiff base Π bond in CS-MR and CS-GTEMR, which induces a weak association between the -OH hydrogen and the Π electron cloud, resulting in a red shift of the hydroxyl stretching vibration peak [31]. Both CS-MR and CS-GTEMR exhibit characteristic absorption peaks of aldehyde groups around 1716 cm^{-1}, indicating the presence of free aldehyde groups from glutaraldehyde that did not react with amino groups [32]. Although the proteins and amino acids, which are the main components of green tea extract in CS-GTEMR, contain free amino groups that can react with glutaraldehyde to form Schiff bases, their low concentrations are insufficient to completely eliminate the pendant aldehyde groups, only reducing their amount. The disappearance of the amino bending vibration absorption peak at 1650 cm^{-1} and the emergence of a Schiff base absorption peak at 1558 cm^{-1} in CS-GTEMR indicate the formation of a -C=N bond through the cross-linking reaction between glutaraldehyde and amino groups. This finding aligns with previous reports stating that Schiff bases primarily form between 1540 and 1590 cm^{-1} [33]. Since the proteins and amino acids in CS-GTEMR contain free amino groups that can react with glutaraldehyde to form Schiff bases, the amide II peak in the FTIR spectrum of CS-GTEMR remains unchanged. After the chitosan cross-linking reaction, the intensity of the ν(C-O) stretching vibration absorption peak of their C_6-OH groups is significantly reduced. The possible reason is that the chitosan molecules undergo a structural transformation from a long-chain structure to a coiled and curly spatial structure after the cross-linking reaction. The primary hydroxyl groups may form hydrogen bonds with nitrogen or hydrogen atoms. Additionally, the high reactivity of the primary hydroxyl groups allows them to bond with the aldehyde group of glutaraldehyde. The characteristic absorption peak of β-D-glucopyranoside in CS-GTEMR has shifted to 916 cm^{-1}, indicating that the cross-linking reaction did not result in the ring-opening of the β-D-glucopyranose ring. Instead, it is likely that the amino groups in CS-GTEMR formed hydrogen bonds with the hydroxyl groups of chitosan, strengthening the force between glucose units in chitosan and leading to a change in the peak position. The presence of a characteristic absorption peak of phenolic hydroxyl groups, ν(H-O), at 1230 cm^{-1} in CS-GTEMR suggests the existence of polyphenolic compounds within the material [34].

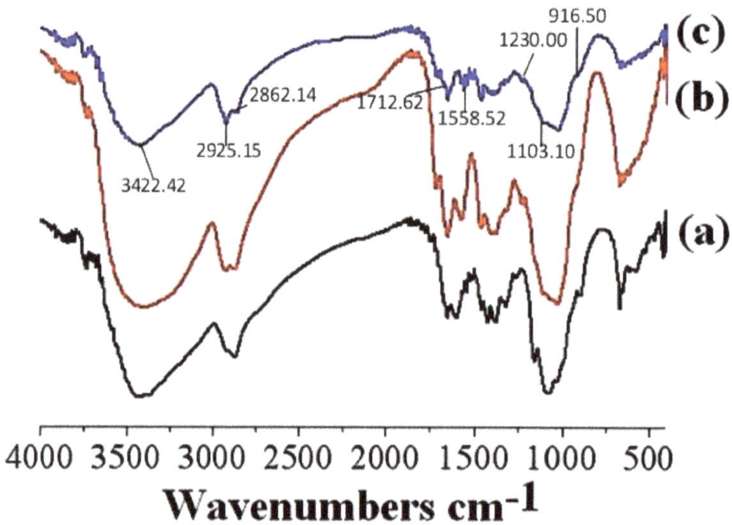

Figure 3. FTIR spectrum (**a**) chitosan powder, (**b**) CS-MR, and (**c**) CS-GTEMR.

Table 2. Main wave numbers of chitosan powder, CS-MR, and CS-GTEMR.

Peaks Attribution	Chitosan Powder (cm^{-1})	CS-MR (cm^{-1})	CS-GTEMR (cm^{-1})
ν(O-H) and ν(N-H)	3444.86	3420.21	3422.42
ν(-CH_3)	2917.82	2925.88	2925.15
ν(-CH_2)	2875.65	2875.98	2862.14

Table 2. *Cont.*

Peaks Attribution	Chitosan Powder (cm^{-1})	CS-MR (cm^{-1})	CS-GTEMR (cm^{-1})
ν(-CHO)	-	1715.07	1712.62
amide I (ν(C=O) and ν(C-N))	1649.84	1650.65	1650.48
δ($-NH_2$)	1597.87	-	-
ν(C=N)	-	1573.36	1558.52
amide II (δ(N-H) and ν(C-N))	1540.99	-	1541.28
amide III (ν(C-N) and δ(N-H))	1259.77	1227.50	1227.18
ν(C-O-C)	1154.80	1102.21	1103.10
C_6-OH (ν(C-O))	1087.55	weaker	weaker
C_3-OH (ν(C-O))	1029.13	1031.00	1024.82
characteristic absorption peak of β-D-glucopyranoside	897.47	weaker	916.50
phenolic hydroxyl group (ν(H-O))	-	-	1230.00

3.1.3. Differential Scanning Calorimetry (DSC) Analysis

Figure 4 displays the DSC curves of chitosan powder, CS-MR, and CS-GTEMR. The first heating stage curve reveals that the onset temperature of the endothermic peak for CS-GTEMR is 42.95 °C, with a peak end temperature of 91.34 °C and an endothermic enthalpy ΔH of 42.02 J/g. In comparison, the experimental data for chitosan powder and CS-MR are 57.06 °C, 82.88 °C, and 40.49 J/g, and 39.93 °C, 87.74 °C, and 63.41 J/g, respectively. The endothermic peak observed in this stage is primarily attributed to the dehydration of chitosan macromolecules [35]. The maximum endothermic peak of CS-GTEMR is greater than that of chitosan powder and CS-MR, while its endothermic enthalpy lies between that of chitosan powder and CS-MR. Most of the water molecules in chitosan powder are bonded to amino groups, forming hydrogen bonds, which require less energy to break. When chitosan is prepared into spherical cross-linked resin, some of its amino groups react with glutaraldehyde to form Schiff bases. The water molecules originally bonded with amino groups are then bonded with hydroxyl groups. This requires more energy to break the hydrogen bonds formed between water molecules and hydroxyl groups, thus explaining why the endothermic enthalpy of CS-MR is higher than that of chitosan powder. In CS-GTEMR, besides bonding with hydroxyl groups, water molecules also bond with the free amino groups of proteins and amino acids. The energy required to break the hydrogen bonds formed between amino groups and water molecules is less than that required to break the hydrogen bonds between water molecules and hydroxyl groups. Therefore, the endothermic enthalpy of CS-GTEMR is greater than that of chitosan powder but less than that of CS-MR.

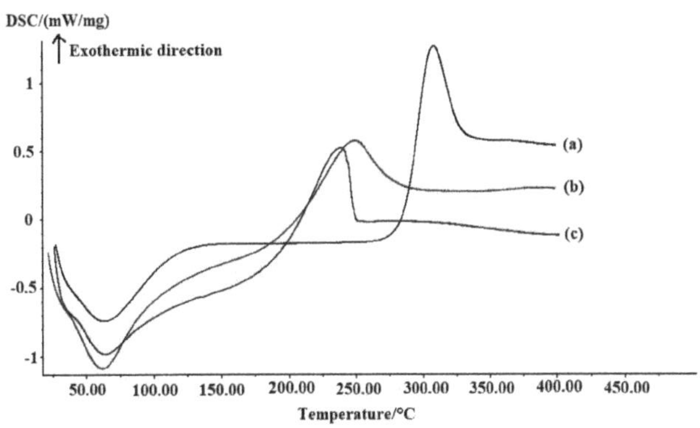

Figure 4. DSC of (**a**) CS-GTEMR, (**b**) CS-MR, and (**c**) chitosan powder.

The second heating stage revealed that the exothermic peak of CS-GTEMR initiated at 202.69 °C and terminated at 247.40 °C, with an exothermic enthalpy of 85.96 J/g. The experimental data for chitosan powder and CS-MR were as follows: 288.58 °C, 326.09 °C, 124.20 J/g, and 215.36 °C, 276.68 °C, 85.63 J/g, respectively. The exothermic peak in this stage is associated with the thermal and oxidative decomposition of chitosan, the thermal decomposition of polysaccharides, polyphenolic compounds, proteins, and amino acids, as well as the volatilization and elimination of volatile products. The thermal degradation temperatures decrease in the order of chitosan powder, CS-MR, and CS-GTEMR. Chitosan exhibits two types of intramolecular hydrogen bonds: one between the hydroxyl group at carbon 3 and the oxygen atom at carbon 5, and the other between the hydroxyl group at carbon 6 and the nitrogen atom of the amino group. Due to its good crystallinity, chitosan demonstrates excellent thermal stability [36]. However, when chitosan is prepared into cross-linked spherical resins, the formation of Schiff bases between amino groups and glutaraldehyde disrupts the intramolecular hydrogen bonds within the chitosan chains. The decrease in the crystallinity of chitosan affects its thermal stability, leading to the conclusion that the thermal stability of CS-MR is lower than that of chitosan powder. When green tea extract is added during the cross-linking process, the hydroxyl and amino groups present in the extract further disrupt the hydrogen bonds within the chitosan chains, resulting in the thermal stability of CS-GTEMR being even lower than that of CS-MR. Additionally, DSC curve data reveal the glass transition temperature ranges for chitosan powder, CS-MR, and CS-GTEMR to be 290.05~300.60 °C, 219.59~223.22 °C, and 192.78~211.18 °C, respectively. These results confirm that the thermal stability of the three substances, from highest to lowest, is chitosan powder, CS-MR, and CS-GTEMR.

3.2. Results and Analysis of Swelling Properties of CS-GTEMR

3.2.1. Results and Analysis of CS-GTEMR Swelling Properties in Water

The swelling ratio of CS-GTEMR in distilled water at different times is shown in Figure 5. CS-GTEMR exhibits swelling properties in distilled water, with an equilibrium swelling ratio of 229.7%. Zhu et al. investigated the swelling behavior of tannic acid-modified keratin/sodium alginate/carboxymethyl chitosan biocomposite hydrogels [37]. The results indicated that the modified hydrogels exhibited a three-dimensional microporous structure with a swelling ratio of 1541.6%. However, the CS-GTEMR prepared in this study appeared spherical in shape and had a more compact texture, resulting in a lower swelling ratio compared to the aforementioned findings. The main components of green tea extract contained in CS-GTEMR mostly possess hydroxyl and amino groups, which easily bond with water molecules, thereby increasing the water absorption and swelling ratio. This result aligns with the high water absorption capacity and porous physical properties of CS-GTEMR. As illustrated in Figure 5, no statistically significant difference ($p > 0.05$) was observed in the swelling ratio of CS-GTEMR between 72 h and 96 h, indicating that the swelling equilibrium had been achieved by 72 h. Extending the swelling time further will not increase the swelling ratio. This phenomenon indicates that the microsphere resin prepared by reversed-phase suspension cross-linking polymerization technology has a tough structure and will not rupture due to excessive swelling in water. This suggests that CS-GTEMR can be applied in aqueous environments.

3.2.2. Results and Analysis of CS-GTEMR Swelling Properties in Different pH Solutions

The equilibrium swelling ratios of CS-GTEMR in buffer solutions with different pH values after 96 h of swelling are presented in Figure 6. Upon completion of the swelling process, no discoloration or microsphere damage was observed in the CS-GTEMR samples across all pH buffers, indicating that CS-GTEMR can maintain a stable structural state in solutions with pH values ranging from 1 to 13, and is capable of absorbing water and swelling under these conditions. Notably, the equilibrium swelling ratio of CS-GTEMR reached a minimum value of 170.2% in a Tris-HCl buffer at pH 9.0 ($p < 0.05$). The maximum equilibrium swelling ratio of the material in a disodium hydrogen phosphate–citric acid buffer solution at pH 3.0 is 267.5% ($p < 0.05$). It should be noted that no significant difference was observed in swelling ratios between the pH 7 and pH 11 groups ($p > 0.05$). However, statistically significant differences were identified between the swelling ratios at pH 13 versus those at both pH 7 and pH 11 ($p < 0.05$). The active hydroxyl and amino groups of polyphenolic compounds, polysaccharides, proteins, and amino acids from green tea extract in CS-GTEMR can bond with water molecules under different pH conditions, resulting in varying equilibrium swelling ratios. These findings suggest that CS-GTEMR can also be applied in solutions with a pH range of 1–13.

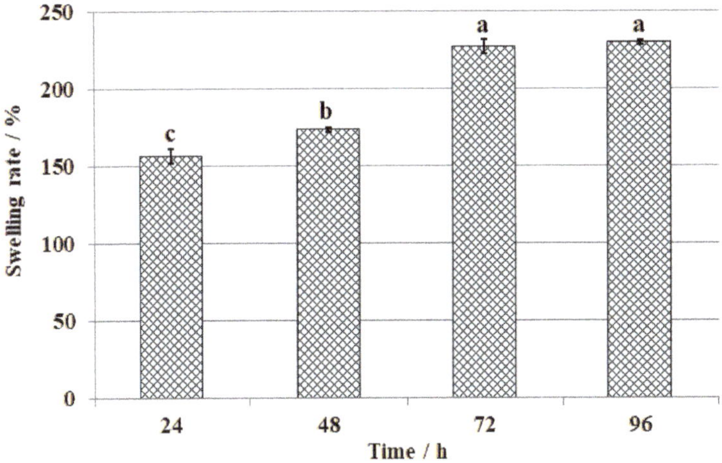

Figure 5. Swelling kinetic curves of CS-GTEMR in water (different letters in each indicator mean significant differences at 0.05 level ($p < 0.05$, $n = 3$)).

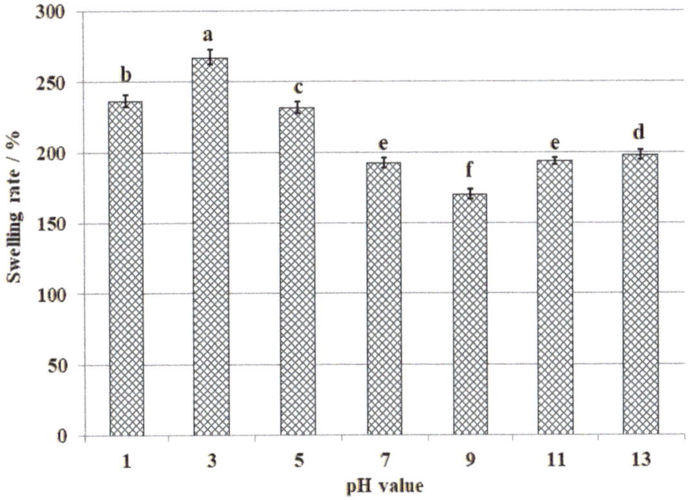

Figure 6. Swelling rates of CS-GTEMR in different pH solutions (different letters in each indicator mean significant differences at 0.05 level ($p < 0.05$, $n = 3$)).

3.3. *Evaluation of the Polyphenolic Compounds in CS-GTEMR*

3.3.1. Content of Polyphenols in CS-GTEMR

The content of polyphenolic compounds encapsulated in CS-GTEMR was determined using the Folin–Ciocalteu method, and the measured value was 50.485 ± 0.840 μg/g. Chitosan is a linear polymer composed of multiple linked glucosamine units. Under acidic pH conditions, it exhibits a high positive charge density due to the presence of a positive charge on each glucosamine unit. Polyphenolic compounds with specific molecular weights and sizes can serve as complexing agents. The coordination interaction between polyphenolic compounds and chitosan can be either reversible or irreversible. The reversible coordination of polyphenolic compounds and chitosan occurs in two stages. In the first stage, polyphenolic compounds and chitosan form soluble complexes through non-covalent bonding, achieving an equilibrium state in solution. In the second stage, the equilibrium state is disrupted, leading to the aggregation and precipitation of these soluble complexes from the solution. This entire process is typically reversible, and under appropriate conditions, the precipitated complexes can be redissolved. In this study, prior to the preparation of CS-GTEMR, a mixture of chitosan and green tea extract was dissolved in an acetic acid solution. During the dissolution process, feather-like aggregates were observed in the mixed solution of chitosan and green tea extract, indicating the formation of reversible complexes between chitosan and the polyphenolic compounds present in the green tea extract. In the emulsification step of the microsphere preparation process, the complexes precipitated from the polyphenolic compounds and chitosan were redissolved to form soluble complexes. After undergoing a secondary cross-linking reaction, polyphenolic compounds are encapsulated within chitosan microspheres in a stable coordination structure. Evidently, this reversible complex dissolution process facilitates the encapsulation of a greater number of polyphenolic compounds within the chitosan microspheres. Furthermore, the enhanced stability of the polyphenolic compounds' binding state within the microspheres is beneficial for their controlled release and the exertion of their active effects during application. During the preparation of chitosan-encapsulated olive leaf extract microspheres using the spray drying method, a soluble complex is formed between the olive leaf extract and the chitosan matrix. If the interaction between the extract and chitosan mixture does not progress to the second stage of aggregation during the spray drying process, the resulting microspheres will contain a low amount of olive leaf extract.

3.3.2. Results and Analysis of the Polyphenolic Compounds Release Rate in Simulated Gastric and Intestinal Fluid

When exposed to simulated gastric fluid and simulated intestinal fluid, the release rates of polyphenolic compounds from CS-GTEMR were 24.934 ± 0.168% and 3.375 ± 0.134%, respectively. It can be inferred that the release rate of polyphenolic compounds is higher in simulated gastric fluid compared to simulated intestinal fluid. The swelling ratios of CS-GTEMR at various pH values obtained in this study indicate that within the pH range of 1–3, which corresponds to simulated gastric fluid, the swelling ratio of CS-GTEMR is significantly higher than at other pH values. In this highly acidic environment, the porous structure of CS-GTEMR exhibits a high swelling ratio, resulting in the exposure of a significant amount of encapsulated polyphenolic compounds to the simulated gastric fluid. Under these conditions, a large number of hydrogen ions may competitively displace the amino groups of chitosan that were originally covalently coordinated with the polyphenolic compounds. Based on the reversible coordination mechanism between polyphenolic compounds and chitosan (as described in Section 3.3.1), under acidic conditions, some of the aggregated and precipitated complexes redissolve and subsequently dissociate into the ligand (chitosan) and the coordinating reagent (polyphenolic compounds). Consequently, a higher release rate of polyphenolic compounds is observed in the simulated gastric fluid. In simulated small intestinal fluid, the release rate of polyphenolic compounds from CS-GTEMR was relatively low, reaching 3.375%. This finding is consistent with the research results obtained by Hameed et al. [38]. They prepared cross-linked chitosan microspheres encapsulating antiviral drugs using a copolymerization technique with polysaccharides extracted from pomegranate peel and chitosan. The swelling experiments demonstrated that the cross-linked chitosan microspheres could control the release of the encapsulated drug in a pH 7.4 solution, while the drug release rate was maximized at pH 1.3. In this study, within the pH range of 7–8, which simulates small intestinal fluid, the swelling ratio of CS-GTEMR reached its minimum value. In this weakly alkaline environment, where only a limited number of hydroxide ions are present, the amino groups of chitosan in CS-GTEMR remain unaffected by hydrogen ions. Consequently, the complexes formed between chitosan and polyphenolic compounds do not dissociate but instead maintain a stable existence. This protective mechanism ensures that the majority of polyphenolic compounds encapsulated within CS-GTEMR are effectively preserved and retained within the system. The cumulative release rate of polyphenolic compounds in simulated gastric fluid and simulated small intestinal fluid was 28.309%, indicating that nearly 72% of the polyphenolic compounds in CS-GTEMR can be further released and exert their active effects. In the large intestine environment, microorganisms can degrade or decompose the chitosan matrix of CS-GTEMR, allowing the retained polyphenolic compounds in CS-GTEMR to be completely released into the large intestine environment. There, they can exert their biological activities, such as antioxidant properties, to protect the health of the organism. Several studies have demonstrated that polyphenolic compounds, upon absorption, are widely distributed in tissues throughout the body, particularly exhibiting high concentrations in the esophagus, small intestine, and large intestine [39–41]. Consequently, a higher quantity of polyphenolic compounds in the large intestine corresponds to a greater content of these compounds in the colon, which is beneficial for the protective effects of polyphenols on colonic health.

3.4. Results and Analysis of Antioxidant Activity

Oxygen free radicals, including O_2^-, ·OH, ·OR, etc., are normal metabolites of human body metabolism. Under normal circumstances, the production and elimination of oxygen free radicals in the body are balanced. The presence of a small amount of oxygen free radicals in the human body can also promote cell proliferation and accelerate the bactericidal and anti-inflammatory effects of cells. However, once the production of oxygen free radicals in the body becomes excessive or the antioxidant system malfunctions, the metabolism of oxygen free radicals will be unbalanced, becoming an important factor causing aging and many diseases. For example, heart disease, hypertension, chronic pneumonia, etc., are all related to oxygen free radicals. Polyphenolic compounds exhibit antioxidant activity due to their ability to scavenge hydroxyl radicals and superoxide anions. Many plant extracts contain polyphenolic compounds, which can serve as natural food antioxidants, protecting the human body from the harmful effects of free radicals [42,43]. Alternatively, these compounds can be prepared into microspheres through coordination with some biopolymers and administered as a prophylactic agent via the digestive tract. This approach can increase the level of polyphenolic compounds in the digestive system, thereby playing a role in disease prevention.

The DPPH radical method is a rapid approach widely used to evaluate the ability of antioxidants to scavenge free radicals within a relatively short period of time. DPPH radical is a highly stable nitrogen-centered radical. If the tested compound can scavenge it, this indicates that the compound has the effect of reducing the effective concentration of hydroxyl radicals, alkyl radicals, or peroxy radicals, or interrupting the lipid peroxidation chain reaction. The DPPH radical ethanol solution is deep purple and exhibits a strong absorption peak near 517 nm. When a radical scavenger is added to the DPPH radical solution, the DPPH radical accepts electrons or hydrogen from the tested substance, transforming into a stable diamagnetic molecule. Consequently, the solution color changes from purple to yellow, and the absorbance value decreases. The magnitude of this decrease in absorbance is linearly related to the extent of radical scavenging. From this relationship, the IC_{50} value of the scavenger can be determined. Therefore, it can be used to detect the scavenging of DPPH free radicals, thereby evaluating the antioxidant properties of the sample. The smaller the IC_{50} value, the greater the scavenging rate and the stronger the antioxidant activity is. After adding CS-GTEMR to the DPPH solution, the optical density at 517 nm decreased rapidly, and the degree of color removal indicated the ability of CS-GTEMR to scavenge DPPH free radicals. Simultaneously, the scavenging abilities of crude tea polyphenols and ascorbic acid for DPPH free radicals were determined, and the results are shown in Table 3. CS-GTEMR has the ability to scavenge DPPH free radicals, and its effect is similar to that of crude tea polyphenols. Sathiyaseelan et al. encapsulated *Melaleuca alternifolia* oil within chitosan–sodium alginate microspheres and investigated the antioxidant, antibacterial, and wound healing properties of these microspheres [44]. The results indicated that the microspheres exhibited DPPH and ABTS radical scavenging abilities. Furthermore, the microspheres demonstrated negligible cytotoxicity and were found to promote the proliferation of NIH3T3 cells in an in vitro scratch assay. This suggests that the polyphenolic compounds encapsulated in CS-GTEMR are very stable and maintain good activity.

Table 3. Results of scavenging DPPH free radicals.

Peaks Attribution	CS-GTEMR	Crude Tea Polyphenol	Ascorbic Acid
IC_{50}	0.16 g/mL	61.49 μg/mL	70.66 μg/mL
scavenging rate/%	59.42 ± 3.99	64.78 ± 1.46	40.35 ± 4.44

Pyrogallol undergoes auto-oxidation and decomposition in a weakly alkaline environment (pH 8.2) to produce $O^{-}{}_{2}\cdot$. As the reaction progresses, $O^{-}{}_{2}\cdot$ accumulates in the system, resulting in a linear increase in the absorbance value of the reaction solution at a wavelength of 325 nm over time. By determining the rate of change in absorbance of the antioxidant reaction solution with time, the ability of the antioxidant to inhibit the accumulation of $O^{-}{}_{2}\cdot$ can be obtained. The kinetic curves for the auto-oxidation of pyrogallol, crude tea polyphenols, ascorbic acid, and CS-GTEMR in scavenging superoxide anion radicals are shown in Figure 7. The kinetic curves for the auto-oxidation of pyrogallol, crude tea polyphenols, and ascorbic acid in scavenging superoxide anion radicals all show a linear upward trend. However, the kinetic curve for CS-GTEMR in scavenging superoxide anion radicals flattens out after 240 s. No significant difference was observed in the absorbance values of the CS-GTEMR curve between 270 s and 300 s ($p > 0.05$). Furthermore, within the 0–300 s time range, significant differences were detected at all time points for the curves of pyrogallol, tea polyphenols, and ascorbic acid ($p < 0.05$). At the end of the reaction, the scavenging rate of CS-GTEMR for superoxide anion radicals was 52.893%, while the scavenging rate of ascorbic acid was 34.238%. This indicates that CS-GTEMR has good scavenging activity for superoxide anion radicals.

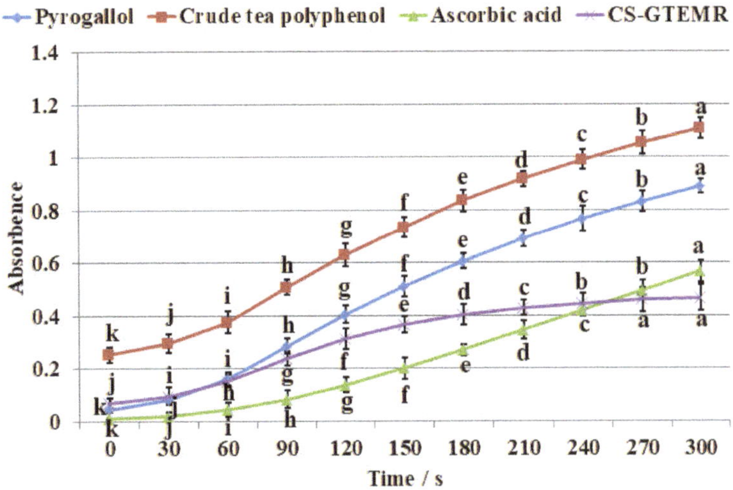

Figure 7. Scavenging superoxide anion radicals activities (different letters in each indicator in the same substance mean significant differences at 0.05 level ($p < 0.05$, $n = 3$)).

Hydroxyl radicals are known as the strongest oxidant, capable of undergoing reactions such as dehydrogenation, addition, and electron transfer. These radicals can react with substances like amino acids, proteins, nucleic acids, and fats, causing oxidative damage to biological organisms. This damage can lead to aging and disease, posing significant harm to the body. Consequently, research on both the oxidative damage caused by hydroxyl radicals and their scavengers has garnered attention. The scavenging effects of ascorbic acid, crude tea polyphenols, and CS-GTEMR on hydroxyl radicals are illustrated in Figure 8. Compared to ascorbic acid and crude tea polyphenols, CS-GTEMR exhibits better scavenging activity against hydroxyl radicals ($p < 0.05$). Furthermore, crude tea polyphenols demonstrated slightly higher hydroxyl radical scavenging activity than ascorbic acid ($p < 0.05$). Zhu et al. prepared catechin-grafted chitosan and investigated its antioxidant activity [45]. Their findings indicated that at a concentration of 1 mg/mL, the catechin-grafted chitosan exhibited a reducing power of 0.51, a hydroxyl radical scavenging rate of 46.81%, and a DPPH radical scavenging rate of 67.08%. This study untangled that the antioxidant activity of catechin-grafted chitosan originates from the phenolic hydroxyl groups on the catechin molecules. This suggests that green tea extract encapsulated in chitosan microspheres can still retain its antioxidant activity.

The determination of hydrogen peroxide scavenging activity is capable of assessing the ability of antioxidants to reduce peroxidizing agents. Although hydrogen peroxide itself does not exhibit high reactivity, its combination with superoxide anion can damage many cellular components. The results of hydrogen peroxide scavenging by ascorbic acid, crude tea polyphenols, and CS-GTEMR are presented in Figure 9. CS-GTEMR demonstrates the strongest effect on hydrogen peroxide, performing better than both ascorbic acid and crude tea polyphenols ($p < 0.05$). Furthermore, both ascorbic acid and crude tea polyphenol demonstrated superior absorbance variation trends compared to hydrogen peroxide ($p < 0.05$).

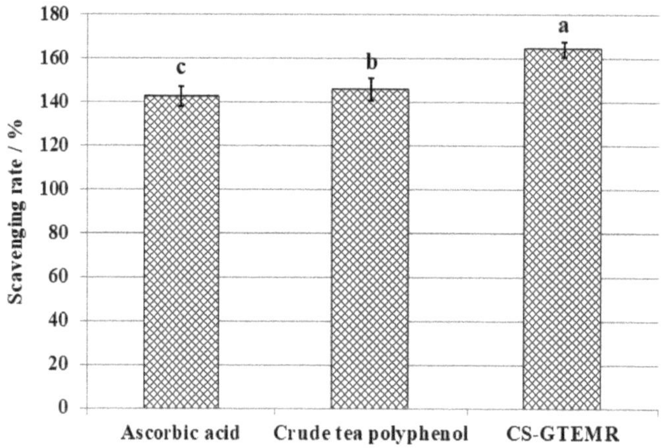

Figure 8. Scavenging hydroxyl radical activities (different letters in each substance mean significant differences at 0.05 level ($p < 0.05$, $n = 3$)).

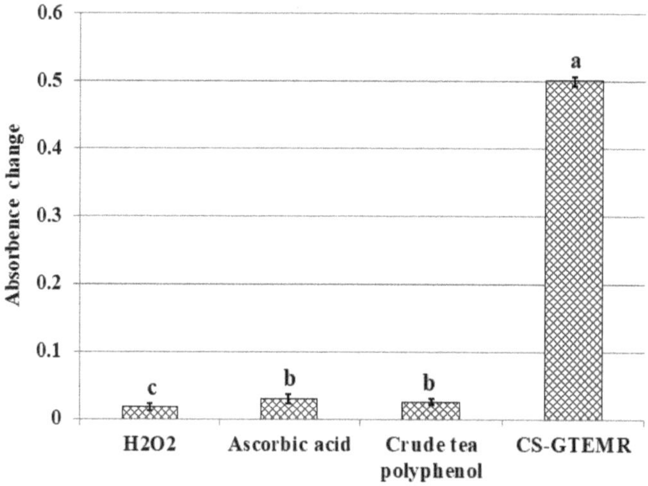

Figure 9. Scavenging hydrogen peroxide activities (different letters in each substance mean significant differences at 0.05 level ($p < 0.05$, $n = 3$)).

Reducing power is a crucial indicator that represents the electron-providing capability of antioxidants. Antioxidants deactivate free radicals by donating electrons through their reducing action, thereby converting these radicals into stable molecules. Polyphenolic compounds, as antioxidants, possess the ability to donate electrons, and their antioxidant efficacy is closely linked to their reducing power. The stronger the reducing power, the greater the antioxidant activity. Figure 10 illustrates the results of iron-reducing power for ascorbic acid, crude tea polyphenols, and CS-GTEMR. Due to the presence of numerous aromatic ring hydroxyl groups, crude tea polyphenols exhibit a higher electron-donating capacity than ascorbic acid, resulting in a higher absorbance value ($p < 0.05$). As for CS-GTEMR, the absorbance value presented is obtained after diluting the reaction solution six times, indicating that its iron-reducing power significantly exceeds that of both ascorbic acid and crude tea polyphenols ($p < 0.05$). The phosphorus molybdenum blue method is based on the reduction of Mo(VI) to Mo(V) by antioxidant compounds, forming a green Mo(V) complex that has maximum absorption at 695 nm. Antioxidants block free radical chain reactions by donating hydrogen atoms. The results are shown in Figure 10. The absorbance value reflecting the molybdenum-reducing capacity of CS-GTEMR was significantly higher than that of crude tea polyphenols and ascorbic acid ($p < 0.05$). In conclusion, CS-GTEMR demonstrates superior reducing power activities.

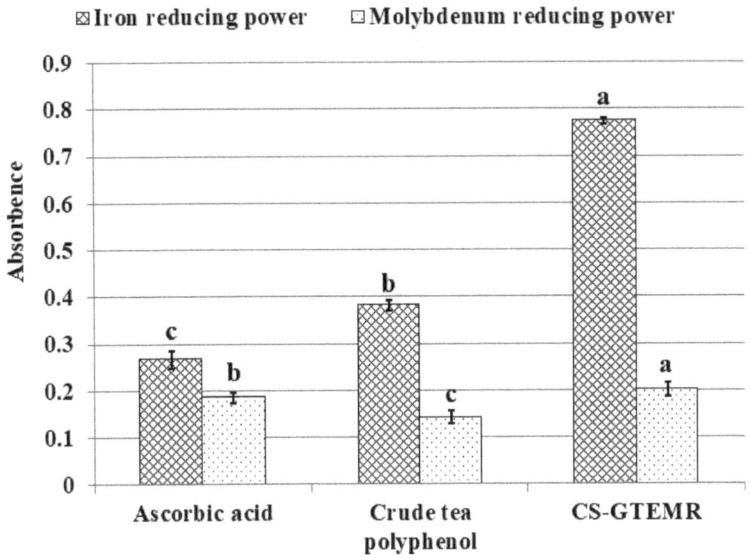

Figure 10. Iron and molybdenum reducing power activities (different letters in each substance in the same indicator mean significant differences at 0.05 level ($p < 0.05$, $n = 3$)).

4. Conclusions

This paper investigates the physical properties, structural characteristics, swelling behavior, polyphenol content, and antioxidant activity of CS-GTEMR prepared using the reversed-phase suspension cross-linking polymerization technique. The results indicated that CS-GTEMR exhibited a reddish-brown spherical shape with a porous, dense, and smooth surface. The average particle size is approximately 316.106 μm, and it contains a total polyphenol content of 50.485 μg/g. The presence of characteristic absorption peaks of phenolic hydroxyl groups in the FTIR spectrum suggests the existence of polyphenolic compounds within CS-GTEMR. The DSC results indicate that the thermal stability of CS-GTEMR is lower than that of chitosan powder and CS-MR. CS-GTEMR exhibits excellent swelling properties in both distilled water and solutions with pH ranging from 1.0 to 13.0. The cumulative release rate of polyphenolic compounds from CS-GTEMR in simulated gastric fluid and simulated intestinal fluid is only 28.309%, suggesting that the polyphenolic compounds within CS-GTEMR have the potential for further release and can exert antioxidant activities, such as scavenging free radicals and reducing power. The aforementioned results demonstrate that the successful encapsulation of green tea extract into the CS-GTEMR system was confirmed through quantitative determination of polyphenolic compounds and identification of the characteristic phenolic hydroxyl peak in FTIR spectroscopy. This confirms the successful preparation of chitosan microsphere materials with developmental potential for green tea extract protection. Furthermore, experimental data on swelling characteristics, controlled release performance in simulated gastrointestinal fluids, and antioxidant activity indicate that CS-GTEMR maintains stability across a wide pH range and achieves controlled release of polyphenolic compounds. Based on these findings, CS-GTEMR demonstrates potential for development as both a directly ingestible nutraceutical and pharmaceutical agent to safeguard human health. Furthermore, in food processing applications such as meat products, baked goods, and fruit/vegetable beverages, CS-GTEMR may serve as a promising natural antioxidant due to its controlled release of polyphenolic compounds. This property effectively inhibits lipid peroxidation, thereby contributing to extended food shelf life. In the future, various functional groups such as flavonoids, pigments, polysaccharides, and peptides could be grafted onto CS-GTEMR, enabling the development of chitosan microspheres with diverse functional activities. Subsequently, their applications may be extended to areas like encapsulation carriers for natural products, sustained-release agents for active substances, and clarification agents for fruit juices and beverages. The development and application of this range of CS-GTEMR products hold promise for playing a significant role in the food and beverage industry, the production of functional food ingredients, and the pharmaceutical industry. Therefore, it is crucial to further explore the activity and applications of the CS-GTEMR series of functional products in the future.

Author Contributions: Conceptualization, L.Y., J.B. and M.W.; data curation, L.Y., S.F., Y.S. and L.W.; formal analysis, L.Y., Y.G., C.J. and J.B.; funding acquisition, S.F., Y.G., C.J. and M.W.; investigation, L.Y., L.W., C.J. and M.W.; methodology, L.Y., S.F., Y.S. and L.W.; project administration, Y.G., Y.S. and C.J.; resources, S.F., Y.G., L.W. and C.J.; validation, L.Y., Y.S., J.B. and C.J.; writing—original draft, L.Y., J.B. and Y.S.; writing—review and editing, L.Y., Y.S. and M.W. All authors have read and agreed to the published version of the manuscript.

Funding: This research was funded by the Department of Science and Technology of Shandong Province of China, under the Key R&D Program of Shandong Province, grant number 2023TZXD074; the Shandong Academy of Agricultural Sciences of China, through the Innovation Project of Agricultural Science and Technology, grant number CXGC2025C19; the Shandong Science and Technology Commissioners of China, through the Typical Case (New Technologies, New Varieties, New Models) of Rural Revitalization, grant number 2022DXAL0102; and the Muping Bureau of Science and Technology of China, through the Technology Innovation Development Special Plan.

Institutional Review Board Statement: Not applicable.

Data Availability Statement: The original contributions presented in this study are included in the article. Further inquiries can be directed to the corresponding author.

Conflicts of Interest: The authors declare no conflicts of interest.

References

1. Righi, D.C.B.F.; Boeira, C.P.; Nora, F.M.D.; Schlesner, S.K.; Kaufmann, A.I.; Barin, J.S.; Machado, P.G.; Ballus, C.A.; do Nascimento, V.R.; Bizzi, C.A.; et al. Green technologies applied to extraction processes and analysis of polyphenolic compounds from *Inga laurina* Willd. *Food Chem.* **2025**, *471*, 142827. [CrossRef] [PubMed]
2. Mashau, M.E.; Kgatla, T.E.; Makhado, M.V.; Mikasi, M.S.; Ramashia, S.E. Nutritional composition, polyphenolic compounds and biological activities of marula fruit (*Sclerocarya birrea*) with its potential food applications: A review. *Int. J. Food Prop.* **2022**, *25*, 1549–1575. [CrossRef]
3. Castaldo, L.; Lombardi, S.; Gaspari, A.; Rubino, M.; Izzo, L.; Narváez, A.; Ritieni, A.; Grosso, M. In vitro bioaccessibility and antioxidant activity of polyphenolic compounds from spent coffee grounds-enriched cookies. *Foods* **2021**, *10*, 1837. [CrossRef]
4. Li, J.; Ma, J.; Zhang, Y.; Zheng, L. Determination of 19 polyphenolic compounds in tea by ultra-high performance liquid chromatography combined with quadrupole-time of flight mass spectrometry. *Food Sci. Hum. Wellness* **2022**, *11*, 719–726. [CrossRef]
5. Alqathama, A.; Abdelhady, M.I.S.; Al-Omar, M.S.; Barghash, M.F.; Barghash, M.F. Antioxidant, anti-inflammatory and cytotoxic activity of *Schinus terebinthifolia* fruit and isolation of a new immunomodulatory polyphenolic compound. *Pharmacogn. Mag.* **2023**, *19*, 13–22. [CrossRef]
6. Liu, S.; Fang, Z.; Ng, K. Incorporating inulin and chitosan in alginate-based microspheres for targeted delivery and release of quercetin to colon. *Food Res. Int.* **2022**, *160*, 111749. [CrossRef]

7. Yu, X.; Chen, Y.; Tan, M. ROS-responsive carboxymethyl chitosan nanoparticles loaded with astaxanthin for alleviating oxidative damage in intestinal cells. *Colloids Surf. B* **2024**, *239*, 113960. [CrossRef]
8. Hossain, R.; Kongchain, A.; Chatatikun, M.; Klangbud, W.K.; Yupanqui, C.T.; Majima, H.J.; Indo, H.P.; Sompol, P.; Sekeroglu, N.; Phongphithakchai, A. Green tea pressurized hot water extract in Atherosclerosis: A multi-approach study on cellular, animal, and molecular mechanisms. *Antioxidants* **2025**, *14*, 404. [CrossRef]
9. De la Fuente-Muñoz, M.; Román-Carmena, M.; Amor, S.; de la Cruz, M.C.; Martorell, P.; Guilera-Bermell, S.; Bou, R.G.; Inarejos-García, A.M.; García-Villalón, A.L.; Granado, M. Supplementation with standardized green/black or white tea extracts attenuates hypertension and ischemia-reperfusion-induced myocardial damage in mice infused with Angiotensin II. *Antioxidants* **2025**, *14*, 47. [CrossRef]
10. Sulaimani, N.; Rosbotham, E.J.; Warnock, R.; Polzella, L.; Judowski, R.; Nicolotti, L.; Houghton, M.J.; Williamson, G.; Bonham, M.P. Time-of-day-dependent effects of a green tea extract on postprandial glycemia and insulinemia in healthy adults: A randomized, controlled, double-blind, cross-over intervention. *Food Funct.* **2025**, *16*, 4122–4133. [CrossRef]
11. Joo, Y.; Lee, H.Y.J.; Jeong, H.; Suh, C.; Hong, H.J.; Kim, Y.; Yu, S.Y.; Lee, C.R.; Shim, Y.; Yoon, S. Effects of Heat-Treated Green Tea Extract on Memory Function and Default Mode Network Connectivity in Individuals with Subjective Memory Impairment: A Randomized, Double-Blinded, Placebo-Controlled Trial. *J. Med. Food* **2025**. [CrossRef] [PubMed]
12. Moalemi, S.F.S.S.; Safari, F.; Ahvati, H. Suppression of cellular proliferation in PC3 prostate cancer cells by green tea extract through induction of miR-34a expression. *Food Sci. Nutr.* **2025**, *13*, e70215. [CrossRef] [PubMed]
13. Yang, H.; Tao, H.; Xu, C.; Song, J.; Teng, C.; Pan, C.; Wei, S. Selenium-enriched green tea extracts: Chemical constituents and effects on antioxidant and anti-inflammatory factors and four major intestinal flora in mice with intestinal disorders. *J. Sci. Food Agric.* **2025**, *105*, 4472–4482. [CrossRef] [PubMed]
14. Ogundare, T.E.; Kulkarni, R.R.; Omaliko, P.C.; Iwuozo, O.C.; Enenya, I.G.; Orimaye, O.E.; Suberu, S.A.; Jeje, O.; Fasina, Y.O. Effect of green tea (*Camellia sinensis*) extract on growth performance, intestinal health, and immune response of Broiler chickens during subclinical necrotic enteritis. *Pathogens* **2025**, *14*, 260. [CrossRef]
15. Bavi, E.P.; Shakerinasab, E.; Hamidinezhad, H.; Nazifi, E. A green and facile approach for fabrication of biocompatible anti-Parkinson chitosan-gelatin-green tea extract composite particles with neuroprotective and Neurotherapeutic effects: In vitro evaluation. *Int. J. Biol. Macromol.* **2022**, *224*, 1183–1195. [CrossRef]
16. Piran, F.; Khoshkhoo, Z.; Hosseini, S.E.; Azizi, M.H. Controlling the antioxidant activity of green tea extract through encapsulation in chitosan-citrate nanogel. *J. Food Qual.* **2020**, *2020*, 7935420. [CrossRef]
17. Chuysinuan, P.; Chunshom, N.; Kotcharat, P.; Thanyacharoen, T.; Techasakul, S.; Ummartyotin, S. The encapsulation of green tea extract in cyclodextrin and loading into chitosan-based composites: Controlled-release behavior and antioxidant properties. *J. Polym. Environ.* **2021**, *29*, 2628–2638. [CrossRef]
18. Kudłacik-Kramarczyk, S.; Drabczyk, A.; Głąb, M.; Gajda, P.; Jaromin, A.; Czopek, A.; Zagórska, A.; Tyliszczak, B. Synthesis and physicochemical evaluation of bees' chitosan-based hydrogels modified with yellow tea extract. *Materials* **2021**, *14*, 3379. [CrossRef]
19. Yu, L.; Bi, J.; Song, Y.; Wang, M. Isotherm, thermodynamics, and kinetics of methyl orange adsorption onto magnetic resin of chitosan microspheres. *Int. J. Mol. Sci.* **2022**, *23*, 13839. [CrossRef]
20. Yu, L.; Song, Y.; Bi, J.; Gao, Y.; Jiang, C.; Yang, Z.; Qi, H.; Yu, H.; Yang, W.; Gong, Q.; et al. Exploring the potent hydrolytic activity of chitosan-cerium complex microspheres resin for organophosphorus pesticide degradation. *Heliyon* **2024**, *10*, e33642. [CrossRef]
21. Pérez, M.; Dominguez-López, I.; Lamuela-Raventós, R.M. The chemistry behind the Folin-Ciocalteu method for the estimation of (poly)phenol content in food: Total phenolic intake in a mediterranean dietary pattern. *J. Agric. Food Chem.* **2023**, *71*, 17543–17553. [CrossRef] [PubMed]
22. Yu, L.; Bi, J.; Song, Y.; Jiang, C.; Qi, H.; Chi, X.; Yang, W.; Shi, C.; Gong, Q.; Wang, M. Products and properties of components from heat-denatured peanut meal following solid-state fermentation by *Aspergillus oryzae* and *Saccharomyces cerevisiae*. *Fermentation* **2023**, *9*, 425. [CrossRef]
23. Zhang, Q.A.; Wang, X.; Song, Y.; Fan, X.H.; García Martín, J.F. Optimization of pyrogallol autoxidation conditions and its application in evaluation of superoxide anion radical scavenging capacity for four antioxidants. *J. AOAC Int.* **2016**, *99*, 504–511. [CrossRef] [PubMed]
24. Falcone, E.; Vigna, V.; Schueffl, H.; Stellato, F.; Vileno, B.; Bouraguba, M.; Mazzone, G.; Proux, O.; Morante, S.; Heffeter, P.; et al. When metal complexes evolve, and a minor species is the most active: The case of bis(phenanthroline)copper in the catalysis of glutathione oxidation and hydroxyl radical generation. *Angew. Chem. Int. Ed. Engl.* **2025**, *64*, e202414652. [CrossRef]
25. Porcher, A.; Guérin, V.; Montrichard, F.; Lebrec, A.; Lothier, J.; Vian, A. Ascorbate glutathione-dependent H_2O_2 scavenging is an important process in axillary bud outgrowth in rosebush. *Ann. Bot.* **2020**, *126*, 1049–1062. [CrossRef]
26. Uma, T.K.N.N.; Szewczyk, P.K.; Karbowniczek, J.E.; Polak, M.; Knapczyk-Korczak, J.; Stachewicz, U. Improving stability and mechanical strength of electrospun chitosan-polycaprolactone scaffolds using genipin cross-linking for biomedical applications. *Macromol. Rapid Commun.* **2024**, 2400869. [CrossRef]

27. Salah, M.; Huang, J.; Zhu, C.; Sobhy, M.; Farag, M.A.; Fang, Y.; Sobhy, R.; Walayat, N.; Khalifa, I.; Maqsood, S.; et al. Chitosan dual gel-like functionalized with flavonoid extract and cinnamaldehyde oil using dual cross-linking agents: Characterization, antioxidant, and antimicrobial effects. *Curr. Res. Food Sci.* **2024**, *9*, 100826. [CrossRef]
28. Yang, Y.; Deng, Z.; Chen, Z.; Li, P.; Du, B.; Li, L. Modification of Inca peanut albumin-polyphenol conjugates by chitosan through laccase catalysis: Structural, interfacial, and functional properties. *Int. J. Biol. Macromol.* **2025**, *289*, 138898. [CrossRef]
29. Miyauchi, M. Water adsorption on hydrophilic fibers and porous and deliquescent materials: Cellulose, polysaccharide, silica, inorganic salt, sugar alcohol, and amino acid. *ACS Omega* **2023**, *8*, 44212–44220. [CrossRef]
30. Çelikçi, N.; Zıba, C.A.; Tümer, M. Chitosan-Based Schiff Base Compounds: Synthesis, Chemical Characterization and Antibacterial Properties. *J. Fluoresc.* **2025**, *23*, 839–1106. [CrossRef]
31. Chang, X.; Mubarak, N.M.; Karri, R.R.; Tan, Y.; Khalid, M.; Dehghani, M.H.; Tyagi, I.; Khan, N.A. Insights into chitosan-based cellulose nanowhiskers reinforced nanocomposite material via deep eutectic solvent in green chemistry. *Environ. Res.* **2023**, *219*, 115089. [CrossRef] [PubMed]
32. Tohamy, H.S. Microwaved schiff base dialdehyde cellulose-chitosan hydrogels for sustained drug release with DFT calculations. *BMC Chem.* **2025**, *19*, 114. [CrossRef] [PubMed]
33. Mehmood, S.; Akhtar, N.; Arshad, M.; Azhar, U.; Ullah, S.; Waris, T.S.; Jabbar, F.; Hasan, A.; Iqbal, F.; Chaudhry, A.A.; et al. A novel methodology for stabilization of silver nanoparticles on cotton, nylon and cotton/nylon fabrics using chitosan and triethyl orthoformate for enhanced and elongated antibacterial performance. *Int. J. Biol. Macromol.* **2024**, *267*, 129256. [CrossRef] [PubMed]
34. Chinarak, K.; Wongnen, C.; Chaijan, M.; Tamman, A.; Donlao, N.; Cheong, L.Z. Unveiling the transformative influence of sonochemistry on formation of whey protein isolate and green tea extract (WPI-GTE) conjugates. *Ultrason. Sonochem.* **2024**, *110*, 107037. [CrossRef]
35. Ata, T.; Al-Ani, I.; Karameh, N.; Atta, M.R.; Dayyih, W.A. Alectinib-loaded chitosan-alginate nanoparticles: A novel synthesis method with in vitro and in vivo evaluations. *Pharmaceutics* **2025**, *17*, 492. [CrossRef]
36. Kadam, D.; Lele, S.S. Cross-linking effect of polyphenolic extracts of Lepidium sativum seedcake on physicochemical properties of chitosan films. *Int. J. Biol. Macromol.* **2018**, *114*, 1240–1247. [CrossRef]
37. Zhu, L.; Ouyang, F.; Fu, X.; Wang, Y.; Li, T.; Wen, M.; Zha, G.; Yang, X. Tannic acid modified keratin/sodium alginate/carboxymethyl chitosan biocomposite hydrogels with good mechanical properties and swelling behavior. *Sci. Rep.* **2024**, *14*, 12864. [CrossRef]
38. Hameed, A.R.; Majdoub, H.; Jabrail, F.H. Effects of surface morphology and type of cross-lnking of chitosan-pectin microspheres on their degree of swelling and favipiravir release behavior. *Polymers* **2023**, *15*, 3173. [CrossRef]
39. Cheng, B.; Feng, H.; Li, C.; Jia, F.; Zhang, X. The mutual effect of dietary fiber and polyphenol on gut microbiota: Implications for the metabolic and microbial modulation and associated health benefits. *Carbohydr. Polym.* **2025**, *358*, 123541. [CrossRef]
40. Wang, X.; Qi, Y.; Zheng, H. Dietary polyphenol, gut microbiota, and health benefits. *Antioxidants* **2022**, *11*, 1212. [CrossRef]
41. Wan, M.L.Y.; Co, V.A.; El-Nezami, H. Dietary polyphenol impact on gut health and microbiota. *Crit. Rev. Food Sci. Nutr.* **2021**, *61*, 690–711. [CrossRef] [PubMed]
42. Gerasopoulos, A.; Mantzouridou, F.T.; Nenadis, N. Physicochemical properties of olive leaf powders and incorporation in chitosan-based edible films for improved functionality. *Food Hydrocoll.* **2025**, *160*, 110748. [CrossRef]
43. Carrapiso, A.I.; Pimienta, M.; Martín, L.; Cardenia, V.; Andrés, A.I. Effect of a chitosan coating enriched with an olive leaf extract on the characteristics of pork burgers. *Foods* **2023**, *12*, 3757. [CrossRef]
44. Sathiyaseelan, A.; Zhang, X.; Wang, M.H. Enhancing the antioxidant, antibacterial, and wound healing effects of melaleuca alternifolia oil by microencapsulating it in chitosan-sodium alginate mcrospheres. *Nutrients* **2023**, *15*, 1319. [CrossRef]
45. Zhu, W.; Zhang, Z. Preparation and characterization of catechin-grafted chitosan with antioxidant and antidiabetic potential. *Int. J. Biol. Macromol.* **2014**, *70*, 150–155. [CrossRef]

polymers

MDPI

Article

Flufenamic Acid-Loaded Electrospun Nanofibers Based on Chitosan/Poly(vinyl alcohol) Polymeric Composites for Drug Delivery in Biomedical Applications

Kuppu Sakthi Velu [1,†], Mohammad Aslam [1,†], Ramachandran Srinivasan [2], Prathap Somu [3,*] and Sonaimuthu Mohandoss [1,4,*]

[1] School of Chemical Engineering, Yeungnam University, Gyeongsan 38541, Republic of Korea; sakthi.velu4@yu.ac.kr (K.S.V.); mohammadaslam13@yu.ac.kr (M.A.)
[2] Centre for Ocean Research, Sathyabama Research Park, Sathyabama Institute of Science and Technology, Chennai 600119, Tamil Nadu, India; srinivasan.cor@sathyabama.ac.in
[3] Department of Biotechnology and Chemical Engineering, School of Engineering, Faculty of Science, Technology and Architecture (FoSTA), Manipal University Jaipur, Dehmi Kalan, Off. Jaipur-Ajmer Expressway, Jaipur 303007, Rajasthan, India
[4] Centre of Molecular Medicine and Diagnostics, Saveetha Dental College and Hospitals, Saveetha Institute of Medical and Technical Sciences, Saveetha University, Chennai 600077, Tamil Nadu, India
* Correspondence: prathap.somu@jaipur.manipal.edu (P.S.); drsmohandoss@yu.ac.kr (S.M.)
† These authors contributed equally to this work.

Abstract: Nanostructured drug-delivery systems with enhanced therapeutic potential have gained attention in biomedical applications. Here, flufenamic acid (FFA)-loaded chitosan/poly(vinyl alcohol) (CHS/PVA; CSPA)-based electrospun nanofibers were fabricated and characterized for antibacterial, anticancer, and antioxidant activities. The FFA-loaded CSPA (FCSPA) nanofibers were characterized by scanning electron microscopy, Fourier-transform infrared spectroscopy, X-ray diffraction (XRD), and differential scanning calorimetry to evaluate their formation process, functional group interactions, and crystallinity. Notably, the average diameter of FCSPA nanofibers decreased with increasing CSPA contents (CSPA-1 to CSPA-3), indicating that FFA addition to CSPA-3 significantly decreased its diameter. Additionally, XRD confirmed the dispersion of FFA within the CSPA amorphous matrix, enhancing drug stability. FCSPA nanofibers exhibited a high swelling ratio (significantly higher than that of the CSPA samples). Biodegradation studies revealed that FCSPA exhibited accelerated weight loss after 72 h, indicating its improved degradation compared with those of other formulations. Furthermore, it exhibited a significantly high drug-encapsulation efficiency, ensuring sustained release. FCSPA nanofibers exhibited excellent antibacterial activity, inhibiting *Staphylococcus aureus* and *Escherichia coli*. Regarding anticancer activity, FCSPA decreased HCT-116 cell viability, highlighting its controlled drug-delivery potential. Moreover, FCSPA demonstrated superior antioxidation, scavenging DPPH free radicals. These findings highlight FCSPA nanofibers as multifunctional platforms with wound-healing, drug-delivery, and tissue-engineering potential.

Keywords: chitosan; poly(vinyl alcohol); electrospinning; drug-release; antibacterial; anticancer; antioxidant

1. Introduction

Electrospinning is a powerful and adaptable technology for fabricating nano-to-micrometer-scale fibers. It involves the alteration of various operational, process, and formulation parameters [1]. The fabrication of fibers through electrospinning of polymer solutions has been widely studied, with key parameters including applied voltage, tip-to-collector distance, solution feed rate, and solution characteristics [2]. The resultant fiber mats can exhibit a nonwoven and/or aligned fiber structure, depending on the type of collector used. Electrospinning has been used to fabricate nanofibers with extraordinary properties, such as extremely high surface area/volume ratios and high porosity owing to extremely small pore sizes. However, electrospun drug-carrying fibers exhibit high surface area/volume ratios, making them suitable for diffusion- and disintegration-based drug delivery [3]. Furthermore, electrospun nanofibers can encapsulate considerable amounts of drugs (up to 40%) and facilitate prolonged, multiple-phase, or rapid drug delivery [4]. Consequently, nonwoven mats comprising electrospun nanofibers have considerable potential for several uses, including biomedical applications such as drug delivery, wound dressing, and antibacterial uses [5].

Chitosan (CHS), a (1-4)-linked 2-amino-2-deoxy-D-glucopyranose, is a derivative of chitin, which is a common naturally occurring polysaccharide [6]. Because of its biocompatible, renewable, biodegradable, wound-healing, antimicrobial, and antitumor properties, CHS has attracted significant interest for its application in wound dressing, wound healing, drug delivery, and tissue engineering [7]. Therefore, CHS-based nanofibers have emerged as valuable biological materials owing to their biocompatibility, biodegradability, and antibacterial characteristics. However, the amine (NH_2) group in the backbone of CHS saccharide and the strong hydrogen bonding between NH_2 and hydroxyl (–OH) groups prevent continuous fiber formation in this material [8]. However, its polycationic behavior in solution limits the feasibility of directly synthesizing CHS through electrospinning [9]. Numerous studies have attempted to increase the electrospinning capabilities of CHS by combining it with other polymers, such as poly(vinyl pyrrolidone) [10], poly(vinyl alcohol) (PVA) [11], poly(ethylene oxide) [12], silk fibroin [13], and zein [14].

Among them, PVA is a synthetic, linear, semicrystalline polymer composed of a carbon chain and an OH group as its backbone and functional component, respectively [15,16]. PVA exhibits numerous essential characteristics, such as easy accessibility, water solubility, high film-forming capacity, and thermostability [17]. Furthermore, PVA exhibits superior fiber-forming capacity and hydrophilicity, and fibers electrospun from PVA have been commercialized since the 1950s [18]. Furthermore, PVA can improve fiber spinning by reducing the repulsive forces in charged polymer solutions. Notably, CHS can be integrated with PVA for improved electrospinning, considering their relative miscibility and comparable electrospinning behaviors, e.g., coagulation, orientation, and cross-linking [18]. Integrating PVA and CHS imparts the solution with the features of each component and with additional valuable properties, such as good electrospinnability and mechanical capabilities. Thus, CHS/PVA (CSPA) is a candidate for antimicrobial applications [19]. Additionally, electrospun CSPA nanofibers have been extensively investigated in recent years because of their biocompatibility, biodegradability, and antibacterial capabilities in tissue engineering [20–23].

Several nano-delivery systems, including nanofibers, nanoparticles, nanocapsules, nanoliposomes, and dendrimers, have been investigated for antibacterial drug delivery [24]. The encapsulation of drugs in electrospun nanofibers offers several advantages, including high surface area/volume ratios and porosities [25,26]. Compared to other nanocarriers used for antibacterial drug delivery, nanofibers demonstrate high drug-loading capacity, excellent encapsulation efficiency (EE%), minimal systemic toxicity, and enable both sustained and controlled release profiles. However, no study has experimentally attempted the electrospinning-based loading of CSPA with flufenamic acid (FFA), i.e., the electrospinning-based preparation of FFA-loaded CSPA (FCSPA). FFA, a non-steroidal anti-inflammatory drug, has garnered attention for its drug-delivery potential as well as for its antibacterial, cytotoxic, and anticancer applications [27,28]. FFA exhibits remarkable antibacterial and antibiofilm properties; it significantly interrupts biofilm formation and limits bacterial growth, making it a good alternative for fighting resistant diseases [29,30]. Furthermore, assessing the cytotoxic and anticancer characteristics of this composite structure may provide new avenues for drug-delivery systems targeting cancer cells [27,31].

In the present study, FFA-loaded CHS/PVA nanofibers were prepared by electrospinning, and the properties of CHS- and PVA-based composite nanofibers in different ratios were investigated. The prepared CSPA and FFA-loaded CSPA (FCSPA) nanofibers were characterized by scanning electron microscopy (SEM), Fourier-transform infrared (FTIR) spectroscopy, X-ray diffraction (XRD), and differential scanning calorimetry (DSC) to evaluate their surface morphologies, functional group contents, surface areas, and thermal properties, respectively. Furthermore, the effects of various CSPA ratios on the swelling ratio, in vitro biodegradation, EE%, and release behavior of FCSPA nanofibers were investigated. The antibacterial activities of CSPA and FCSPA nanofibers against gram-negative *Escherichia coli* and gram-positive *Staphylococcus aureus* were determined by colony counting. Additionally, the anticancer and antioxidant activities of CSPA and FCSPA were extensively investigated. Our findings demonstrate that FCSPA nanofibers are versatile platforms with potential for wound healing, drug delivery, and tissue engineering.

2. Materials and Methods

2.1. Materials

CHS (molecular weight [MW]: 50–190 kDa; 95% deacetylation degree) and FFA were obtained from Sigma-Aldrich (St. Louis, MO, USA). PVA (MW: 146–186 kDa, 89% deacetylation degree) was procured from Daejung Company Ltd., Korea (Busan, Republic of Korea), and deionized (DI) water was used as the solvent for the solutions. All chemicals and solvents were used without further purification. The human colon cancer cell line (HCT-116) was purchased from the Korean Cell Line Bank (Seoul, Republic of Korea).

2.2. Preparations of Chitosan/Poly(vinyl alcohol) and Flufenamic Acid-Loaded Chitosan/Poly(vinyl alcohol) Solutions

To prepare the CSPA and FCSPA polymer solutions, 0.3 g of CHS (3% w/v) was dissolved in 10 mL of 1% (v/v) acetic acid and stirred at ambient temperature for 3 h until a clear solution was obtained. Concurrently, 2 g of PVA (10% w/v) was dissolved in 20 mL of distilled water and stirred at 50 °C using a magnetic heater-stirrer for 4 h to ensure complete dissolution. Thereafter, the CHS and PVA solutions were mixed at different weight ratios (w/w): 30/70, 50/50, and 70/30 for CSPA-1, CSPA-2, and CSPA-3, respectively, followed by stirring at ambient temperature for 1 h to achieve homogeneity (Table 1). For the FCSPA solutions, the optimized CSPA sample, i.e., CSPA-3 ratio (70/30 w/w), was selected. FFA was dissolved in the mixing solution at a concentration of 25 mg in 5 mL (Table 1). Subsequently, the mixture was continuously stirred until full dissolution was achieved, ensuring uniform dispersion of all components.

Table 1. Compositions of chitosan (CHS)/poly(vinyl alcohol) (PVA), CSPA, and flufenamic acid-loaded CSPA (FCSPA) nanofiber solutions.

Sample ID	CHS (3% w/v)	PVA (10% w/v)	CHS:PVA Ratio (w/w)	FFA Content
CSPA-1	0.9 g in 30 mL	7 g in 70 mL	30:70	-
CSPA-2	0.9 g in 30 mL	7 g in 70 mL	50:50	-
CSPA-3	0.9 g in 30 mL	7 g in 70 mL	70:30	-
FCSPA	0.9 g in 30 mL	7 g in 70 mL	70:30 (CSPA-3)	25 mg in 5 mL

2.3. Electrospinning Procedure

The prepared CSPA-1, CSPA-2, CSPA-3, and FCSPA solutions were loaded into a 5 mL syringe with a 22 G needle and secured using a syringe pump. The flow rate was set to 1 mL/h, with a maintained needle-collector distance of 14 cm. Next, they were subjected to a high voltage of 30 kV, and the room humidity was controlled at 60% to facilitate uniform fiber formation (Scheme 1). Thereafter, the electrospun fibers were collected on an aluminum foil-wrapped collector, washed multiple times with distilled water to remove the residual acetic acid, and dried overnight in a desiccator at room temperature.

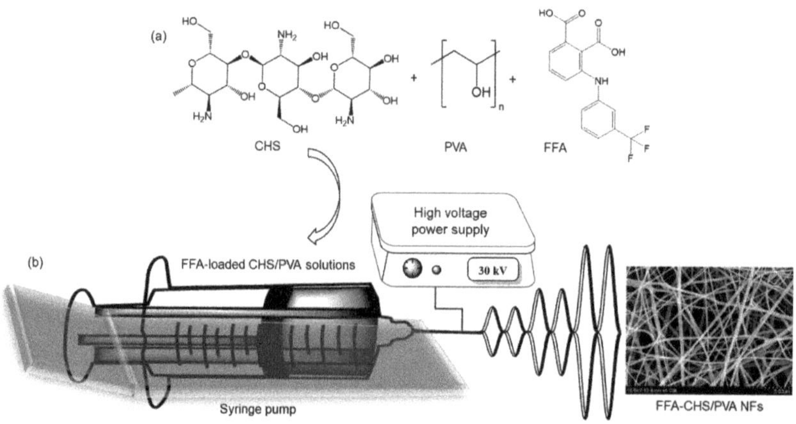

Scheme 1. (**a**) Chemical structures of CHS, PVA, and FFA. (**b**) Schematic of the prepared FCSPA nanofibers via electrospinning.

2.4. Characterizations

Before the electrospinning procedure, the shear viscosities at 100 s^{-1}; solution conductivities; and surface tensions of the CSPA-1, CSPA-2, CSPA-3, and FCSPA solutions were tested. Their viscosities were analyzed using a Brookfield LVT viscometer with a small-sample thermostated adapter, spindle, and chamber SC4-18/13R at 25 ± 0.1 °C. Further, their conductive characteristics were measured using an Orion 162 conductivity meter at room temperature, and their surface tensions were measured using the pendant drop method and a tensiometer (OCA20, Dataphysics Instruments, Filderstadt, Germany). For all in vitro assays, CSPA-1, CSPA-2, CSPA-3, and FCSPA nanofiber suspensions were prepared at a standardized concentration (10 mg) per well to ensure consistent exposure and account for the dose-dependent nature of the observed biological responses.

2.4.1. Scanning Electron Microscopy

The surface morphologies and fiber structures of the CSPA-1, CSPA-2, CSPA-3, and FCSPA electrospun nanofibers were analyzed by SEM (FESEM JSM-7600F, JEOL, Tokyo, Japan). The samples were sputter-coated with gold using an auto fine coater before imaging to enhance their conductivities. SEM imaging was performed at an accelerating voltage of 200 kV, and the fiber diameters were measured using ImageJ software (1.54j); 50 fiber diameters were randomly selected for statistical analyses.

2.4.2. Fourier-Transform Infrared Spectroscopy

FTIR spectroscopy was performed to determine the chemical functional groups in the samples and to examine the molecular interactions in the pure compounds, i.e., CSPA-3, and FCSPA nanofibers. The FTIR spectra were verified within the 4000–400 cm^{-1} range using an attenuated total reflectance module–equipped FTIR spectrometer (Perkin-Elmer, Waltham, MA, USA).

2.4.3. X-Ray Diffraction

XRD (Shimadzu Corporation, Kyoto, Japan) was performed to investigate the crystalline structures of the pure molecules, CSPA-3, and FCSPA nanofibers. The measurements were performed using a Cu–Kα radiation source at 40 kV in a 2θ range of 10–80°.

2.5. Swelling Ratio

The swelling properties of the CSPA-1, CSPA-2, CSPA-3, and FCSPA nanofibers were evaluated. To achieve this, the samples were immersed in phosphate-buffered saline (PBS) solution at pH 7.4 and room temperature. The electrospun CSPA-1, CSPA-2, CSPA-3, and FCSPA nanofibers were cut into uniform pieces (1 cm × 1 cm) to ensure consistency during testing. The initial dry weight (W_0) of each sample was recorded using a high-precision analytical balance. Subsequently, the samples were fully immersed in the PBS solution for predetermined intervals (0, 1, 2, 3, 4, 5, 6, 9, 12, 15, 18, 21, and 24 h) while ensuring complete submersion and minimal interference. At each interval, the samples were carefully removed using tweezers, and excess surface water was gently blotted using filter paper to prevent structural deformation. The swollen weight (W_1) was immediately measured and recorded.

The swelling percentage was calculated as follows:

$$\text{Swelling ratio } (\%) = [(W_1 - W_0)/W_0] \times 100, \quad (1)$$

where W_0 is the initial dry weight and W_1 is the swollen weight after immersion.

2.6. Degradation Profile

The in vitro degradation behaviors of the CSPA-1, CSPA-2, CSPA-3, and FCSPA nanofibers were evaluated by immersing them in PBS (pH 7.4) containing 0.15% (*w*/*v*) lysozyme to simulate physiological enzymatic degradation. The nanofiber mats (~10 mg) were cut into uniform pieces (1 cm × 1 cm) and placed in 10 mL PBS solution at 37 °C inside a shaking incubator set at 100 rpm. At specified intervals (24, 48, and 72 h), the samples were carefully removed, rinsed with DI water to remove the remaining salts, and dried in a vacuum at 40 °C until a stable weight was achieved.

The degradation rate was determined as follows:

$$\text{Degradation } (\%) = [(W_0 - W_t)/W_0] \times 100, \quad (2)$$

where W_0 is the initial dry weight and W_t is the remaining dry weight at each point.

2.7. Drug-Encapsulation Efficiency and Drug Release

To determine the drug EE% of the CSPA-3 electrospun nanofibers, we dissolved a determined mass of each sample in 1 mL of acetic acid. Thereafter, the FFA concentration of the solution was measured using high-performance liquid chromatography (Agilent Technology, Santa Clara, CA, USA) at a maximum wavelength (λ_{max}) of 254 nm. Subsequently, we developed a conventional calibration curve for FFA, and the EE% was computed as follows:

$$\text{EE\%} = \text{Total amount of encapsulated FFA/Theoretical amount of FFA} \times 100 \quad (3)$$

The release profile of FFA from the CSPA-3 nanofibers was evaluated in PBS (pH 7.4) at 37 °C under dynamic conditions. The electrospun FCSPA-3 nanofiber mats were cut into 1 cm × 1 cm sections and precisely weighed (~10 mg). The nanofiber mats were immersed in 50 mL PBS (pH 7.4) in sealed glass vials. These vials were incubated in a shaking water bath or an orbital shaker at 37 ± 0.5 °C with gentle agitation at 100 rpm. To maintain the sink conditions, 1 mL of the release medium was withdrawn and replaced with an equal volume of fresh PBS at predetermined intervals (0, 1, 2, 3, 4, 5, 6, 9, 12, 15, 18, 21, and 24 h). Thereafter, the obtained samples were filtered using 0.45 μm membrane filters, and the concentration of the released FFA was measured by ultraviolet–visible (UV–Vis) spectrophotometry at λ_{max} = 254 nm. The calibration curve for FFA in PBS was developed to assess the drug concentration, after which the cumulative drug release (%) was estimated.

2.8. Bacterial Culture

The antibacterial activities of the FFA, CSPA-1, CSPA-2, CSPA-3, and FCSPA nanofibers were assessed against *S. aureus* and *E. coli* using the colony-counting method. To begin, bacterial suspensions were prepared by inoculating fresh bacterial cultures in Mueller–Hinton broth (MHB) using an incubation time of 24 h at 37 °C until they reached a concentration of 10^6 CFU/mL. The electrospun nanofiber samples (3.5 cm^2) were sterilized under UV light for 30 min before immersion in separate tubes containing 10 mL of bacteria-inoculated MHB. The tubes were incubated at 37 °C with continuous shaking at 100 rpm, and at predetermined intervals (0, 2, 4, 6, 12, and 24 h), 100 μL aliquots of the bacterial suspension were withdrawn, serially diluted in sterile PBS, and spread onto Mueller–Hinton agar plates. After 24 h of incubation at 37 °C, the colony-forming units were counted and compared with those of the control group (ciprofloxacin and vancomycin), consisting of bacteria without nanofiber treatment. The percentage reduction in bacterial viability was subsequently calculated as follows:

$$\text{Bacterial reduction (\%)} = [(\text{CFU\_control} - \text{CFU\_treated})/\text{CFU\_control}] \times 100 \quad (4)$$

The bacterial cultures (initial optical density at 600 nm [OD_{600}]: ~0.05) were treated with FFA-loaded nanofibers, free FFA, or a standard antibiotic (vancomycin and ciprofloxacin) in 96-well plates. The cultures were incubated with shaking at 37 °C, and the OD_{600} was measured at 0, 4, 8, 12, and 24 h using a microplate reader. All treatments were performed in triplicate to evaluate the time-dependent antibacterial effects.

2.9. Cytotoxicity Assay

The cytotoxicities of the FFA, CSPA-1, CSPA-2, CSPA-3, and FCSPA nanofibers were assessed on HCT-116 cell lines using the 3-[4,5-dimethylthiazol-2-yl]-2,5 2,5-diphenyl tetrazolium bromide (MTT) assay. The HCT-116 cells were cultured in Dulbecco's modified Eagle's medium (DMEM) supplemented with 10% fetal bovine serum, 100 U/mL penicillin, and 100 μg/mL streptomycin and maintained at 37 °C in an incubator containing 5% CO_2. Electrospun nanofiber samples were cut into 6 mm diameter discs, sterilized under UV light for 30 min, and placed in 96-well culture plates. The cells were seeded at a density of 5×10^3 cells per well and incubated for 24 h to facilitate cell attachment. Following incubation, the culture medium was replaced with fresh medium containing FFA, CSPA-1, CSPA-2, CSPA-3, and FCSPA nanofibers, which were prepared via immersion in DMEM for 48 h at 37 °C. Next, 20 μL of MTT solution (5 mg/mL in PBS) was added to each well, followed by incubation for 4 h. Formazan crystals were immersed in 150 μL of dimethyl sulfoxide, and the absorbance was measured at 570 nm using a microplate reader. Thereafter, the cell viability was estimated using the following equation:

$$\text{Cell viability (\%)} = (\text{Absorbance of test sample}/\text{Absorbance of control}) \times 100 \quad (5)$$

2.10. Cell Imaging

To investigate the cell adhesion and morphology of the samples, HCT-116 cells were seeded onto a control as well as onto FFA, CSPA-1, CSPA-2, CSPA-3, and FCSPA nanofiber scaffolds at a density of 5×10^3 cells per well in 24-well plates. Next, the cells were incubated at 37 °C in a 5% CO_2 environment for 48 h to facilitate attachment and proliferation. Subsequently, they were fixed with 4% paraformaldehyde for 15 min, washed three times with PBS, and permeated with 0.1% Triton X-100 for 5 min. Next, the nuclei were counterstained with 4′,6-diamidino-2-phenylindole (DAPI) and propidium iodide (PI) for 10 min. Following staining, the samples were rinsed with PBS and observed under a fluorescence microscope to assess cell morphology.

2.11. Antioxidant Activity

The antioxidant activities of free FFA, CSPA-1, CSPA-2, CSPA-3, and FCSPA nanofibers were determined using a 2,2-diphenyl-1-picrylhydrazyl (DPPH) free radical scavenging assay. Each 10 mg sample was separately immersed in 1 mL of DPPH ethanolic solution (10^{-4} mol/L) and incubated in the dark at room temperature for varied durations (6, 12, 18, and 24 h). Additionally, a DPPH solution containing FFA was used as the control. The percentage of DPPH-scavenging activity was calculated as follows:

$$\text{DPPH-scavenging activity (\%)} = A_0 - A_i/A_0 \times 100 \tag{6}$$

where A_0 is the absorbance of the control solution (DPPH) and A_i is the absorbance of the solution containing the corresponding free FFA, CSPA-1, CSPA-2, CSPA-3, and FCSPA nanofibers.

2.12. Statistical Analysis

The studies were performed in triplicate, and the data were reported as mean ± standard deviation. SPSS version 27.0 was used to perform a one-way analysis of variance, followed by Tukey's test. A p-value < 0.05 was considered statistically significant.

3. Results and Discussion

3.1. Nanofiber Morphologies

The surface morphologies and physical properties of the electrospun scaffolds were analyzed by SEM. Figure 1 shows the effects of varying the CSPA ratios (30/70 [CSPA-1], 50/50 [CSPA-2], and 70/30 [CSPA-3]), along with the FFA-loaded CSPA-3 fiber mats, designated as FCSPA. The SEM images revealed that the electrospun CSPA-1, CSPA-2, and CSPA-3 nanofibers were bead-free, continuous, and randomly oriented [32]. A comparative analysis of Figure 1a,c,e demonstrates that increasing the CHS content caused a substantial increase in the average diameter of the nanofiber, induced noticeable morphological alterations, and caused an increase in the random-alignment degree [32]. Specifically, at a CSPA ratio of 70/30 (w/w), the nanofibers appeared to be smooth and without bead defects (Figure 1e), making this composition optimal for further investigation. Among the electrospun nanofibers, the FCSPA variant exhibited reduced average fiber diameter with increasing CSPA content (Figure 1g) [33]. Notably, FFA incorporation caused a significant decrease in the fiber diameter as the size of the CSPA-3 nanofiber decreased from 347 ± 61 nm to 81 ± 27 nm, following FFA-drug loading (Figure 1i–l) [34]. The variations in the CSPA content affected the fiber diameter and indirectly affected the FFA-loading efficiency of the CSPA-3 nanofibers. Notably, CHS addition increased the average fiber diameter, as observed with the increasing CHS contents of the CSPA-1, CSPA-2, and CSPA-3 formulations. In contrast, the presence of FFA in the CSPA-3 formulation (FCSPA) caused a reduction in the average fiber diameter, likely due to the enhanced surface charge and electrostatic repulsion, which facilitated the formation of finer nanofibers.

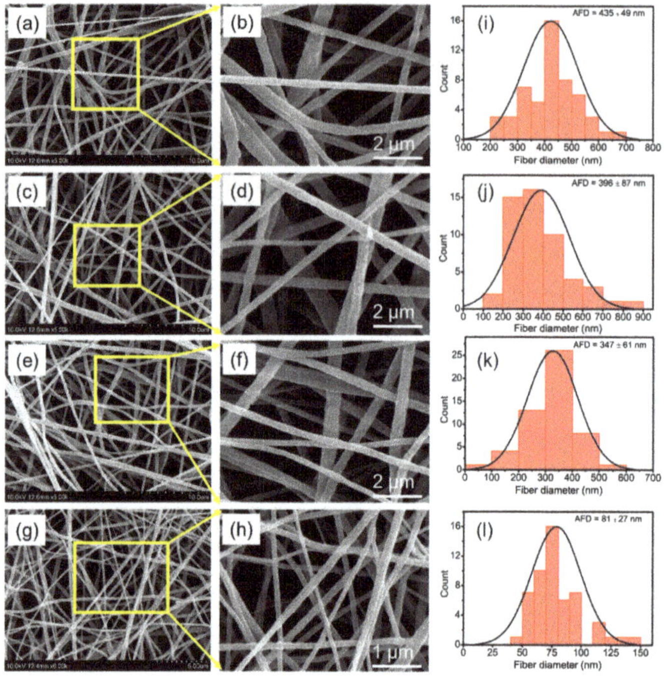

Figure 1. Field-emission scanning electron microscopy (FESEM) micrographs of (**a**,**b**) CSPA-1, (**c**,**d**) CSPA-2, (**e**,**f**) CSPA-3, and (**g**,**h**) FCSPA nanofibers. (**i**–**l**) Distributions of the diameters of the CSPA-1, CSPA-2, CSPA-3, and FCSPA nanofibers.

3.2. Surface Properties of Nanofibers

The FTIR spectra of CHS, PVA, CSPA (*w*/*w*: 70/30), CSPA-3 nanofibers, FFA, FFA-loaded CSPA (*w*/*w*: 70/30), and FCSPA nanofibers are shown in Figure 2. In the FTIR spectrum of CHS, the peaks at 1649 and 1547 cm^{-1} corresponded to the C=O vibration and N–H bending, respectively (Figure 2a). The band at 1380 cm^{-1} was associated with CH_2 deformation, and the absorption band at 1077 cm^{-1} corresponded to the stretching of the C–O–C bond [15]. The FTIR spectrum of PVA showed a broad O–H stretching band at 3200–3600 cm^{-1}, indicating the presence of –OH groups and hydrogen bonding, whereas C–H stretching occurred at 2800–3000 cm^{-1}, indicating methylene (–CH_2) vibrations (Figure 2b). The C=O stretching at 1659 cm^{-1} corresponded to residual acetate groups, and the O–H bending at 1592 cm^{-1} was related to the absorbed water [16]. Furthermore, C–H bending was observed at 1400–1500 cm^{-1}, and C–O stretching was observed around 1141 cm^{-1} and was associated with the polymer crystallinity. The FTIR spectrum of the CSPA-3 nanofibers showed broad O–H and N–H stretching peaks at 3278–3326 cm^{-1}, indicating hydrogen bonding between –OH and NH_2 (Figure 2c). Furthermore, C–H stretching was observed at 2800–2950 cm^{-1}, corresponding to –CH_2 vibrations, whereas the C=O stretching band at 1640–1690 cm^{-1} was attributed to the presence of polymer interactions or residual acetate groups [21]. The C–O–C and C–N stretching between 1085 and 1144 cm^{-1} reflected the polysaccharide structures of CHS and PVA.

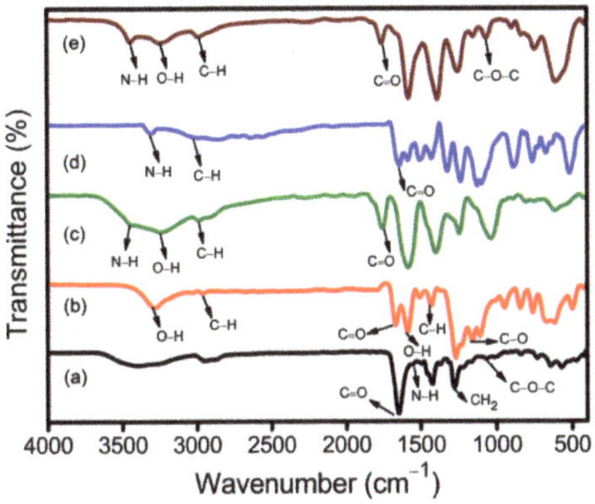

Figure 2. Fourier-transform infrared (FTIR) spectra of (a) CHS, (b) PVA, (c) CSPA-3 nanofibers, (d) FFA, and (e) FCSPA nanofibers.

A broad N–H stretching band was observed in the FTIR spectrum of FFA at 3321 cm^{-1}, and it corresponded to the presence of a secondary NH_2 group. The C=O stretching vibration at 1652 cm^{-1} corresponded to the carbonyl group, whereas the C–H stretching between 2800 and 3100 cm^{-1} corresponded to the aromatic and aliphatic C–H vibrations (Figure 2d). The C=C stretching at 1500–1600 cm^{-1} represented vibrations of the aromatic rings, and the C–F stretching between 1100 and 1400 cm^{-1} was attributed to the presence of a trifluoromethyl group [35]. The FTIR spectrum of FCSPA nanofibers exhibited broad O–H and N–H stretching bands at 3200–3500 cm^{-1}, highlighting hydrogen bonding from -OH and NH_2 (Figure 2e). The C=O stretching peak at 1650 cm^{-1} indicated FFA incorporation into the CSPA-3 matrix. The C–H stretching vibrations at 2800–3000 cm^{-1} were attributed to the CH_2 groups, and the C–O–C and C–N stretching bands between 1085 and 1145 cm^{-1} indicated the formation of ether and NH_2 bonds, representing the FCSPA nanofibers. The incorporation of FFA largely shifted the peak positions with changes in intensity, highlighting the interactions between the FFA drug and the CSPA-3 matrix. These spectral characteristics confirmed the integration of FFA into the CSPA-3 nanofiber matrix.

The XRD patterns of the crystalline and amorphous structures of pure CHS, PVA, CSPA (*w*/*w*; 70/30), CSPA-3 nanofibers, FFA, FFA-loaded CSPA (*w*/*w*; 70/30), and FCSPA nanofibers were analyzed (Figure 3). The XRD patterns of CHS and PVA exhibited broad, diffuse peaks. CHS exhibited a distinct peak at approximately $2\theta = 20.3°$ (Figure 3a), whereas PVA exhibited a peak around $2\theta = 19.2°$ (Figure 3b), characteristic of an amorphous phase [32,33]. In contrast, the XRD spectrum of the electrospun CSPA (*w*/*w*; 70/30) nanofibers (Figure 3c) exhibited decreased intensity for the peaks associated with CHS and PVA, indicating the formation of intermolecular and intramolecular hydrogen bonds between CHS and PVA during electrospinning [32,33]. The XRD pattern of FFA revealed sharp, well-defined peaks at 2θ of 13.4°, 14.6°, 17.8°, 18.3°, 19.2°, 22.7°, 24.3°, 26.1°, 30.1°, and 31.0°, indicating its crystalline nature (Figure 3d) [36]. However, the XRD pattern of the FFA-loaded CSPA (*w*/*w*: 70/30) nanofibers did not show the sharp crystalline peaks of FFA. The peaks were replaced by broad peaks. This change in the diffraction pattern indicates disruption of the crystalline structure of FFA owing to the dissolution of FFA in water and its blending with the CSPA matrix during electrospinning (Figure 3e). Consequently, the absence of distinct diffraction peaks for FFA may be attributed to its low concentration in the composite fibers, which probably falls below the detection threshold of the instrument. Additionally, FFA may be present in amorphous or molecularly dispersed states within the CSPA matrix, further contributing to the lack of identifiable crystalline features [37].

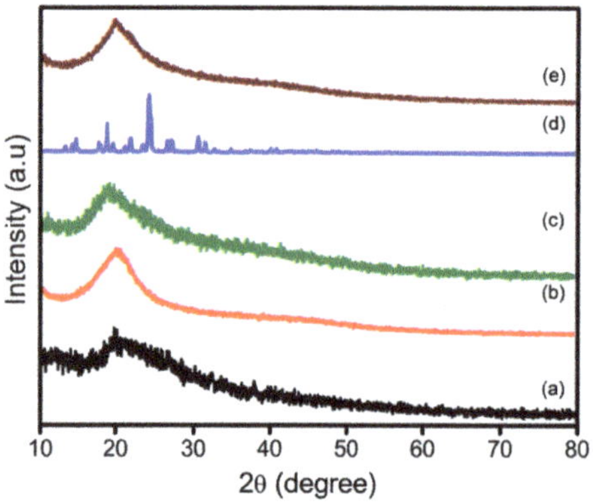

Figure 3. X-ray diffraction (XRD) patterns of (a) CHS, (b) PVA, (c) CSPA-3 nanofibers, (d) FFA, and (e) FCSPA nanofibers.

3.3. Thermal Properties of Nanofibers

Figure 4 depicts the DSC analyses of CHS, PVA, CSPA (*w/w*; 70/30), CSPA-3 nanofibers, FFA, FFA-loaded CSPA (*w/w*; 70/30), and FCSPA nanofibers. The DSC study of pure CHS and PVA revealed a rather large and highly endothermic curve with peaks at 196 °C and 192 °C (Figure 4) [20,22]. Moreover, the CSPA-3 nanofiber exhibited a peak shift toward lower temperatures, resulting in the development of a broad endothermic curve at 164 °C. The melting-point reduction of PVA in the blends indicated that there was slight mixing between PVA and CHS [22,23]. Furthermore, the decrease in the endothermic heat was attributable to the presence of amorphous CHS, which damaged the crystalline structure of PVA in the polymer blends. The pure FFA drug exhibited a pronounced endothermic peak at 160 °C, indicating crystal melting. Similarly, the peaks of the endothermic curve of FCSPA nanofibers migrated to higher temperatures at 212 °C, following the addition of FFA-loaded CSPA-3 nanofibers. Upon the addition of FFA, the endothermic curves of the FCSPA nanofibers became less noticeable and eventually disappeared. These results indicated substantial interactions between the functional groups of CSPA and FFA. The DSC results exhibited good agreement with the XRD pattern, indicating consistent thermal and structural properties of the material.

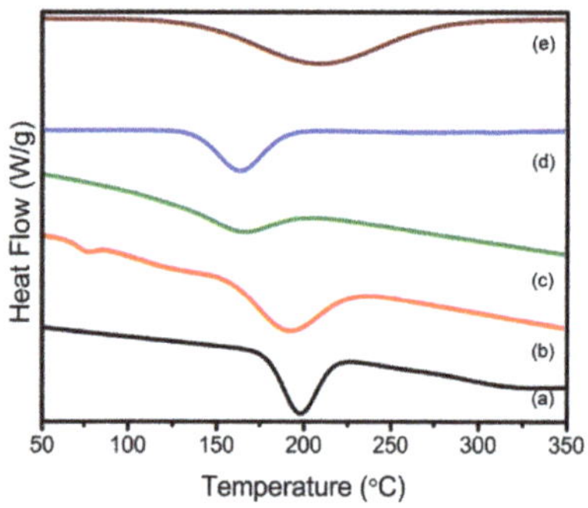

Figure 4. Differential scanning calorimetry (DSC) analyses of the (a) CHS, (b) PVA, (c) CSPA-3 nanofibers, (d) FFA, and (e) FCSPA nanofibers.

3.4. Swelling Evaluation

The water-absorption capacities of the scaffolds were evaluated by measuring their swelling in a PBS solution. The results demonstrated that all nanofibers facilitated effective water uptake, which increased with prolonged exposure to the aqueous environment. To determine the swelling percentages of the CSPA-1, CSPA-2, CSPA-3, and FCSPA nanofiber mats (Figure 5a), we determined their equilibrium swelling ratios at 24 h, obtaining approximately 169 ± 8.4%, 187 ± 6.3%, 245 ± 12.2%, and 302 ± 15.1%, respectively. As shown in Figure 5a, differences were observed in the swelling behaviors of the scaffolds, with the FCSPA nanofiber mats exhibiting the highest water uptake compared with those of CSPA-1, CSPA-2, and CSPA-3 [38]. The increase in the swelling percentage may be attributed to the enhancement of the –OH groups in the FCSPA nanofibers during blending. Additionally, the higher swelling capacity of the FCSPA nanofibers indicated that the presence of FFA in the scaffolds increased their surface area, making them more suitable for cell adhesion and infiltration. Overall, the FFA-loaded CSPA-3 nanofibers exhibited high absorption capacities (>300%), maintaining a stable swollen state [39]. This property is beneficial for preserving moisture conditions, making them promising for wound-healing applications as well as ensuring controlled and sustained drug release in drug-delivery systems. The SEM graphs (Figure 5b,c) show that the FCSPA nanofibers maintained their integrity after being immersed in PBS for 24 h, thereby indicating their stability in water solution.

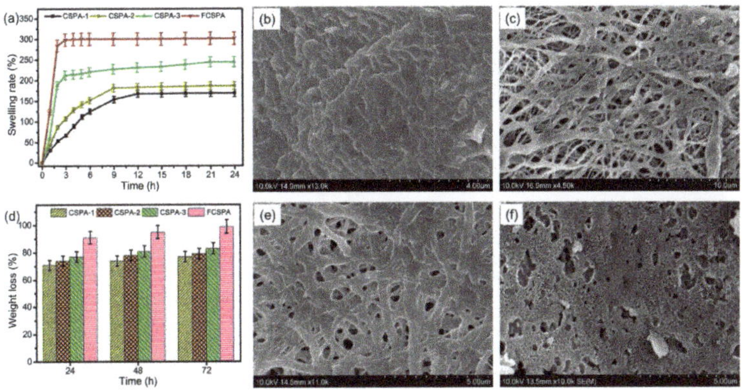

Figure 5. (**a**) Swelling ratios of CSPA-1, CSPA-2, CSPA-3, and FCSPA nanofibers at different incubation times. Scanning electron microscopy (SEM) images of the degrees of swelling of the FCSPA nanofibers (**b**) before and (**c**) after immersion in PBS for 24 h. (**d**) In vitro biodegradation of CSPA-1, CSPA-2, CSPA-3, and FCSPA nanofibers at different incubation times. SEM images of in vitro degradation of the FCSPA nanofibers (**e**) before and (**f**) after 72 h of immersion in PBS (n = 3; $p < 0.05$).

3.5. In Vitro Degradation

Biodegradation is a key factor in determining the suitability of nanofibers for biomedical applications. The CSPA-1, CSPA-2, CSPA-3, and FCSPA nanofiber mats exhibited progressive weight loss over time (Figure 5d). Notably, the degradation rate of the FCSPA nanofiber mat was higher than that of CSPA-1, CSPA-2, and CSPA-3. This may be due to the incorporation of FFA into the CSPA-3 formulation, which disrupts polymer chain entanglement and accelerates the degradation process [40]. The highest degradation rate was observed in the FCSPA nanofibers, likely due to the increased content of the highly hydrophilic polymer PVA. After 72 h, the degradation rates for CSPA-1, CSPA-2, and CSPA-3 were 77.2 ± 2.38%, 79.0 ± 3.07%, and 83.6 ± 3.19%, respectively. In contrast, the FCSPA scaffolds exhibited significantly higher degradation rates (98.12 ± 3.31%), highlighting their excellent biodegradability [40]. The degradation behaviors of the FCSPA nanofibers before and after PBS immersion for 24 h were investigated by SEM. In Figure 5e, the morphology of the FCSPA nanofibers appeared to have degraded; however, their fibrous structure was still visible before 72 h. In contrast, Figure 5f shows that after 72 h, the surface leveled off, with some pores present on the surface. This change was attributed to the excellent water solubility of PVA in the blend, which ensured moisture retention and facilitated the biodegradation of the FCSPA nanofibers.

3.6. Loading Efficiency and In Vitro Drug Release

The absence of thermal reactions during electrospinning is advantageous for the encapsulation of thermolabile bioactive substances. As illustrated in Figure 6a, the EE% of FFA in CSPA-1, CSPA-2, and CSPA-3 nanofibers were 77.3 ± 1.86%, 83.4 ± 2.31%, and 91.2 ± 0.95%, respectively. In all FFA-loaded nanofibers, the EE% exceeded 75%, demonstrating the effective dispersion of FFA within the CSPA solution, where it remained stably encapsulated after electrospinning [41]. A notable increase in EE% was observed when the CSPA ratio was adjusted to 70/30 (*w*/*w*). This enhancement was attributed to the increased CHS concentration, which facilitated the formation of additional hydrogen bonds between CHS and PVA. Consequently, this interaction likely weakened the binding capacity of FFA with the CSPA matrix, resulting in an enhanced EE%.

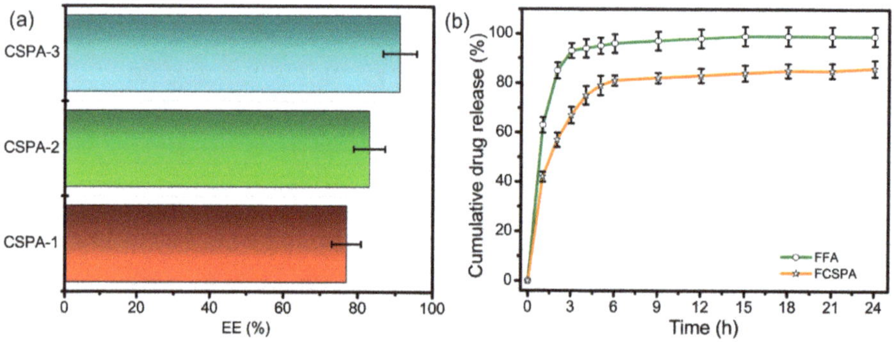

Figure 6. (**a**) Drug-encapsulation efficiencies (EE%) of the CSPA-1, CSPA-2, and CSPA-3 nanofibers. (**b**) Drug-release profiles of the FCSPA nanofibers in PBS (pH 7.4) at various intervals (n = 3; $p < 0.05$).

UV–Vis spectroscopy was performed to analyze the release profile of FFA from the CSPA-3 nanofiber mats, providing insights into the structure–function relationship of the electrospun FCSPA nanofibers in PBS within 24 h (Figure 6b). The release behavior of free FFA was used as a control. After 24 h, the release rate of FFA from the FCSPA nanofibers reached 93.27 ± 2.31%, whereas free FFA exhibited a release rate of 91.34 ± 0.98% within 9 h [41]. Additionally, free FFA demonstrated a rapid release of 96.1 ± 1.20% within the first 7 h, which was significantly higher than that of FCSPA nanofibers, which released 88.14 ± 1.37% of the loaded FFA drug. Within the first 3 h, only 68.4% of the drug was released from the FCSPA nanofibers, a considerably lower rate than that of pure FFA. Moreover, the higher water-uptake capacity of the FCSPA nanofibers may have contributed to the accelerated release of FFA. Burst release, a phenomenon frequently observed in various drug-delivery systems, was evident. This effect can benefit specific applications, such as wound treatment and targeted drug delivery, where an initial high drug concentration is beneficial.

3.7. Antibacterial Activity

The antibacterial properties of FFA, CSPA-1, CSPA-2, CSPA-3, and FCSPA nanofibers against *S. aureus* (gram-positive) and *E. coli* (gram-negative) were evaluated using the colony-counting method (Figure 7a). Among them, FCSPA demonstrated a significantly higher inhibition rate than the blank CSPA-1, CSPA-2, and CSPA-3 nanofibers, highlighting the strong antibacterial performance of the FFA drug [42]. As shown in Figure 7b,c, the antibacterial efficacy of the FFA, CSPA-1, CSPA-2, CSPA-3, and FCSPA nanofibers ranged from 39.2% to 98.5% and 39.8% to 96.7% against *S. aureus* and *E. coli*, respectively. The inhibition rate of the FCSPA nanofibers reached 98.5 ± 3.89% against *S. aureus* and 96.7 ± 2.17% against *E. coli*, which were significantly higher than those of FFA (39.2 ± 1.75% and 39.8 ± 2.09%), CSPA-1 (54.3 ± 1.21% and 49.8 ± 2.39%), CSPA-2 (68.7 ± 2.18% and 63.6 ± 1.99%), and CSPA-3 (79.7 ± 2.06% and 76.5 ± 1.28%), respectively (Figure 7b,c) [10]. Interestingly, the antibacterial activity was more pronounced against *S. aureus* than against *E. coli*, which may be attributed to structural differences in the cell walls of gram-positive and gram-negative bacteria [23,33]. These findings suggest that the incorporation of FFA into CSPA-3 nanofibers significantly enhances their antimicrobial properties against both bacterial strains. Bacterial growth kinetics revealed that the standard antibiotics (vancomycin and ciprofloxacin) maintained the lowest OD_{600} values (0.05–0.1), indicating strong inhibition. FCSPA nanofibers showed moderate suppression with OD_{600} reaching 0.25 at 24 h, whereas free FFA exhibited the least effect with OD_{600} increasing to 0.89 (Figure 7d,e). These results highlight the enhanced and sustained antibacterial activity of the FCSPA nanofiber–based delivery system compared with that of free FFA.

3.8. Anticancer Analysis

Figure 8a shows the percentage of viable cells over 24 and 48 h of incubation. Compared with the control, the nanofibers demonstrated increased cell viability, with FFA obtaining 64.93 ± 2.14% and 68.69 ± 3.18%, CSPA-1 showing 61.37 ± 1.29% and 59.81 ± 2.47%, CSPA-2 at 56.31 ± 0.98% and 53.29 ± 2.07%, and CSPA-3 at 68.19 ± 2.62% and 71.25 ± 1.93% viability over the 24 and 48 h incubation periods, respectively [14]. Furthermore, the FCSPA nanofiber scaffolds exhibited significantly enhanced cell viability, reaching 73.02 ± 2.17% and 74.21 ± 1.67% at 24 and 48 h, respectively. However, a slight reduction in cell viability was observed in the FCSPA nanofibers, suggesting that FFA incorporation influenced the biocompatibility of the CSPA-3 nanofiber mats. This may be attributed to the decreased density of amino groups due to FFA grafting onto the CSPA backbone [37]. FFA appears to create a more conducive environment for cell attachment, differentiation, and proliferation owing to its inherent biocompatibility and biodegradability. These properties make it a promising candidate for biomedical applications, particularly in tissue engineering and regenerative medicine. Figure 8b illustrates the cellular behavior of HCT-116 cells cultured on the nanofiber scaffolds after 48 h of incubation. Cellular internalization was assessed using DAPI and PI staining, and the resulting images were captured using confocal laser scanning microscopy. The fluorescence images demonstrated that, compared with the control, the nanofiber scaffolds supported significantly higher cellular activity [43]. A greater number of HCT-116 cells adhered to and spread across the surfaces of FFA, CSPA-1, CSPA-2, CSPA-3, and FCSPA nanofibers, indicating enhanced cell–substrate interactions [44].

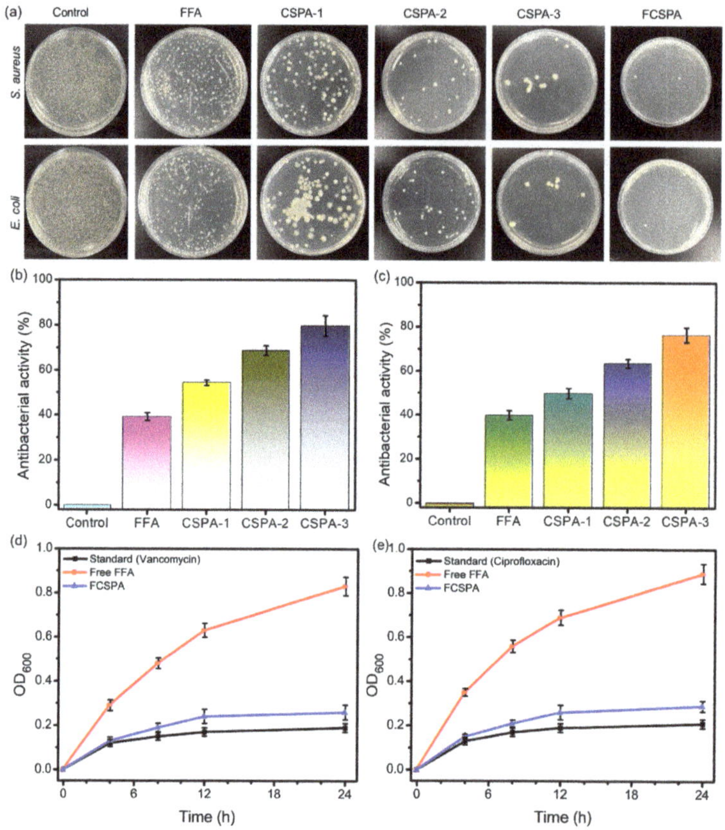

Figure 7. (**a**) Antimicrobial activities of FFA, CSPA-1, CSPA-2, CSPA-3, and FCSPA nanofibers against *S. aureus* and *E. coli*. Inhibition rates of FFA, CSPA-1, CSPA-2, CSPA-3, and FCSPA nanofibers against (**b**) *S. aureus* and (**c**) *E. coli*. (**d**,**e**) Bacterial growth kinetics (OD_{600}) over time for FCSPA nanofibers, free FFA, and a standard antibiotic (n = 3; $p < 0.05$).

3.9. Antioxidant Analysis

The radical scavenging activity of newly fabricated materials is typically assessed using the DPPH radical entrapment method to evaluate their antioxidant properties in biological systems. The DPPH free radical scavenging activities of free FFA, CSPA-1, CSPA-2, CSPA-3, and FCSPA nanofibers exhibited time-dependent responses, with the scavenging rate increasing over time (6, 12, 18, and 24 h), indicating the sustained release of FFA from the CSPA (FCSPA) matrixes (Figure 9). Notably, the antioxidant activity of the FFA-loaded CSPA-3 nanofiber formulation (86.21 $\pm$ 2.36%) was significantly higher than that of the free FFA (48.5 $\pm$ 1.92%), CSPA-1 (39.8 $\pm$ 1.24%), CSPA-2 (61.7 $\pm$ 0.99%), and CSPA-3 (73.14 $\pm$ 3.17%) at 24 h. This enhancement can be attributed primarily to the antioxidant properties of CHS and the incorporation of FFA into the nanofiber structure [40,45]. The increased scavenging activity of FCSPA nanofibers underscores the significant contribution of FFA to its antioxidant performance. The notable antioxidant activity of the FFA-loaded CSPA-3 nanofibers was likely attributable to the high concentration of antioxidants present in the FFA. As the immersion time in the DPPH solution increased, more FFA was released from the nanofibers, resulting in a corresponding increase in the DPPH free radical scavenging activity [10,40,45]. The antioxidant activity of FFA is attributed to the presence of an –OH group at the carbon position. These findings indicate that the incorporation of

FFA into the CSPA-3 nanofiber substantially enhances its antioxidant capacity, making it a promising candidate for biomedical applications requiring sustained antioxidant effects.

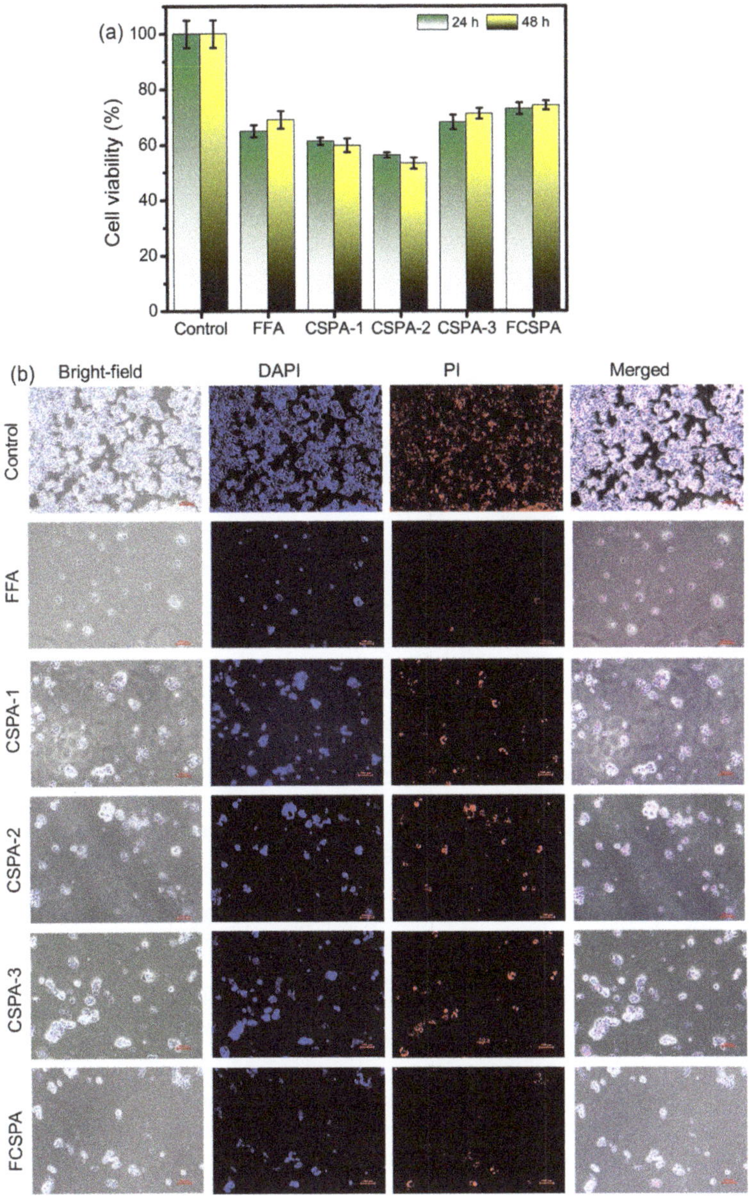

Figure 8. (**a**) MTT assay for the HCT-116 cancer cells of control as well as the FFA, CSPA-1, CSPA-2, CSPA-3, and FCSPA nanofibers after 24 and 48 h. (**b**) HCT-116 cancer cells treated with control as well as FFA, CSPA-1, CSPA-2, CSPA-3, and FCSPA nanofibers using bright-field, fluorescence–DAPI, PI staining, and merged images after 48 h. Scale bar: ~100 μm. (n = 3; $p < 0.05$).

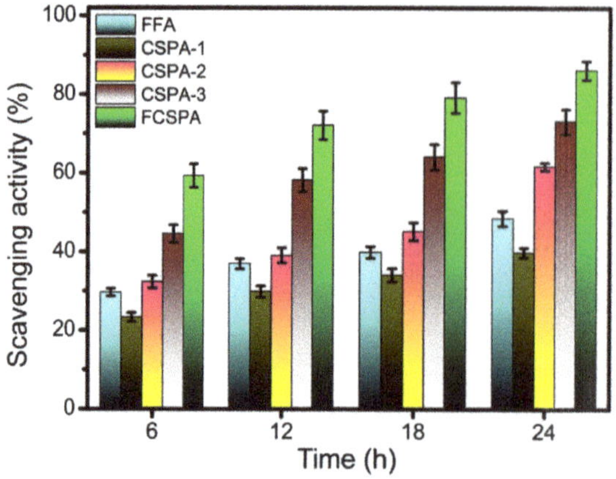

Figure 9. Radical scavenging activities of the free FFA, CSPA-1, CSPA-2, CSPA-3, and FCSPA nanofibers at different intervals (6, 12, 18, and 24 h) (n = 3; $p < 0.05$).

As presented in Table 2, the FFA-loaded CSPA nanofibers developed in this study demonstrated superior EE (91.2%) compared to other bioactive compound–loaded nanofiber systems reported previously [46–49]. Moreover, the sustained release of FFA over 72 h significantly exceeds the release durations typically observed for compounds like quercetin, ibuprofen, and resveratrol, which commonly exhibit burst or moderate-duration release profiles. The pronounced antioxidant activity and morphological integrity of the FCSPA nanofibers further highlight the effectiveness of the CSPA matrix in stabilizing and delivering FFA. These findings underscore the enhanced performance and potential applicability of the FFA nanofiber system in controlled drug delivery and bioactive wound-healing platforms.

Table 2. Comparative performances of FCSPA nanofibers and other reported bioactive compound–encapsulating nanofibers.

Drug	Polymer Matrix	EE (%)	Release Duration (h)	Bioactivity	Ref.
FFA	CSPA	91.2	72	Antioxidant	This study
Curcumin	PCL/PEG	70	48	Anti-inflammatory	[46]
Quercetin	PVA/CHS	78	12	Antioxidant	[47]
Ibuprofen	PLA	65–75	8–24	Anti-inflammatory	[48]
Resveratrol	Gelatin	80–90	48	Antioxidant	[49]

4. Conclusions

In the present study, we encapsulated FFA in CSPA nanofibers via electrospinning, yielding FCSPA. The morphologies, chemical compositions, surface characteristics, and thermal properties of the prepared CSPA and FCSPA nanofibers were examined by SEM, FTIR, XRD, and DSC, respectively. The FESEM images of the CSPA and FCSPA nanofibers exhibited a narrow distribution, uniform structures, and a smooth morphology. The FCSPA nanofibers exhibited superior swelling capacity (302 ± 15.1% at 24 h), accelerated biodegradation (98.12 ± 3.31% at 72 h), and high EE% (91.2 ± 0.95%). The controlled drug-release experiments indicated the stability of FFA drugs that were released within 24 h (93.27 ± 2.31%). The FCSPA nanofibers inhibited *E. coli* (96.7 ± 2.17%) and *S. aureus* (98.5 ± 3.89%), and their antibacterial effects against gram-positive bacteria were better than those against gram-negative bacteria. Furthermore, their cell viabilities and anticancer activities indicated that the FCSPA nanofibers exhibited low toxicity and good anticancer activity against the HCT-116 cell line. The FCSPA nanofibers delivered the highest antioxidant performance among the tested samples, achieving an 86.21 ± 2.36% DPPH-scavenging activity in 24 h. These results highlight the fabricated FCSPA nanofiber as a promising candidate for wound-healing, drug-delivery, and tissue-engineering applications. However, because we have not comprehensively validated the aforementioned biomedical applications of our FCSPA nanofibers, we anticipate that further in vivo studies will focus on validating their clinical potential.

Author Contributions: Conceptualization, K.S.V. and M.A.; methodology, K.S.V., M.A., and R.S.; software, K.S.V., M.A., S.M., and P.S.; validation, K.S.V., M.A., S.M., and P.S.; formal analysis, K.S.V., M.A. and R.S.; investigation, K.S.V., M.A., S.M., and P.S.; resources, K.S.V., M.A., and R.S.; data curation, K.S.V., M.A., and R.S.; writing—original draft preparation, K.S.V. and M.A.; writing—review and editing, S.M. and P.S.; visualization, K.S.V., R.S., M.A., S.M., and P.S.; supervision, S.M. and P.S. All authors have read and agreed to the published version of the manuscript.

Funding: This research received no external funding.

Institutional Review Board Statement: Not applicable.

Data Availability Statement: The original contributions presented in this study are included in the article. Further inquiries can be directed to the corresponding authors.

Conflicts of Interest: The authors declare no conflict of interest.

Abbreviations

FFA	Flufenamic acid
CHS	Chitosan
PVA	Poly(vinyl alcohol)
CSPA	CHS/PVA
FCSPA	FFA-loaded CHS/PVA
SEM	Scanning electron microscopy
FTIR	Fourier-transform infrared
XRD	X-ray diffraction
DSC	Differential scanning calorimetry
FESEM	Field-emission scanning electron microscopy
E. coli	*Escherichia coli*
S. aureus	*Staphylococcus aureus*
DPPH	2,2-Diphenyl-1-picrylhydrazyl
MW	Molecular weight
DI	Deionized
NF	Nanofiber
PBS	Phosphate-buffered saline
W_0	Initial dry weight
W_1	Swollen weight
EE%	Drug-encapsulation efficiency
MHB	Mueller–Hinton broth
CFU	Colony-forming units
OD_{600}	Optical density at 600 nm
MTT	3-[4,5-dimethylthiazol-2-yl]-2,5 diphenyl tetrazolium bromide
DMEM	Dulbecco's modified Eagle's medium
DAPI	4′,6-Diamidino-2-phenylindole
PI	Propidium iodide

References

1. Shabani, A.; Al, G.A.; Berri, N.; Castro-Dominguez, B.; Leese, H.S.; Martinez-Hernandez, U. Electrospinning Technology, Machine Learning, and Control Approaches: A Review. *Adv. Eng. Mater.* **2025**, *27*, 2401353. [CrossRef]
2. Al-Abduljabbar, A.; Farooq, I. Electrospun Polymer Nanofibers: Processing, Properties, and Applications. *Polymers* **2023**, *15*, 65. [CrossRef] [PubMed]
3. Ahmadi Bonakdar, M.; Rodrigue, D. Electrospinning: Processes, Structures, and Materials. *Macromol* **2024**, *4*, 58–103. [CrossRef]
4. Emerine, R.; Chou, S.-F. Fast Delivery of Melatonin from Electrospun Blend Polyvinyl Alcohol and Polyethylene Oxide (PVA/PEO) Fibers. *AIMS Bioeng.* **2022**, *9*, 178–196. [CrossRef]
5. Ekrami, E.; Khodabandeh Shahraky, M.; Mahmoudifard, M.; Mirtaleb, M.S.; Shariati, P. Biomedical Applications of Electrospun Nanofibers in Industrial World: A Review. *Int. J. Polym. Mater. Polym. Biomater.* **2023**, *72*, 561–575. [CrossRef]
6. Mawazi, S.M.; Kumar, M.; Ahmad, N.; Ge, Y.; Mahmood, S. Recent Applications of Chitosan and Its Derivatives in Antibacterial, Anticancer, Wound Healing, and Tissue Engineering Fields. *Polymers* **2024**, *16*, 1351. [CrossRef]
7. Qasim, S.B.; Zafar, M.S.; Najeeb, S.; Khurshid, Z.; Shah, A.H.; Husain, S.; Rehman, I.U. Electrospinning of Chitosan-Based Solutions for Tissue Engineering and Regenerative Medicine. *Int. J. Mol. Sci.* **2018**, *19*, 407. [CrossRef]
8. Taokaew, S.; Chuenkaek, T. Developments of Core/Shell Chitosan-Based Nanofibers by Electrospinning Techniques: A Review. *Fibers* **2024**, *12*, 26. [CrossRef]
9. Ibrahim, M.A.; Alhalafi, M.H.; Emam, E.A.M.; Ibrahim, H.; Mosaad, R.M. A Review of Chitosan and Chitosan Nanofiber: Preparation, Characterization, and Its Potential Applications. *Polymers* **2023**, *15*, 2820. [CrossRef]
10. Liu, X.; Wang, S.; Ding, C.; Zhao, Y.; Zhang, S.; Sun, S.; Zhang, L.; Ma, S.; Ding, Q.; Liu, W. Polyvinylpyrrolidone/Chitosan-Loaded Dihydromyricetin-Based Nanofiber Membrane Promotes Diabetic Wound Healing by Anti-Inflammatory and Regulating Autophagy-Associated Protein Expression. *Int. J. Biol. Macromol.* **2024**, *259*, 129160. [CrossRef]
11. Kaur, H.; Singh, S.; Rode, S.; Chaudhary, P.K.; Khan, N.A.; Ramamurthy, P.C.; Gupta, D.N.; Kumar, R.; Das, J.; Sharma, A.K. Fabrication and Characterization of Polyvinyl Alcohol-Chitosan Composite Nanofibers for Carboxylesterase Immobilization to Enhance the Stability of the Enzyme. *Sci. Rep.* **2024**, *14*, 19615. [CrossRef] [PubMed]
12. Sarac, B.; Gürbüz, R.; Soprunyuk, V.; Yüce, E.; Rezvan, A.; Schranz, W.; Eckert, J.; Ozcan, A.; Sarac, A.S. Chitosan-Containing Electrospun Poly(Ethylene Oxide)-Polybutadiene-CNT Fibers. *Polym. Adv. Technol.* **2024**, *35*, e6403. [CrossRef]
13. Deng, S.; Huang, Y.; Hu, E.; Ning, L.J.; Xie, R.; Yu, K.; Lu, F.; Lan, G.; Lu, B. Chitosan/Silk Fibroin Nanofibers-Based Hierarchical Sponges Accelerate Infected Diabetic Wound Healing via a HClO Self-Producing Cascade Catalytic Reaction. *Carbohydr. Polym.* **2023**, *321*, 121340. [CrossRef]
14. Zidar, A.; Zupančič, Š.; Kristl, J.; Jeras, M. Impact of Polycaprolactone, Alginate, Chitosan and Zein Nanofiber Physical Properties on Immune Cells for Safe Biomedical Applications. *Int. J. Biol. Macromol.* **2024**, *282*, 137029. [CrossRef]
15. Gopakumar, A.N.; Ccanccapa-Cartagena, A.; Bell, K.; Salehi, M. Development of Crosslinked Polyvinyl Alcohol Nanofibrous Membrane for Microplastic Removal from Water. *J. Appl. Polym. Sci.* **2024**, *141*, e55428. [CrossRef]
16. Liu, H.; Chen, R.; Wang, P.; Fu, J.; Tang, Z.; Xie, J.; Ning, Y.; Gao, J.; Zhong, Q.; Pan, X.; et al. Electrospun Polyvinyl Alcohol-Chitosan Dressing Stimulates Infected Diabetic Wound Healing with Combined Reactive Oxygen Species Scavenging and Antibacterial Abilities. *Carbohydr. Polym.* **2023**, *316*, 121050. [CrossRef]
17. Gautam, L.; Warkar, S.G.; Ahmad, S.I.; Kant, R.; Jain, M. A Review on Carboxylic Acid Cross-Linked Polyvinyl Alcohol: Properties and Applications. *Polym. Eng. Sci.* **2022**, *62*, 225–246. [CrossRef]
18. Costa-Júnior, E.S.; Barbosa-Stancioli, E.F.; Mansur, A.A.P.; Vasconcelos, W.L.; Mansur, H.S. Preparation and Characterization of Chitosan/Poly(Vinyl Alcohol) Chemically Crosslinked Blends for Biomedical Applications. *Carbohydr. Polym.* **2009**, *76*, 472–481. [CrossRef]
19. Salleh, N.A.M.; Afifi, A.M.; Zuki, F.M.; SalehHudin, H.S. Enhancing Mechanical Properties of Chitosan/PVA Electrospun Nanofibers: A Comprehensive Review. *Beilstein J. Nanotechnol.* **2025**, *16*, 286–307. [CrossRef]
20. Olvera Bernal, R.A.; Olekhnovich, R.O.; Uspenskaya, M.V. Chitosan/PVA Nanofibers as Potential Material for the Development of Soft Actuators. *Polymers* **2023**, *15*, 2037. [CrossRef]
21. Menazea, A.A.; Ahmed, M.K. Wound Healing Activity of Chitosan/Polyvinyl Alcohol Embedded by Gold Nanoparticles Prepared by Nanosecond Laser Ablation. *J. Mol. Struct.* **2020**, *1217*, 128401. [CrossRef]
22. Jia, Y.T.; Gong, J.; Gu, X.H.; Kim, H.Y.; Dong, J.; Shen, X.Y. Fabrication and Characterization of Poly (Vinyl Alcohol)/Chitosan Blend Nanofibers Produced by Electrospinning Method. *Carbohydr. Polym.* **2007**, *67*, 403–409. [CrossRef]
23. Hang, A.T.; Tae, B.; Park, J.S. Non-Woven Mats of Poly(Vinyl Alcohol)/Chitosan Blends Containing Silver Nanoparticles: Fabrication and Characterization. *Carbohydr. Polym.* **2010**, *82*, 472–479. [CrossRef]
24. Patra, J.K.; Das, G.; Fraceto, L.F.; Campos, E.V.R.; Rodriguez-Torres, M.D.P.; Acosta-Torres, L.S.; Diaz-Torres, L.A.; Grillo, R.; Swamy, M.K.; Sharma, S.; et al. Nano Based Drug Delivery Systems: Recent Developments and Future Prospects. *J. Nanobiotechnol.* **2018**, *16*, 71. [CrossRef]

25. Martínez, E.K.T.; Bravo, J.M.C.; Medina, A.S.; González, G.L.P.; Gómez, L.J.V. A Summary of Electrospun Nanofibers as Drug Delivery System: Drugs Loaded and Biopolymers Used as Matrices. *Curr. Drug Deliv.* **2018**, *15*, 1360–1374. [CrossRef]
26. Gaydhane, M.K.; Sharma, C.S.; Majumdar, S. Electrospun Nanofibres in Drug Delivery: Advances in Controlled Release Strategies. *RSC Adv.* **2023**, *13*, 7312–7328. [CrossRef]
27. Chi, Y.; Li, K.; Yan, Q.; Koizumi, S.; Shi, L.; Takahashi, S.; Zhu, Y.; Matsue, H.; Takeda, M.; Kitamura, M.; et al. Nonsteroidal Anti-Inflammatory Drug Flufenamic Acid Is a Potent Activator of AMP-Activated Protein Kinase. *J. Pharmacol. Exp. Ther.* **2011**, *339*, 257–266. [CrossRef]
28. Lazar, A.D.; Dinescu, S.; Albu-Kaya, M.G.; Gharbia, S.; Hermenean, A.; Costache, M. Release of the Non-Steroidal Anti-Inflammatory Drug Flufenamic Acid by Multiparticulate Delivery Systems Promotes Adipogenic Differentiation of Adipose-Derived Stem Cells. *Materials* **2020**, *13*, 1550. [CrossRef]
29. Zhang, Y.; Han, Y.; Wang, L.; Kong, J.; Pan, W.; Zhang, X.; Chen, L.; Yao, Z.; Zhou, T.; Cao, J. Flufenamic Acid, a Promising Agent for the Sensitization of Colistin-Resistant Gram-Negative Bacteria to Colistin. *Microbiol. Spectr.* **2023**, *11*, e0405222. [CrossRef]
30. Chavez-Dozal, A.A.; Jahng, M.; Rane, H.S.; Asare, K.; Kulkarny, V.V.; Bernardo, S.M.; Lee, S.A. In Vitro Analysis of Flufenamic Acid Activity against Candida Albicans Biofilms. *Int. J. Antimicrob. Agents* **2014**, *43*, 86–91. [CrossRef]
31. Liu, X.; Li, Z.; Liu, H.; Zhu, Y.; Xia, D.; Wang, S.; Gu, R.; Zhang, P.; Liu, Y.; Zhou, Y. Flufenamic Acid Inhibits Adipogenic Differentiation of Mesenchymal Stem Cells by Antagonizing the PI3K/AKT Signaling Pathway. *Stem Cells Int.* **2020**, *2020*, 1540905. [CrossRef] [PubMed]
32. Liu, Q.; Ouyang, W.C.; Zhou, X.H.; Jin, T.; Wu, Z.W. Antibacterial Activity and Drug Loading of Moxifloxacin-Loaded Poly(Vinyl Alcohol)/Chitosan Electrospun Nanofibers. *Front. Mater.* **2021**, *8*, 36. [CrossRef]
33. Cheng, Y.; Farasati Far, B.; Jahanbakhshi, M.; Bahrami, S.; Tamimi, P.; Sedaghat, M.; Ghazizadeha, E. Exploring the Potential of a Polyvinyl Alcohol/Chitosan-Based Nanofibrous Matrix for Erythromycin Delivery: Fabrication, in Vitro and in Vivo Evaluation. *RSC Adv.* **2023**, *13*, 18450–18460. [CrossRef] [PubMed]
34. Sasmal, P.; Datta, P. Tranexamic Acid-Loaded Chitosan Electrospun Nanofibers as Drug Delivery System for Hemorrhage Control Applications. *J. Drug Deliv. Sci. Technol.* **2019**, *52*, 559–567. [CrossRef]
35. Pang, Y.; Buanz, A.; Gaisford, S.; Magdysyuk, O.V.; Williams, G.R. Monitoring Polymorphic Phase Transitions in Flufenamic Acid Amorphous Solid Dispersions Using Hyphenated X-Ray Diffraction-Differential Scanning Calorimetry. *Mol. Pharm.* **2022**, *19*, 1477–1487. [CrossRef]
36. Mohandoss, S.; Velu, K.S.; Wahab, R.; Al-Khedhairy, A.A.; Tamizhselvi, R.; Napoleon, A.A.; Palanisamy, S.; You, S.; Lee, Y.R. Enhanced Solubility and Biological Activities of Flufenamic Acid through β-Cyclodextrin Derivatives Inclusion Complexes: A Comprehensive Study. *J. Mol. Liq.* **2024**, *402*, 124765. [CrossRef]
37. Zhang, W.; Zhao, L.; Ma, J.; Wang, X.; Wang, Y.; Ran, F.; Wang, Y.; Ma, H.; Yu, S. Electrospinning of Fucoidan/Chitosan/Poly(Vinyl Alcohol) Scaffolds for Vascular Tissue Engineering. *Fibers Polym.* **2017**, *18*, 922–932. [CrossRef]
38. Rahnama, S.; Movaffagh, J.; Shahroodi, A.; Jirofti, N.; Fazly Bazzaz, B.S.; Beyraghdari, M.; Hashemi, M.; Kalalinia, F. Development and Characterization of the Electrospun Melittin-Loaded Chitosan Nanofibers for Treatment of Acne Vulgaris in Animal Model. *J. Ind. Text.* **2022**, *52*, 152808372211124. [CrossRef]
39. Wsoo, M.A.; Razak, S.I.A.; Bohari, S.P.M.; Shahir, S.; Salihu, R.; Kadir, M.R.A.; Nayan, N.H.M. Vitamin D3-Loaded Electrospun Cellulose Acetate/Polycaprolactone Nanofibers: Characterization, in-Vitro Drug Release and Cytotoxicity Studies. *Int. J. Biol. Macromol.* **2021**, *181*, 82–98. [CrossRef]
40. Yu, H.; Chen, D.; Lu, W.; Zhang, C.; Wang, H.; Peng, Z.; Jiang, H.; Xiao, C. Characterization of Polyvinyl Alcohol/Chitosan Nanofibers Loaded with Royal Jelly by Blending Electrospinning for Potential Wound Dressings. *Int. J. Biol. Macromol.* **2025**, *307*, 141977. [CrossRef]
41. Iqbal, H.; Khan, B.A.; Khan, Z.U.; Razzaq, A.; Khan, N.U.; Menaa, B.; Menaa, F. Fabrication, Physical Characterizations and in Vitro Antibacterial Activity of Cefadroxil-Loaded Chitosan/Poly(Vinyl Alcohol) Nanofibers against Staphylococcus Aureus Clinical Isolates. *Int. J. Biol. Macromol.* **2020**, *144*, 921–931. [CrossRef] [PubMed]
42. Amiri, N.; Ajami, S.; Shahroodi, A.; Jannatabadi, N.; Amiri Darban, S.; Fazly Bazzaz, B.S.; Pishavar, E.; Kalalinia, F.; Movaffagh, J. Teicoplanin-Loaded Chitosan-PEO Nanofibers for Local Antibiotic Delivery and Wound Healing. *Int. J. Biol. Macromol.* **2020**, *162*, 645–656. [CrossRef] [PubMed]
43. Song, Y.; Hu, Q.; Liu, S.; Wang, Y.; Zhang, H.; Chen, J.; Yao, G. Electrospinning/3D Printing Drug-Loaded Antibacterial Polycaprolactone Nanofiber/Sodium Alginate-Gelatin Hydrogel Bilayer Scaffold for Skin Wound Repair. *Int. J. Biol. Macromol.* **2024**, *275*, 129705. [CrossRef]
44. Karabulut, H.; Xu, D.; Ma, Y.; Tut, T.A.; Ulag, S.; Pinar, O.; Kazan, D.; Guncu, M.M.; Sahin, A.; Wei, H.; et al. A New Strategy for the Treatment of Middle Ear Infection Using Ciprofloxacin/Amoxicillin-Loaded Ethyl Cellulose/Polyhydroxybutyrate Nanofibers. *Int. J. Biol. Macromol.* **2024**, *269*, 131794. [CrossRef]
45. Khasteband, M.; Sharifi, Y.; Akbari, A. Chrysin Loaded Polycaprolactone-Chitosan Electrospun Nanofibers as Potential Antimicrobial Wound Dressing. *Int. J. Biol. Macromol.* **2024**, *263*, 130250. [CrossRef]

46. Zahiri, M.; Khanmohammadi, M.; Goodarzi, A.; Ababzadeh, S.; Sagharjoghi Farahani, M.; Mohandesnezhad, S.; Bahrami, N.; Nabipour, I.; Ai, J. Encapsulation of Curcumin Loaded Chitosan Nanoparticle within Poly (ε-Caprolactone) and Gelatin Fiber Mat for Wound Healing and Layered Dermal Reconstitution. *Int. J. Biol. Macromol.* **2020**, *153*, 1241–1250. [CrossRef]
47. Karuppannan, S.K.; Dowlath, M.J.H.; Ramalingam, R.; Musthafa, S.A.; Ganesh, M.R.; Chithra, V.; Ravindran, B.; Arunachalam, K.D. Quercetin Functionalized Hybrid Electrospun Nanofibers for Wound Dressing Application. *Mater. Sci. Eng. B* **2022**, *285*, 115933. [CrossRef]
48. Mohiti-Asli, M.; Saha, S.; Murphy, S.V.; Gracz, H.; Pourdeyhimi, B.; Atala, A.; Loboa, E.G. Ibuprofen Loaded PLA Nanofibrous Scaffolds Increase Proliferation of Human Skin Cells in Vitro and Promote Healing of Full Thickness Incision Wounds in Vivo. *J. Biomed. Mater. Res. Part B Appl. Biomater.* **2017**, *105*, 327–339. [CrossRef]
49. Yu, F.; Li, M.; Yuan, Z.; Rao, F.; Fang, X.; Jiang, B.; Wen, Y.; Zhang, P. Mechanism Research on a Bioactive Resveratrol– PLA–Gelatin Porous Nano-Scaffold in Promoting the Repair of Cartilage Defect. *Int. J. Nanomed.* **2018**, *13*, 7845–7858. [CrossRef]

polymers

MDPI

Article

Design, Synthesis, and Characterization of Novel Pyrazole Cross-Linked Chitosan Derivatives Modified with Zinc Oxide Nanoparticles for Boosting Their Anticancer Activity

Hanan D. Almutairi [1,*], Marwa Abdel-Motaal [1], Marwa Sharaky [2] and Nadia A. Mohamed [1,*]

[1] Department of Chemistry, College of Science, Qassim University, Buraidah 51452, Saudi Arabia; ma.mohammed@qu.edu.sa
[2] Pharmacology Unit, Cancer Biology Department, National Cancer Institute (NCI), Cairo University, Cairo 11796, Egypt; marwa.sharaky@nci.cu.edu.eg
* Correspondence: 431214217@qu.edu.sa (H.D.A.); na.ahmed@qu.edu.sa (N.A.M.)

Abstract: A new series of chitosan-based pyrazole derivatives was successfully prepared via crosslinking chitosan using either malonopyrazole (MPy-Cs) or thiopyrazole (TPy-Cs). Three derivatives of TPy-Cs were produced based on their content of TPy, namely TPy-Cs1, TPy-Cs2, and TPy-Cs3 of crosslinking degrees of 71, 48, and 29%, respectively. Further, various weight ratios of ZnO nanoparticles were loaded into some of these derivatives to obtain the corresponding ZnONP bio-composites. FTIR, XRD, SEM, and TEM techniques were employed to emphasize the chemical, internal, and morphological structure of these derivatives. Although MPy-Cs derivatives did not show any activity against all the examined cancer cell lines, TPy-Cs derivatives exhibited an appreciable anticancer activity which greatly improved with increasing their TPy content, i.e., from TPy-Cs3 to TPy-Cs1. The TPy-Cs1 displayed IC_{50} (14.4 μg/mL) against the HN9 cell line that was comparable to the Doxorubicin (DOX) standard drug (12.6 μg/mL). Among all the prepared composites, TPy-Cs3/ZnONPs-5% was the most potent anticancer candidate against all the tested cancer cell lines, although it does not exceed the anticancer activity of DOX. Tpy-Cs2 and its ZnONP composites were safe on normal human skin fibroblast (HSF) cell lines. Thus, the inclusion of both TPy and ZnONPs into the chitosan matrix fostered its anticancer efficiency.

Keywords: chitosan; pyrazole; synthesis; ZnO nanoparticles; anticancer activity; cytotoxicity

1. Introduction

Cancer is a leading global cause of death, resulting from abnormal cell proliferation that spreads throughout the body. Chemotherapy drugs play a crucial role in combating this deadly disease [1]. To enhance survival rates and decrease tumor size, numerous cancer patients depend on the use of chemotherapeutic medications [2]. While chemotherapeutic drugs enhance the effectiveness of cancer treatment, certain patients may endure drug-induced side effects [3].

The high-dose requirement, poor bioavailability, low selectivity index, development of drug resistance, and non-specific interactions are major drawbacks of chemotherapeutic drugs. Therefore, there is an urgent need to develop a suitable drug delivery system to reduce the therapeutic dose or frequency, and thereby minimize the toxic effects of the anticancer drugs [4,5].

Biopolymers are utilized as carriers in delivering active pharmaceutical ingredients. They play a crucial role in developing various delivery systems, such as hydrogels, micelles, tablets, capsules, and particulate systems (nanoparticles, beads, and micro) [6–8]. As a carrier, the biopolymer must be non-toxic, biodegradable, and biocompatible. The latter two properties help to remove the carrier after drug administration [9,10]. Some biopolymers used in drug delivery systems include cellulose, alginate, gellan gum, pectin, gum arabic, guar gum, starch, gelatin, chondroitin sulfate, and hyaluronic acid [11]. Among these, chitosan is one of the most widely used biopolymers in the pharmaceutical industry [12–14].

Chitosan is a biocompatible and biodegradable polymer that is derived from chitin [15]. Chitosan shows various biochemical activities such as antiproliferative and antimicrobial activity, immune activation, cholesterol level-lowering activities, production of phytoalexin-eliciting activities, antihypertensive action(s), and neuroprotective, wound-healing, and antiulcer activities [16,17]. Chitosan is frequently utilized as a delivery vehicle for nucleic acids, chemotherapeutic drugs, or anticancer compounds due to its biocompatibility. Additionally, chemically modified chitosan derivatives can be used as antiproliferative agents, either on their own or in conjunction with other anticancer agents [18,19].

Chitosan can cause programmed death of cancer cells, often by activating caspases and other apoptotic pathways [20]. It also helps to inhibit tumor growth and metastasis by reducing inflammation [21]. Chitosan increases the activity of immune cells like macrophages and natural killer (NK) cells, improving the ability of the body to target cancer cells [22]. The biocompatibility of the chitosan enables it to be used as a carrier for anticancer drugs, increasing efficacy while reducing side effects [23]. It also inhibits the formation of new blood vessels, which tumors require to grow and spread [24], and may influence metabolic pathways in cancer cells, disrupting their energy production and viability [25].

Pyrazoles are five-membered heterocycles that form a class of compounds particularly useful in organic synthesis. They are one of the most studied groups of compounds within the azole family. A wide variety of synthesis methods and synthetic analogues have been reported over the years. The presence of the pyrazole nucleus in different structures leads to diverse applications in areas such as technology, medicine, and agriculture. They are described as inhibitors of protein glycation, and have been found to possess antibacterial, antifungal, anticancer, antidepressant, anti-inflammatory, anti-tuberculosis, antioxidant, and antiviral properties [26,27].

Nowadays, pyrazole systems have garnered increased attention as biomolecules due to their intriguing pharmacological properties. This heterocycle can be found in numerous well-established drugs across diverse therapeutic categories [28,29]. Pyrazole derivatives frequently inhibit the proliferation of cancer cells. They can disrupt key signaling pathways that control cell cycle progression, resulting in apoptosis (programmed cell death). Many pyrazole compounds inhibit specific enzymes involved in cancer cell metabolism. For example, some pyrazole derivatives have been shown to inhibit cyclooxygenase (COX) enzymes, which are involved in tumor growth and progression. Pyrazoles can activate apoptotic pathways in cancer cells by regulating the expression of apoptosis-related proteins such as Bcl-2 family proteins and caspases. Certain pyrazole derivatives may inhibit angiogenesis, the formation of new blood vessels required for tumor growth. This may involve inhibiting the vascular endothelial growth factor (VEGF) signaling pathway. Pyrazoles can influence key signaling pathways, including the PI3K/Akt and MAPK pathways, which are frequently dysregulated in cancer. Some pyrazoles can cause oxidative stress in cancer cells, resulting in their death. The increased reactive oxygen species (ROS) levels can damage cellular components, accelerating apoptosis [30,31].

Zink oxide nanoparticles (ZnONPs) can trigger apoptosis in cancer cells via intrinsic pathways. This process involves mitochondrial dysfunction, which results in the release of cytochrome C into the cytosol, activating caspases and initiating apoptotic cascades [32].

ZnONPs have been shown to produce ROS within cancer cells, which cause oxidative stress. Increased ROS levels can cause cellular damage, disrupt redox balance, and activate apoptotic pathways [33,34]. ZnONPs upregulate cyclin-dependent kinase inhibitors, which can result in cell cycle arrest in the G1, S, or G2/M phases. This stops cancer cells from growing and can make other therapies more successful [35]. By preventing the production of blood vessels, ZnONPs may have anti-angiogenic effects by blocking vascular endothelial growth factors. This restricts the access of the tumors to oxygen and nutrition, which stops them from growing [36]. According to some research, ZnONPs have the ability to specifically target cancer stem cells, which are essential for tumor dissemination and recurrence. ZnONPs can enhance overall therapy results by decreasing the population of cancer cells [37]. ZnONPs have the ability to improve the effectiveness of other therapeutic treatments, including radiation and chemotherapy. Better treatment results and fewer negative effects may result from this synergistic impact [38].

Accordingly, in this work and for the first time, the anticancer activity of chitosan was boosted by chemical crosslinking via its primary -NH_2 groups using the bioactive bis-pyrazole moieties. This was achieved by preparing a series of new pyrazole cross-linked chitosan derivatives via reacting chitosan with a bis-pyrazole linked with either a malonic linkage (MPy-Cs) or a thiocarbonyl group (TPy-Cs). The incorporation of nitrogen-rich pyrazole nuclei, substituted with amino, ester, and thiocarbonyl groups, as cross-linkages between the chitosan chains, leads to enhancing the interactions between their positively charged sites and the negatively charged cancer cell membranes. Also, some new ZnONPs/pyrazole cross-linked chitosan composites were prepared by dispersing two different ZnONP concentrations of 3 and 5% based on the weight of MPy-Cs and TPy-Cs derivatives, to improve their anticancer activity. Thus, the amalgamation of bis-pyrazole, ZnONPs, and chitosan in the same structure is considered to be a good strategy to accomplish adequate systems for competing with the traditional anticancer agents. The structure of the prepared derivatives and their ZnONPs/composites was characterized using FTIR, XRD, SEM and TEM. Their efficacy against three different types of cancer cells (human colorectal carcinoma cells (HCT_{116}), human skin carcinoma cells (A375), and human tongue carcinoma cells (HN9)) was investigated. Their cytotoxicity against human skin fibroblasts HSF was also evaluated.

2. Materials and Methods

2.1. Materials

Chitosan (Cs, 1.0–3.0 × 10^5 g mol^{-1} molecular weight, 98% degree of deacetylation) was purchased from Acros Organics (Morris Plains, NJ, USA). Carbon disulfide (99.9%), dimethyl sulfate (99%), dimethyl sulfoxide anhydrous (DMSO, 99.9%), and thiocarbohydrazide (98%) were provided by Sigma-Aldrich (Munich, Germany). Ethyl cyanoacetate (98%) was supplied by Loba Chemie (Mumbai, India). Potassium hydroxide pellets (98%) were obtained from Pan-Reac. AppliChem-ITW Reagent (Darmstadt, Germany). Dimethylformamide and methanol were purchased from Fisher Scientific UK Ltd. (Leicestershire, UK). Acetone was provided by NV (Zedelgem, Belgium). Zinc oxide nanoparticles (ZnONPs) (product number: NCZ4701, CAS number: 1314-13-2, purity > 99.99%, particle size: 20 nm, form: nanopowder) were supplied by Nanochemazodne Inc. (Leduc, AB, Canada). Malonohydrazide and ethyl 2-cyano-3,3-bis(methylthio)acrylate were prepared according to the methods previously described [39,40]. Three human cancer cell lines included human colorectal carcinoma cells (HCT_{116}), human skin carcinoma cells (A375), and human tongue carcinoma cells (HN9), in addition to normal human skin fibroblasts (normal HSF) which were obtained from the American Type Culture Collection (ATCC, Manassas, VA, USA). They were maintained at the National Cancer Institute (NCI), Cairo

University, Cairo, Egypt, in DMEM media containing 10% fetal bovine serum and 1% penicillin-streptomycin, and incubated in 5% CO_2 in a humidified atmosphere at 37 °C.

2.2. Methods

2.2.1. Synthesis of Novel Malonopyrazole Cross-Linked Chitosan (MPy-Cs) Derivative

Step A: A mixture of malonohydrazide (0.1 mmol) and ethyl 2-cyano-3,3-bis (methylthio)acrylate (0.2 mmol) was refluxed in 20 mL of methanol for 2 h. The formed precipitate was filtered off, dried, and crystallized from methanol to afford the pure bis-pyrazole product (MPy, Scheme 1A) as faint brown crystals (yield = 85%, mp = 180–182 °C). IR (KBr) ν_{max} (cm^{-1}): 3446 and 3335 (NH_2), 1721 (C=O ester), 1657 (C=O amide and C=N in the pyrazole rings), 1593 and 1532 (C=C in the pyrazole rings), 947 and 852 (C-S) as shown in Figure 1. ^{1}H-NMR (DMSO-*d6*) *d* (ppm) (Figure 2): 1.26 (t, *J* = 7.03 Hz, 6H, $2CH_3$ ester), 2.36 (s, 6H, $2SCH_3$), 3.3 (s, 2H, CH_2), 4.13 (q, *J* = 7.01 Hz, 4H, $2CH_2$ of ester), 6.04 (s, 4H, $2NH_2$) and ^{13}C-NMR (DMSO-*d6*) δ (ppm) (Figure 3): 11.87(-CH_3, ester group), 14.93 (-CH_3, -SMe), 59.14 (-CH_2, ester), 92.03 (-CO-CH_2-CO-, 147.65, 153.39, 163.8, 164.88, 164.94 (-C-SMe), 165.45, 165.98 (2 -C=O). MS (70 eV, EI, %), m/z = 487 (M+, 19), Anal. Calcd.: %C, 43.40; %H, 4.71; %N, 17.86; %S, 13.63 (470.52): Found: %C, 43.38; %H, 4.69; %N, 17.85; %S, 13.66.

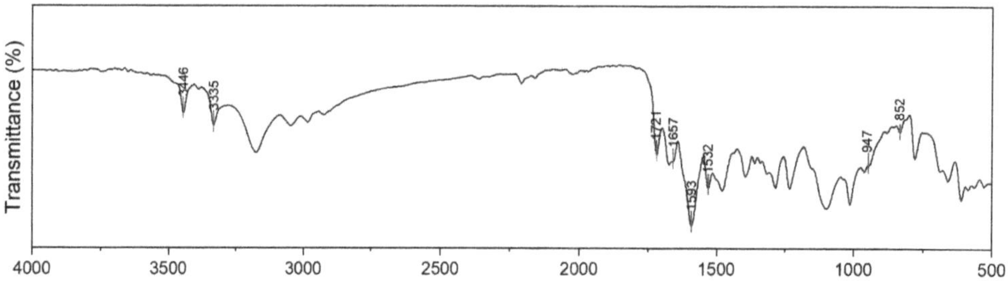

Figure 1. FTIR spectrum of MPy.

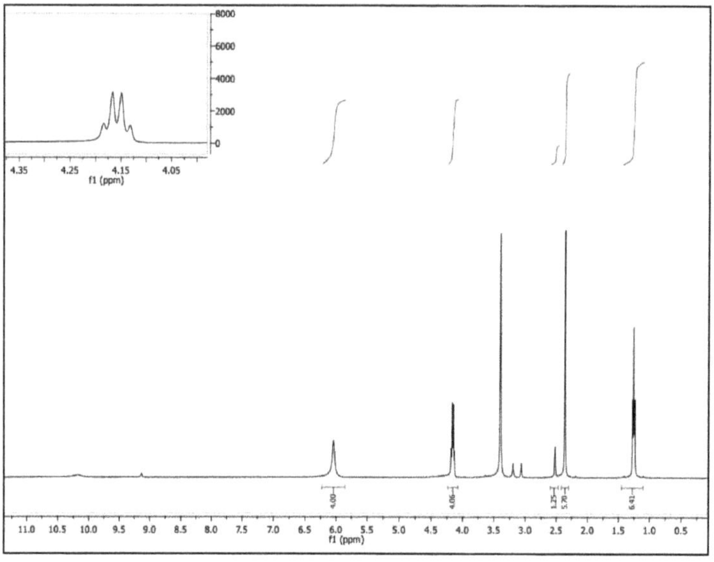

Figure 2. ^{1}H-NMR spectrum of MPy.

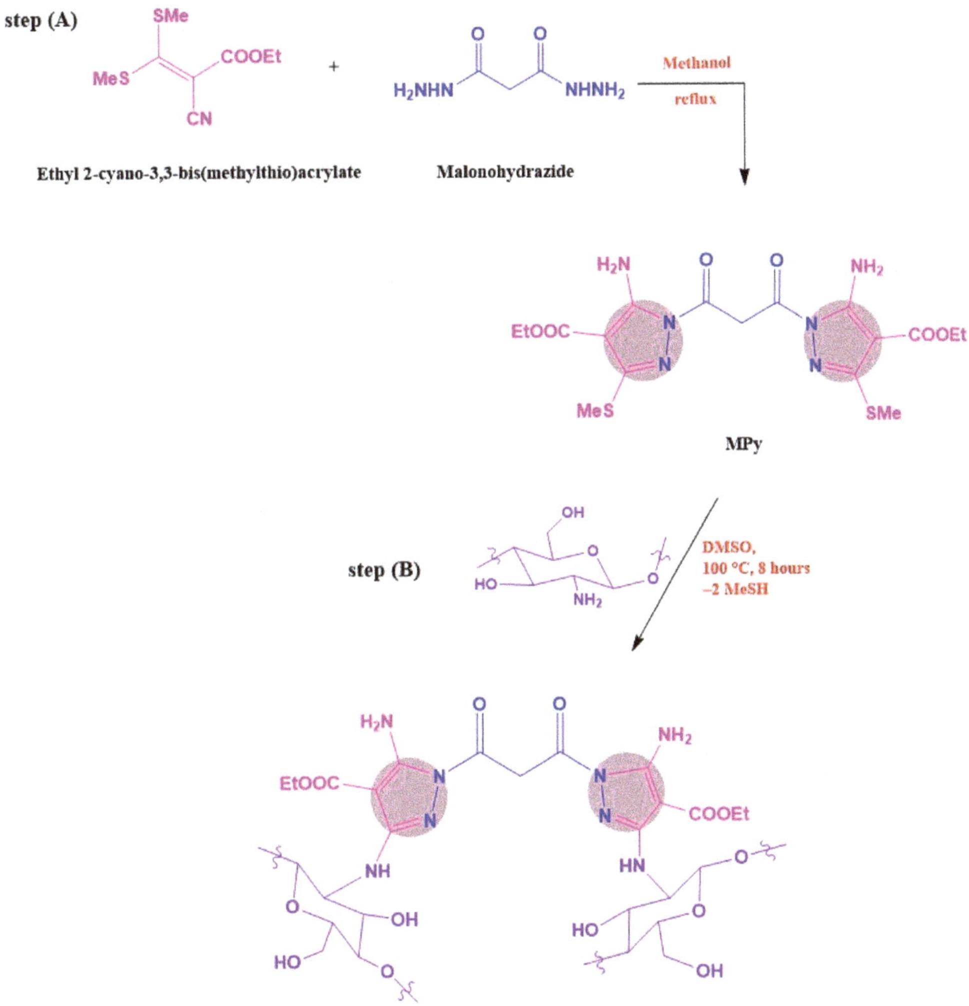

Scheme 1. Synthesis of novel MPy-Cs derivative: (**A**) synthesis of MPy; (**B**) synthesis of MPy-Cs.

Step B: MPy (0.1 mol) was added to chitosan (0.2 mol, already swelled in 50 mL DMSO) and stirred at 100 °C for 8 h, and then at RT for 12 h, and thereafter poured into a beaker containing 400 mL acetone with gentle stirring. The reaction mixture was cooled at 4 °C for 4 h, and the formed MPy-Cs derivative (Scheme 1B) was filtered. Exhaustive extraction using a Soxhlet for 10 h of the MPy-Cs with methanol allowed for its purification from the unreacted MPy. The MPy-Cs after extraction was washed well with hot methanol and dried in an air-circulating oven at 60 °C until reaching a constant weight.

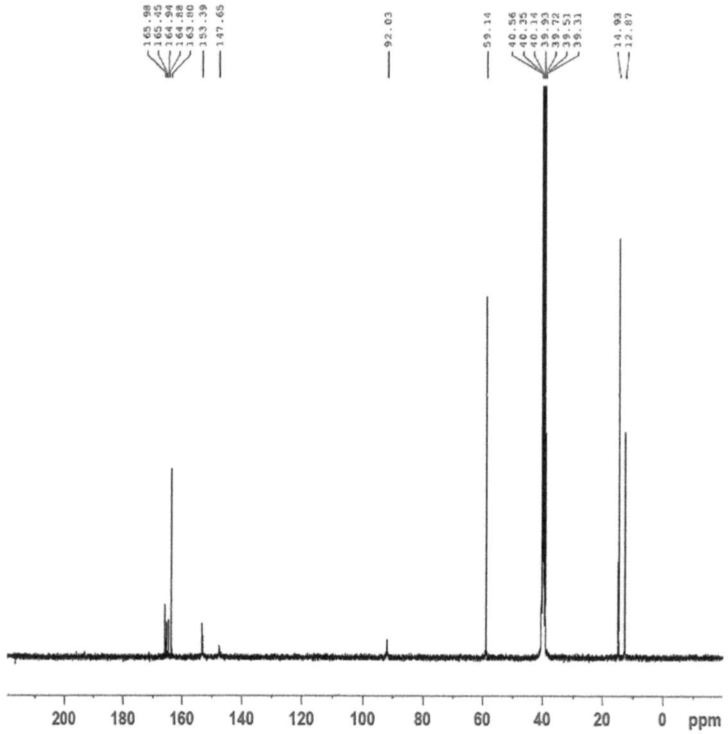

Figure 3. ^{13}C-NMR spectrum of MPy.

2.2.2. Synthesis of Novel Thiopyrazole Cross-Linked Chitosan (TPy-Cs) Derivatives

TPy-Cs derivatives were prepared by adding ethyl 2-cyano-3,3-bis(methylthio)acrylate (0.2 mol) to the suspended chitosan (0.2 mol) in DMSO (30 mL). The mixture was stirred at 100 °C for 4 h, and then at RT for 12 h, obtaining MA-Cs (Scheme 2). After that, three different predetermined quantities of thiocarbohydrazide (TCH, Scheme 2 and Table 1) were added in situ and stirred at 100 °C for 4 h then at RT for 12 h. The reaction mixtures were poured into a beaker containing 400 mL acetone with stirring. After cooling at 4 °C for 4 h, the formed products (TPy-Cs derivatives) were filtered off. The unreacted ethyl 2-cyano-3,3-bis(methylthio)acrylate and thiocarbohydrazide were extracted from the TPy-Cs derivatives with methanol using a Soxhlet apparatus for 10 h. The TPy-Cs derivatives after extraction were washed well with hot methanol and dried in an air-circulating oven at 60 °C until reaching a constant weight. The reactant molar ratio was listed in Table 1 to obtain three novel thiopyrazole cross-linked chitosan (TPy-Cs) derivatives denoted as TPy-Cs1, TPy-Cs2, and TPy-Cs3 of decreasing the TPy moiety content, respectively.

Table 1. Synthesis, elemental analysis and crosslinking degree of novel TPy-Cs derivatives.

Derivative Code	MA-Cs: TCH (Molar Ratio)	MA-Cs (g)	TCH (g)	Elemental Analysis (%)					Crosslinking Degree (%)
				C	H	N	O	S	
Cs	-	-	-	44.79	6.82	8.65	39.74	-	-
TPy-Cs1	1: 0.5	2	0.32	44.80	5.09	16.60	28.79	4.72	71
TPy-Cs2	1: 0.25	2	0.16	44.80	5.66	13.99	32.38	3.17	48
TPy-Cs3	1: 0.125	2	0.08	44.79	6.12	11.88	35.29	1.92	29

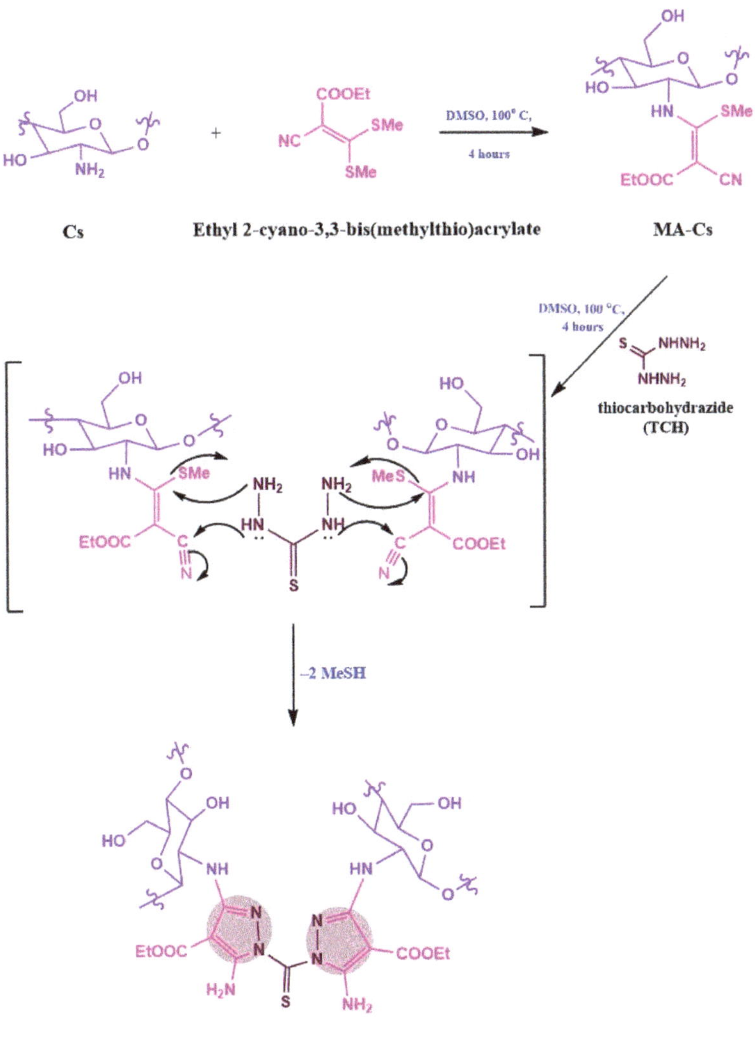

Scheme 2. Synthesis of novel TPy-Cs derivatives.

2.2.3. Synthesis of TPy-Cs/ZnONPs and MPy-Cs/ZnONP Composites

Using a mortar, two different precalculated quantities of ZnONPs were crushed separately with a fixed weight of either the Soxhlet-purified TPy-Cs2, TPy-Cs3, or MPy-Cs (0.25 g), then stirred in 10 mL of acetone for 24 h at room temperature. The ZnONP bio-composites were filtered and dried at 55 °C. The used ZnONP quantities were 3 and 5% based on the weight of the TPy-Cs and MPy-Cs derivatives. Thus, six composites, designated as TPy-Cs2/ZnONPs-3%, TPy-Cs2/ZnONPs-5%, TPy-Cs3/ZnONPs-3%, and TPy-Cs3/ZnONPs-5%, MPy-Cs/ZnONPs-3% and MPy-Cs/ZnONPs-5%, were produced. The interaction of TPy-Cs with ZnONPs in their composites, as a representative example, is shown in Scheme 3.

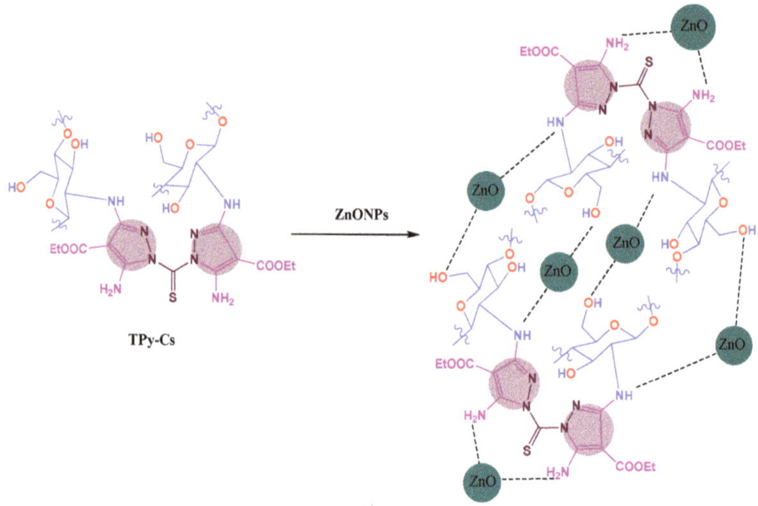

Scheme 3. Presumptive presentation of the interaction between TPy-Cs and ZnONPs in their composites.

2.3. Measurements

2.3.1. FTIR Spectroscopy

FTIR spectroscopy measurements of pyrazole cross-linked chitosan derivatives and their ZnONP composites were recorded on Agilent Technologies FTIR Spectrometer (Cary 600 Series, Santa Clara, CA, USA) in the wavenumber range from 4000 to 400 cm^{-1}.

2.3.2. X-Ray Diffractometry (XRD)

X-ray diffractometer (Joel JDX-8030, Akishima, Japan) was utilized to investigate the internal structure of the pyrazole cross-linked chitosan derivatives and their ZnONP composites at a diffraction angle (2θ) throughout a range of 5 to 90°, with a 5° min^{-1} speed.

2.3.3. Scanning Electron Microscopy (SEM)

The surface morphology of the gold-coated pyrazole cross-linked chitosan derivatives and their ZnONP composites was studied using an ultra-high-resolution Schottky Field Emission SEM device (JEOL 7610F, Tokyo, Japan).

2.3.4. Transmission Electron Microscopy (TEM)

The inspection of the morphology of the suspended solution of the TPy-Cs2/ZnONPs-5% was carried out using an advanced high-resolution Field Emission Transmission Electron Microscope (HR-JEOL, JEM-2100F, Tokyo, Japan).

2.3.5. Anticancer and Cytotoxicity Assay

The activity of MPy-Cs, TPy-Cs derivatives and their ZnONP composites beside the Doxorubicin (DOX, as a reference drug) against HCT_{116} (human colorectal carcinoma cells), A375 (human skin carcinoma cells) and HN9 (human tongue carcinoma cells), and the normal HSF (human skin fibroblasts) cell lines was evaluated by sulphorhodamine-B (SRB) assay [41]. In brief, the cells were seeded at a density concentration of 3×10^3 cells/well in 96-well microtiter plates They were allowed to attach for 24 h before being incubated with the investigated derivatives and their nanocomposites. Consequently, cells were treated with doses 25, 50, 100, and 200 μg/mL of these samples solubilized in DMSO. Five wells were utilized for every concentration with continued incubation for 48 h. Dimethyl sulfoxide (DMSO) was used as the control medium (1% *v*/*v*) with no effect on the experiment. After terminating the incubation, the cells were fixed with 20% trichloroacetic acid and each well was colored with 0.4% SRB dye. Each well's optical density (O.D.) was measured spectrophotometrically at 570 nm using an ELISA microplate reader (TECAN sunrise™, Männedorf, Switzerland). The mean survival percentage of the cells was calculated as follows: O.D. (treated cells)/O.D (control cells) × 100. The Half-maximal Inhibitory Concentration (IC_{50}) value of each sample was calibrated from the graph of the dose–response curve for each concentration using Graph Pad Prizm software (version 8.2).

Further, in this work, we used the outgrowth technique to generate the primary culture of normal human skin fibroblasts (HSF) from young people's skin explants. HSF cells were displayed in samples (up to 200 μg/mL) for 48 h.

3. Results and Discussion

3.1. Synthesis of the MPy-Cs, TPy-Cs Derivatives and Their ZnONP Composites

Novel pyrazole cross-linked chitosan derivatives were synthesized via a chemical modification of chitosan by incorporating a bis-pyrazole linked with either a malonic linkage (MPy-Cs, Scheme 1) or thiocarbonyl group (TPy-Cs, Scheme 2) to enhance its anticancer activity.

The synthesis of MPy-Cs included two steps: first is a condensation of ethyl 2-cyano-3,3-bis(methylthio)acrylate with malonohydrazide (2:1) in boiling methanol with stirring for 2 h, obtaining MPy (Scheme 1A, its structure was affirmed depending on spectral measurements). Secondly, the target MPy-Cs was achieved by heating MPy with chitosan (1:2) in DMSO at 100 °C for 8 h to condense the -SMe groups of the synthesized MPy with the amino groups of chitosan (Scheme 1B).

On the other hand, the synthesis of a novel series of TPy-Cs involved firstly the formation of the intermediate MA-Cs via treating equimolar amounts of chitosan with ethyl 2-cyano-3,3-bis(methylthio)acrylate in DMSO with stirring at 100 °C for 4 h (Scheme 2). Afterward, three different molar ratios of thiocarbohydrazide (TCH) were condensed with MA-Cs (1:2) in situ to afford a series of TPy-Cs derivatives denoted as TPy-Cs1, TPy-Cs2, and TPy-Cs3 (Scheme 2 and Table 1).

Further, some ZnO nanocomposites based on MPy-Cs, TPy-Cs2, and TPy-Cs3 were prepared using two different concentrations of ZnONPs of 3 and 5% (based on the weight of these derivatives), producing MPy-Cs/ZnONPs-3%, MPy-Cs/ZnONPs-5%, TPy-Cs2/ZnONPs-3%, TPy-Cs2/ZnONPs-5%, TPy-Cs3/ZnONPs-3%, and TPy-Cs3/ZnONPs-5%.

3.2. Characterization of MPy-Cs, TPy-Cs Derivatives and Their ZnONP Composites

3.2.1. FTIR Analysis

FTIR spectrum of Cs (Figure 4) showed, between 3700 and 3000 cm^{-1}, a very broad absorption peak, corresponding to the stretching vibration of the hydroxyl groups overlapping with that of -NH_2 groups and their hydrogen bonds. In this wavenumber range, there is a doublet peak at 3356 and 3294 cm^{-1} related to the -NH_2 groups. The symmetric stretching vibration peaks of the -CH and -CH_2 groups in the moieties of pyranose appeared at 2905 and 2867 cm^{-1}, respectively. Further, two weak peaks that appeared at 1648 and 1589 cm^{-1} are attributable to amide I and amide II, respectively, confirming the high Cs degree of deacetylation. The four absorption peaks that appeared at 1150, 1060, 1027, and 895 cm^{-1} confirmed the saccharide moieties of Cs [42].

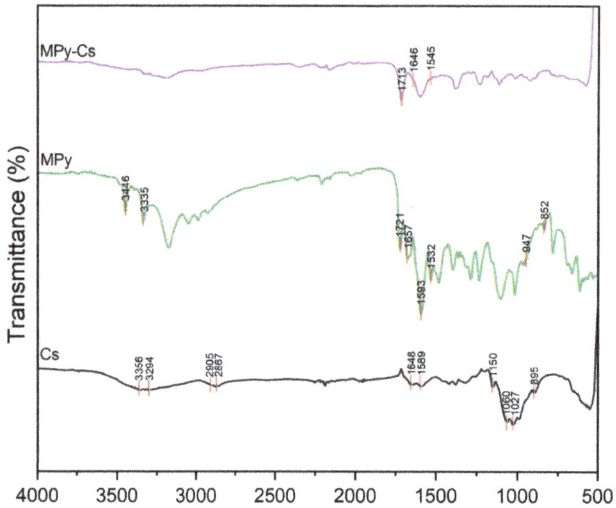

Figure 4. FTIR spectrum of MPy-Cs derivative.

In addition to the characteristic stretching vibration peaks of Cs (Figure 4) and MPy (Figures 1 and 4), the FTIR spectrum of MPy-Cs (Figure 4) showed some additional distinguished peaks; at 1713 cm^{-1} related to the C=O of the ester, a broad peak ranged from 1646 to 1545 cm^{-1} attributed to the overlapped C=O of amide, C=C, and C=N of pyrazole rings.

FTIR spectrum of ethyl 2-cyano-3,3-bis(methylthio)acrylate (Figure 5, Scheme 2) showed some characteristic stretching vibration peaks at 2207 cm^{-1} assigned to CN group, at 1695 cm^{-1} corresponded to C=O of the ester, at 1650 cm^{-1} related to the C=C bond, at 2980 cm^{-1} indicating C-H bonds in the hydrocarbon chain, and at 910 and 857 cm^{-1} attributed to C-S groups [43].

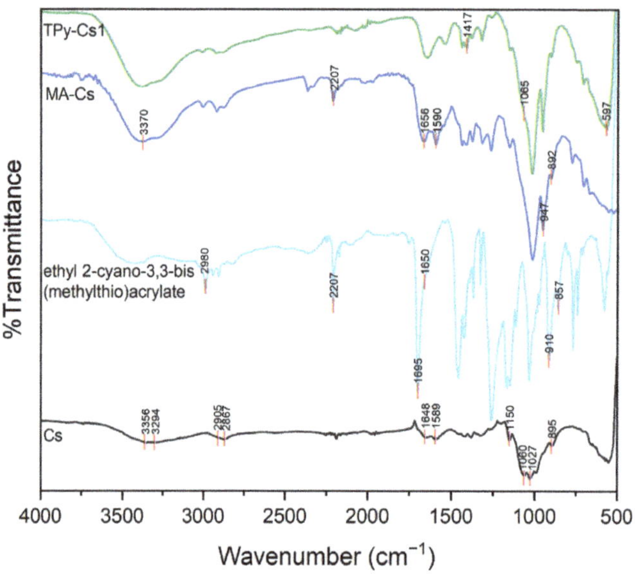

Figure 5. FTIR spectrum of TPy-Cs1 derivative.

FTIR spectrum of MA-Cs (Figure 5) showed stretching vibration peaks at 947 and 892 cm-1 ascribed to the C-S group, indicating the occurrence of the condensation reaction between Cs and ethyl 2-cyano-3,3-bis(methylthio)acrylate with the elimination of MeSH (Scheme 2). This is evidenced by the disappearance of the doublet peak of the primary amine groups of Cs at 3356 and 3294 cm^{-1}, which was completely replaced by a peak at 3370 cm^{-1}. In addition to the appearance of the absorption peaks, 2207, 1656 and 1590 cm^{-1} corresponded to the CN group, C=O group of the ester and C=C bonds, respectively.

FTIR spectra of TPy-Cs1 derivative (Figure 5) showed the disappearance of the peaks of C-S groups at 947 and 892 cm^{-1}, indicating the condensation between the -SMe groups of MA-Cs and the -NH_2 groups of the thiocabohydrazide (TCH) with the elimination of MeSH (Scheme 2). The peak corresponding to CN groups at 2207 cm^{-1} also disappeared. Further, the absorption peak of the C=S group at 1065 cm^{-1} overlapped with that of the C–O group, obtaining a broad, strong peak. The absorption peaks related to N–C–S groups appeared at 1417 and 597 cm^{-1} [44].

FTIR spectra of TPy-Cs2/ZnONPs-3% and TPy-Cs2/ZnONPs-5%, as representative examples for the ZnONP biocomposites, (Figure 6), showed absorption peaks comparable to those of their parent TPy-Cs2. The peaks at around the wavenumber of 566 cm^{-1}, which corresponded to the expected N-ZnO and O-ZnO bonds in these biocomposites, seemed to overlap with those of their parent TPy-Cs2 [45].

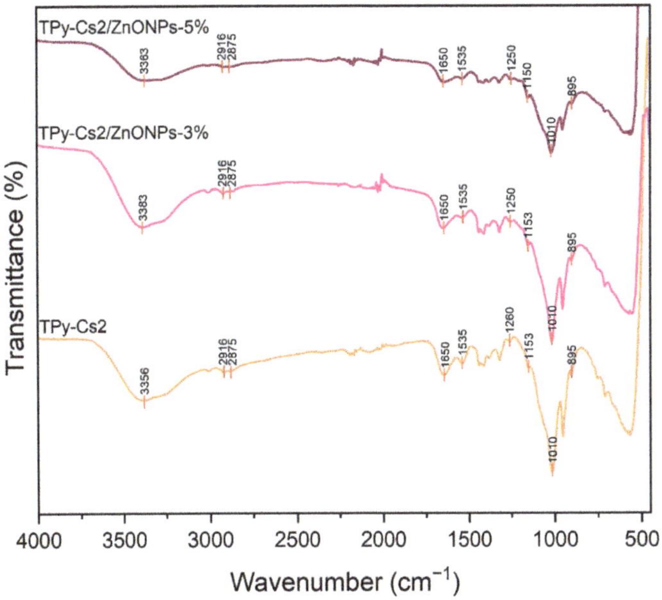

Figure 6. FTIR spectra of TPy-Cs2/ZnONP composites.

3.2.2. XRD Analysis

The virgin chitosan is distinguished by an inner structure involving amorphous and crystalline regions which manifested in its XRD pattern as two peaks near 2θ = 10° and 20°, respectively, as shown in Figure 7 [46]. This is ascribed to its capability to form a considerable number of hydrogen bonds due to its possession of a lot of hydroxyl and primary amine polar groups. Modification of chitosan with MPy and TPy leads to a remarkable reduction in its number of hydrogen bonds due to the consumption of its amino groups during the cross-linking process. This led to a separation of the chitosan chains away from each other, an increment in the amorphous region and a reduction in the crystalline fraction. This is shown from the evanescence of the peak at 2θ = 10° and a broadening and a notable lowering of the intensity of the peak at 2θ = 20° in XRD patterns of MPy-Cs, TPy-Cs1, TPy-Cs2, and TPy-Cs3 (Figure 7).

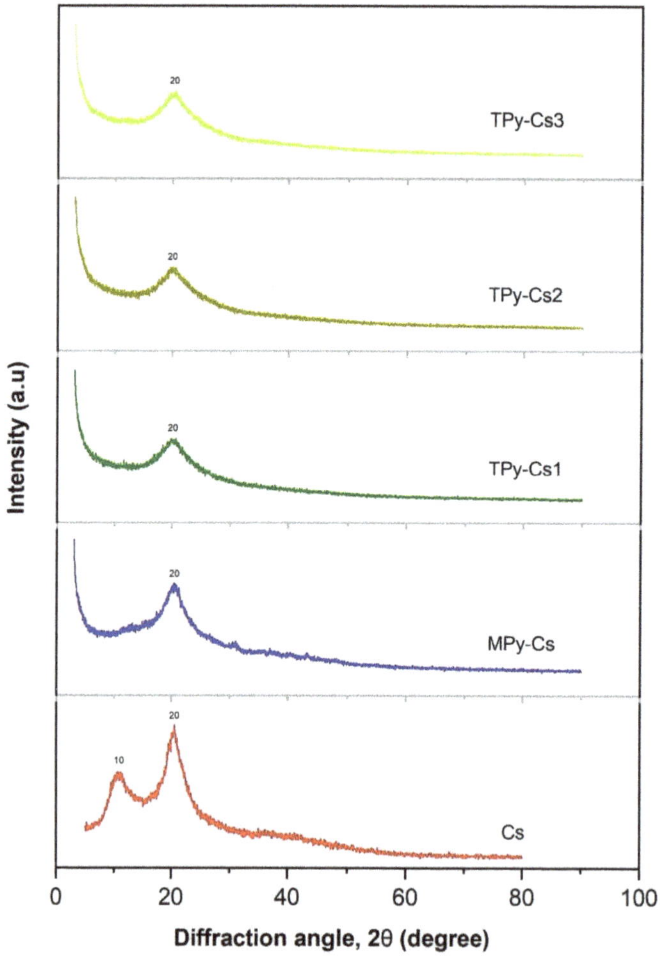

Figure 7. XRD patterns of MPy-Cs and TPy-Cs derivatives.

In order to confirm the incorporation of ZnONPs into the matrices of MPy-Cs and TPy-Cs, XRD measurements were performed (Figure 8). All the patterns of MPy-Cs/ZnONPs and TPy-Cs/ZnONP composites showed seven new peaks, beside the amorphous peak of MPy-Cs and TPy-Cs, near 2θ = 20°, at 2θ = 31.72°, 34.42°, 36.24°, 47.5°, 56.6°, 62.78° and 67.82°, which were indexed to crystal planes of (100), (002), (101), (102), (200), (201), and (202), respectively. These peak and their corresponding crystal planes are in good agreement with those for the pure ZnO according to the standard (JCPDS card no. 36-1451). The intensity of these peaks increased with increasing the ZnONP content in the composites. This evidences the successful preparation of the MPy-Cs/ZnONPs and TPy-Cs/ZnONP composites [47].

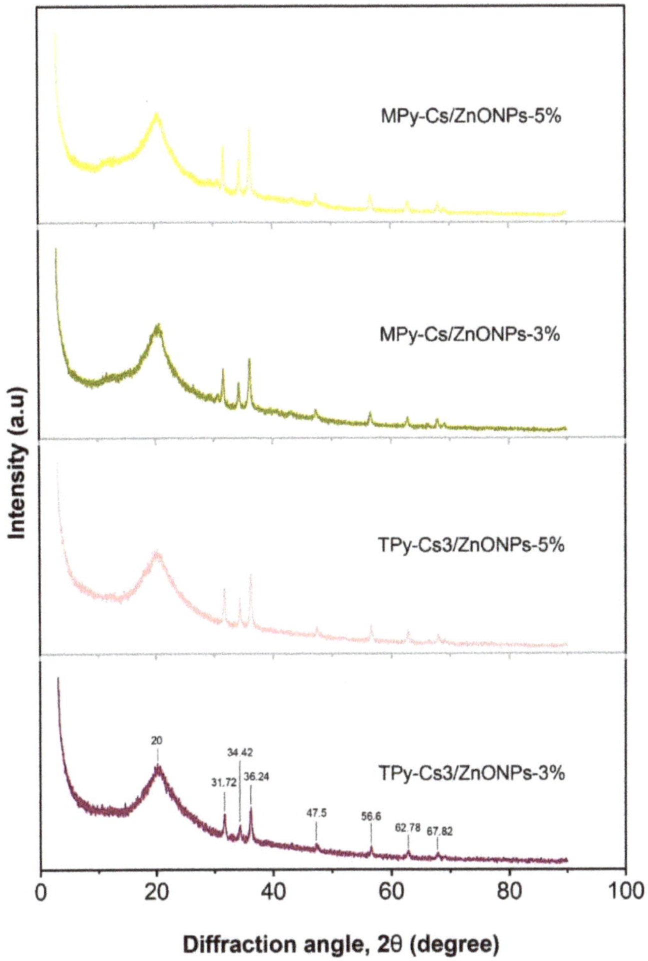

Figure 8. XRD patterns of MPy-Cs and TPy-Cs3-based ZnONP composites.

3.2.3. SEM Analysis

Figure 9 depicts SEM images of topographical features on the surfaces of Cs, MPy-Cs, and TPy-Cs derivatives. The Cs surface was found to be smooth and free of any lumps. Upon cross-linking reaction, a drastic change in the surface appearance of chitosan has been observed. As compared with chitosan, the images confirm that the surfaces of its derivatives are rough and have grooves with a porous structure and high degree of lumps of varying sizes due to the differences in the size of the inserted modifier (MPy and TPy) by the crosslinking process. Further, the intensity of these lumps increased with increasing the crosslinking degree and can be arranged as follows: MPy-Cs = TPy-Cs1 > TPy-Cs2 > TPy-Cs3. It is assumed that their pores are considered to be the water penetration regions and the sites of interaction between the external stimuli and the hydrophilic functional groups of these derivatives. The incorporation of the long chain cross-linker MPy and TPy between the chitosan chains effectively reduces the hydrogen bonds between the chitosan chains resulting in the formation of the microporosity developing to a more open structure. These pores permit these derivatives to absorb a large quantity of water and polar solvents that support the interaction between them and the incoming ingredients. The images show that although the three TPy-Cs derivatives have a comparable surface appearance, and the distribution and the size of their pores are different from TPy-Cs1 to TPy-Cs3. Since the pores became denser with increasing the amount of the crosslinker from TPy-Cs3 to TPy-Cs1. Thus, these derivatives are characterized by a microporous structure with a large surface area which makes them able to be used as good drug delivery systems.

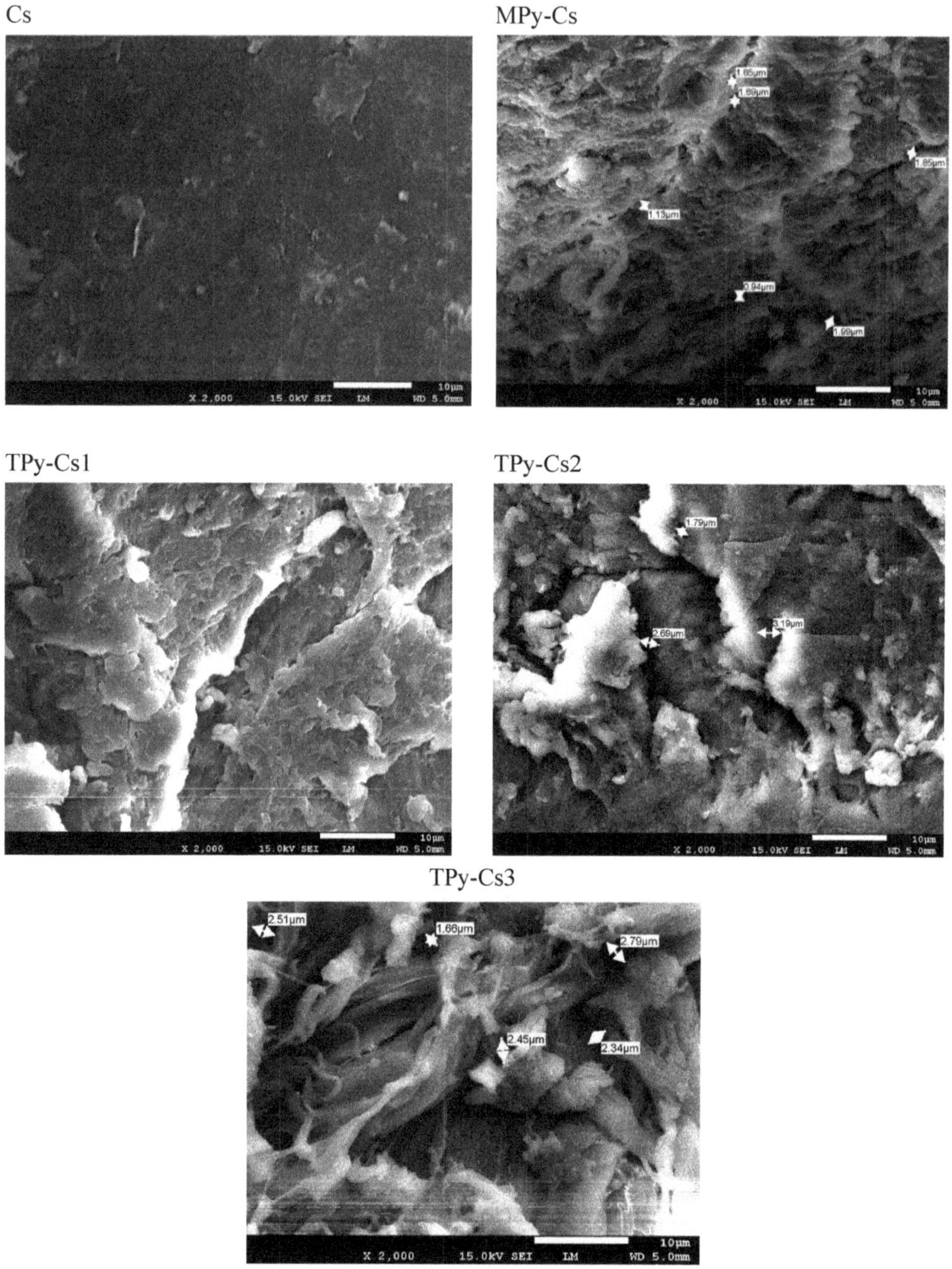

Figure 9. SEM images of MPy-Cs, and TPy-Cs derivatives.

3.2.4. TEM Analysis

TEM images of the TPy-Cs2/ZnONPs-5% composite, at different magnifications, are presented in Figure 10. The uniform dispersion of ZnONPs into the TPy-Cs2 matrix was clearly observed. The ZnONPs appeared as spherical-like shapes, and their sizes ranged from 14.21 to 29.20 nm.

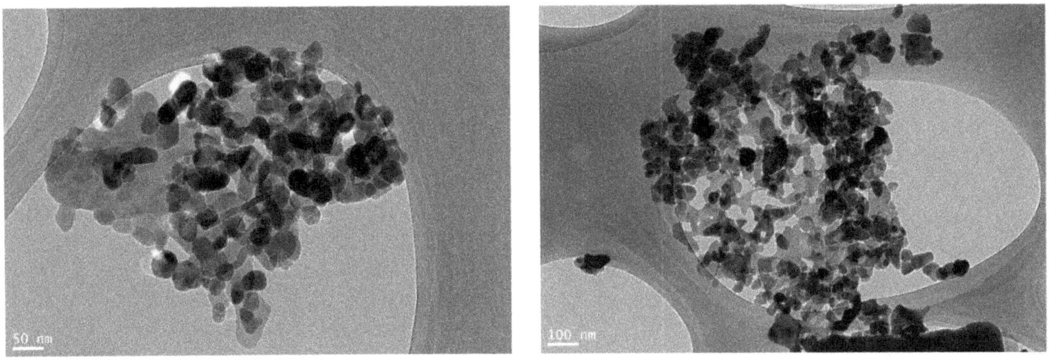

Figure 10. TEM images of TPy-Cs2/ZnONPs-5% composite at different magnifications.

3.3. Anticancer and Cytotoxic Activity

3.3.1. Anticancer Activity

The activity of the newly synthesized samples, including MPy-Cs, TPy-Cs derivatives, and their ZnONP composites against the three cancer (HCT_{116}, A375, and HN9) cell lines using sulphorhodamine-B (SRB) assay was examined. Their ability to cause 50% inhibition of the cell growth (IC_{50}) relative to the DOX reference drug is shown in Figures 11–13 and Table 2. The cancer cells were treated with doses 25, 50, 100, and 200 µg/mL of the investigated samples solubilized in DMSO. The Half-maximal Inhibitory Concentration (IC_{50}) value of each sample was calibrated from the graph of the dose–response curve for each concentration using Graph Pad Prizm software, (version 8.2). The complete data, from which IC_{50} was calculated, were provided via non-published material. In comparison to TPy-Cs derivatives, the MPy-Cs derivative exhibited no signs of activity against all the examined cancer cell lines, suggesting the role of thiocarbonyl group in enhancing the anticancer activity. This was supported by the activity of the TPy-Cs derivatives for inhibiting the tested cancer cell lines that greatly improved with increasing their thiopyrazole content, i.e., from TPy-Cs3 to TPy-Cs1. This was illustrated by their IC_{50} values against all the inspected cancer cell lines. Against the HCT_{116} cell line, the IC_{50} values of TPy-Cs3, TPy-Cs2, and TPy-Cs1 were 98.07, 90.12, and 32.64 µg/mL, respectively (Figure 11 and Table 2). Against the A375 cell line, the IC_{50} values of TPy-Cs3, TPy-Cs2, and TPy-Cs1 were 145.66, 42.53, and 20.10 µg/mL, respectively (Figure 12 and Table 2). Against the HN9 cell line, the IC_{50} values of TPy-Cs3, TPy-Cs2, and TPy-Cs1 were 147.19, 57.87, and 14.40 µg/mL, respectively (Figure 13 and Table 2).

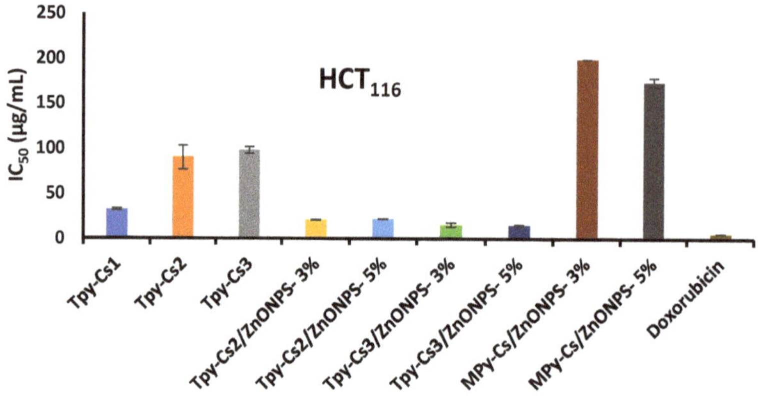

Figure 11. IC_{50} for the investigated MPy-Cs, TPy-Cs derivatives, and their ZnONP composites, beside DOX (as reference drug) against HCT_{116} cancer cell line at 48 h incubation. The results are expressed as the mean ± SD of three separate experiments performed in three replicates.

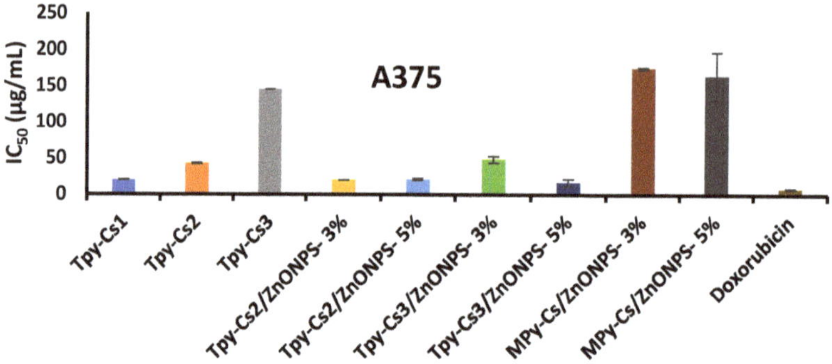

Figure 12. IC_{50} for the investigated MPy-Cs, TPy-Cs derivatives, and their ZnONP composites, beside DOX (as reference drug) against A375 cancer cell line at 48 h incubation. The results are expressed as the mean ± SD of three separate experiments performed in three replicates.

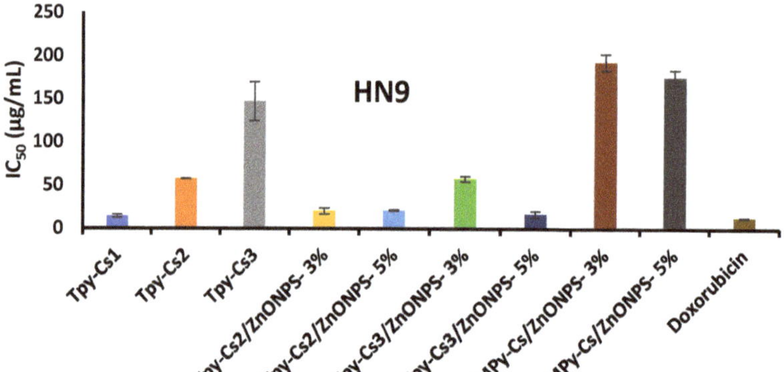

Figure 13. IC_{50} for the investigated MPy-Cs, TPy-Cs derivatives, and their ZnONP composites, beside DOX (as reference drug) against HN9 cancer cell line at 48 h incubation. The results are expressed as the mean ± SD of three separate experiments performed in three replicates.

Table 2. Anticancer activity for the tested MPy-Cs, TPy- Cs derivatives, and their ZnONP composites against HCT116, A375, and HN9 cell lines. The results are expressed as the mean ± SD of three separate experiments performed in three replicates.

Sample	HCT_{116}		A375		HN9	
	IC_{50} (μg/mL)	SD (±)	IC_{50} (μg/mL)	SD (±)	IC_{50} (μg/mL)	SD (±)
TPy-Cs1	32.64	1.08	20.10	0.35	14.40	1.97
TPy-Cs2	90.12	13.21	42.53	0.82	57.87	0.16
TPy-Cs3	98.07	3.66	145.66	0	147.19	22.34
TPy-Cs2/ZnONPs-3%	20.98	0.33	19.67	0.28	20.57	3.52
TPy-Cs2/ZnONPs-5%	21.68	0.28	20.63	1.21	21.60	0.61
TPy-Cs3/ZnONPs-3%	15.21	2.29	48.06	4.59	57.69	2.89
TPy-Cs3/ZnONPs-5%	14.82	0.69	16.14	4.91	16.95	3.74
MPy-Cs	-	-	-	-	-	-
MPy-Cs/ZnONPs-3%	199.35	0	174.38	1.35	192.86	9.49
MPy-Cs/ZnONPs-5%	173.89	4.69	163.79	33.28	175.33	8.63
DOX	5.68	0.37	7.89	0.59	12.60	0.60

In comparison to the activity of the reference Doxorubicin drug (IC_{50} = 12.60 μg/mL) against the HN9 cell line, derivative TPy-Cs1 (IC_{50} = 14.40 μg/mL) showed very close activity to that of this drug. On the other hand, it was found that the level of the anticancer activity of the other two TPy-Cs derivatives (TPy-Cs2 and TPy-Cs3) is greatly lower than that of the reference Doxorubicin drug; thus, they were selected for the creation of some ZnONP bio-composites to enhance their anticancer activity. As would be predicted, incorporating ZnONPs into the matrix of TPy-Cs2 appreciably enhanced its anticancer activity as evidenced by dropping its IC_{50} value from 90.12 to 20.98 and 21.68 μg/mL for TPy-Cs2/ZnONPs-3% and TPy-Cs2/ZnONPs-5%, respectively, against the HCT_{116} cell line (Figure 11 and Table 2). Also, against the A375 cell line, the IC_{50} value of TPy-Cs2 improved from 42.53 to 19.67 and 20.63 μg/mL for TPy-Cs2/ZnONPs-3% and TPy-Cs2/ZnONPs-5%, respectively (Figure 12 and Table 2). Similarly, against the HN9 cell line, the IC_{50} value of TPy-Cs2 lowered from 57.87 to 20.57 and 21.6 μg/mL for TPy-Cs2/ZnONPs-3% and TPy-Cs2/ZnONPs-5%, respectively (Figure 13 and Table 2).

An analogous behavior was observed when ZnONPs were impregnated into the matrix of TPy-Cs3, since its IC_{50} value decreased from 98.07 to 15.21 and 14.82 μg/mL for TPy-Cs3/ZnONPs-3% and TPy-Cs3/ZnONPs-5%, respectively, against the HCT_{116} cell line (Figure 11 and Table 2). Again, its IC_{50} value reduced from 145.66 to 48.06 and 16.14 μg/mL for TPy-Cs3/ZnONPs-3% and TPy-Cs3/ZnONPs-5%, respectively, against A375 cell line (Figure 12 and Table 2). Also, its IC_{50} value decreased from 147.19 to 57.69 and 16.95 μg/mL for TPy-Cs3/ZnONPs-3% and TPy-Cs3/ZnONPs-5%, respectively, against the HN9 cell line (Figure 13 and Table 2). Thus, the TPy-Cs3/ZnONPs-5% was observed to be the most potent anticancer candidate against all the tested cell lines (Figure 14), although it does not exceed the anticancer activity of Doxorubicin.

Although the MPy-Cs derivative has no potency to inhibit the activity of the three inspected cancer cell lines, its ZnONP composites showed a moderate one. IC_{50} of MPy-Cs/ZnONPs-3% and MPy-Cs/ZnONPs-5% were 199.35 and 173.89 μg/mL, respectively, against the HCT_{116} cell line (Figure 11 and Table 2), 174.38 and 163.79 μg/mL, respectively, against the A375 cell line (Figure 12 and Table 2), 192.86 and 175.33 μg/mL, respectively, against the HN9 cell line (Figure 13 and Table 2).

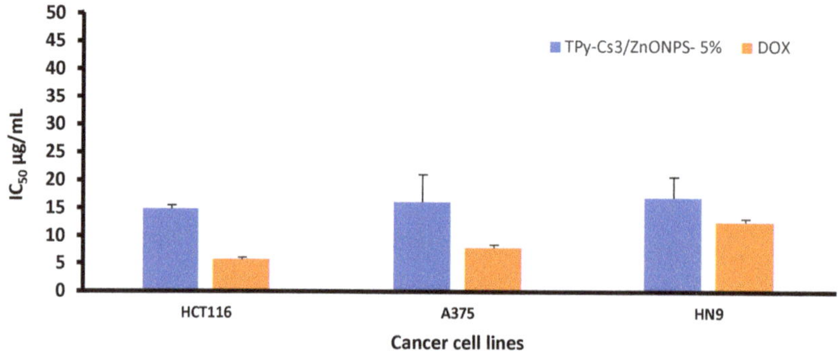

Figure 14. IC_{50} for TPy-Cs3/ZnONPs-5% against the three cancer cell lines at 48 h incubation. The results are expressed as the mean ± SD of three separate experiments performed in three replicates.

3.3.2. Cytotoxic Activity

The toxic action of the TPy-Cs2 derivative and its ZnONP composite (as a representative example of the prepared samples) on normal human skin fibroblast cell line (HSF) was inspected at a concentration range between 25 and 200 μg/mL (Figure 14 and Table 3). The results revealed that at a concentration of 25 μg/mL of the TPy-Cs2 derivative, TPy-Cs2/ZnONPs-3% and TPy-Cs2/ZnONPs-5% composites had a slight impact on the viability of the HSF cells (96.77, 93.55 and 77.67%, respectively), as can be observed in Figure 15 and Table 3. A progressive lowering in the viability was obtained by increasing the concentration of the investigated samples to 200 μg/mL. The viability of the HSF cells at 200 μg/mL concentration of TPy-Cs2 derivative, TPy-Cs2/ZnONPs-3% and TPy-Cs2/ZnONPs-5% composites was 72.44, 63.49 and 57.59%, respectively. Thus, the cytotoxicity for 50% of HSF cells (CC_{50}) was found to be over 200 μg/mL concentration of the three inspected samples. As noted in Figures 10–12 and Table 2, the TPy-Cs2 derivative and TPy-Cs2/ZnONPs-3% and TPy-Cs2/ZnONPs-5% composites demonstrated a better activity against all the inspected cancer cell lines at lower IC_{50} values (ranged from 20.63–90.12 μg/mL) than that which resulted during the assessment of the cytotoxicity (CC_{50} > 200 μg/mL). This illustrates the safe features of these samples and suggests that they can be used as one of the suitable anticancer drugs.

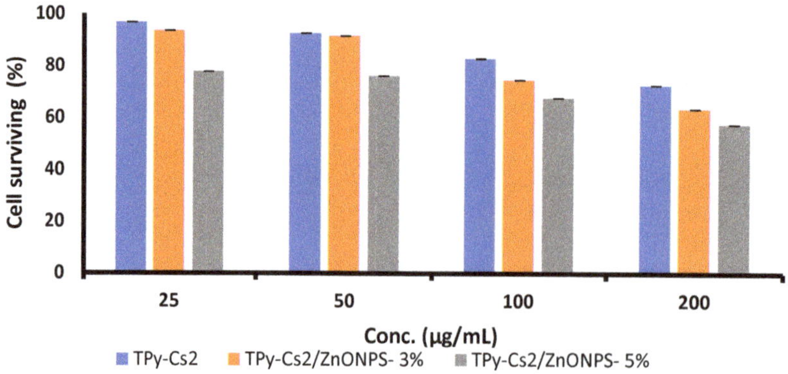

Figure 15. Cytotoxicity of TPy-Cs2 derivative and its ZnONP composites on the normal human skin fibroblast (HSF) cell line. The results are expressed as the mean ± SD of three separate experiments performed in three replicates.

Table 3. Cytotoxic effect of TPy-Cs2 derivative and its ZnONP composites on the normal human skin fibroblast (HSF) cell line. The results are expressed as the mean ± SD of three separate experiments performed in three replicates.

Sample Conc. (μg/mL)	TPy-Cs2		TPy-Cs2/ZnONPs- 3%		TPy-Cs2/ZnONPs- 5%	
	Surviving (%)	SD (±)	Surviving (%)	SD (±)	Surviving (%)	SD (±)
0	100	0.00	100	0.00	100	0.00
25	96.77	0.04	93.55	0.04	77.67	0.04
50	92.57	0.04	91.55	0.07	75.98	0.01
100	82.84	0.02	74.49	0.05	67.59	0.03
200	72.44	0.04	63.49	0.03	57.59	0.01

3.3.3. Structure–Activity Relationship (SAR)

The promising anticancer properties of every component of the synthesized composites, including chitosan, pyrazole, and ZnONPs, prompted us to assess their inhibitory activity to gain insight into the mechanism of the anticancer activity of these composites with various concentrations.

The molecular insight into the anticancer activity of these targets involves a manifold approach, with membrane transport playing a critical role. These composites influence the increased permeability and retention (EPR) effect, which causes tumor tissues to accumulate preferentially. In cellular biology, the term "membrane transport" refers to a collection of mechanisms that control the passage of ions and other small molecules across biological membranes, which are made up of protein-containing lipid bilayers. Passage through the membrane is controlled by selective membrane permeability, a characteristic of biological membranes that enables them to segregate components of different chemical natures.

Drugs are delivered into cells via the influence of the attached function groups. In detail, the unique qualities of promising biopolymer chitosan facilitate cellular damage through interactions between its positively charged amino groups and the negatively charged cancer cell membranes and the hydroxyl-rich chitosan backbone facilitating electrostatic interactions with the negatively charged cell membrane. Furthermore, chitosan stimulates apoptosis pathways in cancer cells and suppresses angiogenesis, which is the formation of blood vessels that feed tumors. Additionally, its distinctive biocompatibility and biodegradability improve the targeted delivery of anticancer drugs to tumor cells precisely [48].

The introduction of substituted pyrazole ligand with its nitrogen-containing heterocycle (which is associated with amino and ester plus the important thione groups) contributes to the composite's ability to intercalate with DNA, disrupting its structure and inhibiting replication via binding with the protein membrane (Figure 16). Additionally, the aforementioned IC_{50} values of the tested pyrazolo-chitosan derivatives affirmed the much greater inhibitory potential of the TPy-Cs derivatives compared to the MPy-Cs derivative. This may be due to the presence of thione (C=S) groups, which enhance their inhibition activity via improvement of the binding between the investigating TPy-Cs derivatives and the amino acid of cancer cell membrane lining protein, leading to the damage of the cancer cell [49].

On the other hand, the pyrazole moiety can coordinate with ZnONPs, influencing their release kinetics and stability. Numerous previous research studies have demonstrated that ZnONPs have therapeutic anticancer potential due to their unique features. This is due to a series of steps that occur in cancer cells, beginning with a loss of inner balance and redox state instability, followed by Zn^{+} created by oxidation of ZnONP internalization facilitated by endocytosis, followed by lysosomal escape, allowing the nanoparticles to be agglomerated in the cytoplasm and membranes of cells, which causes cellular damage and leads to cell death. Also, oxidative stress causes DNA damage, resulting in cell demise [50–52]. Therefore, the loading of 3% and 5% concentrations of ZnONPs into the synthesized pyrazole cross-linked chitosan derivatives played a vital role in the enhancement of the inhibitory activity against the employed cancer cell lines.

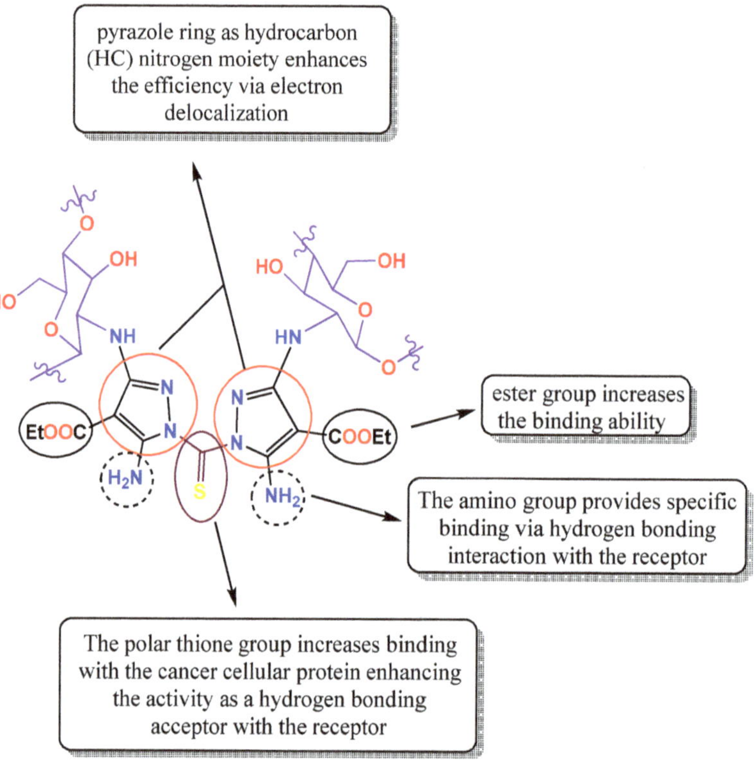

Figure 16. Structure–activity relationship for the TPy-Cs derivatives.

From the displayed results, it was concluded that the ideal concentration among all samples is 5% ZnONPs loaded into TPy-Cs3 (TPy-Cs3/ZnONPs-5%), as it exhibited the most potency against the three examined tumor cells.

The synergy between these functional groups—the cationic chitosan, the DNA-intercalating functionalized pyrazole moiety, and the redox-active ZnONPs—creates a versatile anticancer mechanism. This mechanism involves enhanced cellular uptake, membrane disruption, DNA damage, mitochondrial dysfunction, and cell cycle arrest, finally leading to apoptosis. The strategic combination of these functionalities provides a targeted and potent therapeutic effect, minimizing off-target toxicity and maximizing efficacy.

4. Conclusions

Some new pyrazole derivatives based on biopolymer chitosan were successfully produced. This was achieved through crosslinking of chitosan chains using either malonopyrazole to obtain MPy-Cs derivative or three different concentrations of thiopyrazole to obtain three TPy-Cs derivatives: TPy-Cs1, TPy-Cs2, and TPy-Cs3 with crosslinking degrees of 71, 48, and 29%, respectively. Moreover, impregnation of different amounts of ZnONPs into most of the prepared TPy-Cs and MPy-Cs derivatives was carried out using

3 and 5% ZnONPs based on the weight of these derivatives, attaining the corresponding ZnONP bio-composites. Some analytical techniques (FTIR, XRD, SEM, and TEM) were used to prove the chemical, inner, and morphological structure of these derivatives and their corresponding ZnONP composites. Their anticancer performance was significantly enhanced with an increase in their TPy and ZnONP content. Despite the fact that the MPy-Cs derivative showed no inhibition activity against all the inspected cancer cell lines, TPy-Cs derivatives possessed a considerable anticancer activity which was boosted as a function of their thiopyrazole content, i.e., from TPy-Cs3 to TPy-Cs1. Further, the TPy-Cs1 exhibited IC_{50} (14.4 μg/mL) against the HN9 cell line that was comparable to the standard drug Doxorubicin (12.6 μg/mL). Amongst all the obtained composites, TPy-Cs3/ZnONPs-5% has the most anticancer potency against all the checked cell lines, although it did not surpass the anticancer potency of Doxorubicin. The results indicated the safe features of Tpy-Cs2 and its ZnONP composites on normal human skin fibroblast (HSF) cell lines. Thus, the synthesized Tpy-Cs and their ZnONP composites can be used as one of the suitable anticancer drugs. So, it can be concluded that a combination of TPy, ZnONPs, and chitosan in the same formulation may be deemed as a good approach to improve the anticancer performance of the produced composites.

Author Contributions: Conceptualization, N.A.M. and M.A.-M.; Data curation, N.A.M., M.A.-M. and H.D.A.; Formal analysis, N.A.M. and M.A.-M.; Investigation H.D.A., M.A.-M. and M.S.; Methodology, N.A.M., M.A.-M. and M.S.; Resources, N.A.M., M.A-M. and H.D.A.; Supervision, N.A.M. and M.A.-M.; writing—original draft, N.A.M., H.D.A., M.A.-M. and M.S.; Writing—review and editing, N.A.M. All authors have read and agreed to the published version of the manuscript.

Funding: This research received no external funding.

Institutional Review Board Statement: Not applicable.

Informed Consent Statement: Not applicable.

Data Availability Statement: The data presented in this study are available on request from the corresponding authors.

Conflicts of Interest: The authors declare no conflicts of interest.

References

1. Ward, R.A.; Fawell, S.; Floc'h, N.; Flemington, V.; McKerrecher, D.; Smith, P.D. Challenges and Opportunities in Cancer Drug Resistance. *Chem. Rev.* **2021**, *121*, 3297–3351. [CrossRef]
2. Falzone, L.; Salomone, S.; Libra, M. Evolution of Cancer Pharmacological Treatments at the Turn of the Third Millennium. *Front. Pharmacol.* **2018**, *9*, 1300. [CrossRef]
3. Pérez-Herrero, E.; Fernández-Medarde, A. Advanced targeted therapies in cancer: Drug nanocarriers, the future of chemotherapy. *Eur. J. Pharm. Biopharm.* **2015**, *93*, 52–79. [CrossRef]
4. Senapati, S.; Mahanta, A.K.; Kumar, S.; Maiti, P. Controlled drug delivery vehicles for cancer treatment and their performance. *Signal Transduct. Target. Ther.* **2018**, *3*, 7. [CrossRef] [PubMed]
5. Rayhan, M.A.; Hossen, M.S.; Niloy, M.S.; Bhuiyan, M.H.; Paul, S.; Shakil, M.S. Biopolymer and Biomaterial Conjugated Iron Oxide Nanomaterials as Prostate Cancer Theranostic Agents: A Comprehensive Review. *Symmetry* **2021**, *13*, 974. [CrossRef]
6. Parhi, R. Drug delivery applications of chitin and chitosan: A review. *Environ. Chem. Lett.* **2020**, *18*, 577–594. [CrossRef]
7. Elmehbad, N.Y.; Mohamed, N.A.; El-Ghany, N.A.A. Evaluation of the antimicrobial and anti-biofilm activity of novel salicylhydrazido chitosan derivatives impregnated with titanium dioxide nanoparticles. *Int. J. Biol. Macromol.* **2022**, *205*, 719–730. [CrossRef]
8. Mohamed, N.A. Synthesis, characterization and evaluation of in vitro potential antimicrobial efficiency of new chitosan hydrogels and their CuO nanocomposites. *Int. J. Biol. Macromol.* **2024**, *276*, 133810. [CrossRef]
9. Dass, C.R.; Choong, P.F.M. The use of chitosan formulations in cancer therapy. *J. Microencapsul.* **2008**, *25*, 275–279. [CrossRef]
10. Prabaharan, M. Review Paper: Chitosan Derivatives as Promising Materials for Controlled Drug Delivery. *J. Biomater. Appl.* **2008**, *23*, 5–36. [CrossRef]

11. Hasnain, M.S.; Ahmed, S.A.; Alkahtani, S.; Milivojevic, M.; Kandar, C.C.; Dhara, A.K.; Nayak, A.K. Biopolymers for Drug Delivery. In *Advanced Biopolymeric Systems for Drug Delivery*; Springer: Cham, Switzerland, 2020; pp. 1–29. [CrossRef]
12. Ahmed, T.; Aljaeid, B. Preparation, characterization, and potential application of chitosan, chitosan derivatives, and chitosan metal nanoparticles in pharmaceutical drug delivery. *Drug Des. Dev. Ther.* **2016**, *10*, 483–507. [CrossRef]
13. Mohamed, N.A.; El-Ghany, N.A.A.; Abdel-Aziz, M.M. Synthesis, characterization, anti-inflammatory and anti-Helicobacter pylori activities of novel benzophenone tetracarboxylimide benzoyl thiourea cross-linked chitosan hydrogels. *Int. J. Biol. Macromol.* **2021**, *181*, 956–965. [CrossRef] [PubMed]
14. Elmehbad, N.Y.; Mohamed, N.A. Terephthalohydrazido cross-linked chitosan hydrogels: Synthesis, characterization and applications. *Int. J. Polym. Mater. Polym. Biomater.* **2022**, *71*, 969–982. [CrossRef]
15. Elieh-Ali-Komi, D.; Hamblin, M.R. Chitin and Chitosan: Production and Application of Versatile Biomedical Nanomaterials. *Int. J. Adv. Res.* **2016**, *4*, 411–427. Available online: https://pmc.ncbi.nlm.nih.gov/articles/PMC5094803/ (accessed on 12 April 2025).
16. Prateeksha; Sharma, V.K.; Liu, X.; Oyarzún, D.A.; Abdel-Azeem, A.M.; Atanasov, A.G.; Hesham, A.E.-L.; Barik, S.K.; Gupta, V.K.; Singh, B.N. Microbial polysaccharides: An emerging family of natural biomaterials for cancer therapy and diagnostics. *Semin. Cancer Biol.* **2022**, *86*, 706–731. [CrossRef] [PubMed]
17. Alfuraydi, R.T.; Alminderej, F.M.; Mohamed, N.A. Evaluation of antimicrobial and anti-biofilm formation activities of novel poly(vinyl alcohol) hydrogels reinforced with crosslinked chitosan and silver nano-particles. *Polymers* **2022**, *14*, 1619. [CrossRef]
18. Adhikari, H.S.; Yadav, P.N. Anticancer Activity of Chitosan, Chitosan Derivatives, and Their Mechanism of Action. *Int. J. Biomater.* **2018**, *2018*, 2952085. [CrossRef]
19. Park, B.K.; Kim, M.-M. Applications of Chitin and Its Derivatives in Biological Medicine. *Int. J. Mol. Sci.* **2010**, *11*, 5152–5164. [CrossRef]
20. Shanmuganathan, R.; Edison, T.N.J.I.; LewisOscar, F.; Kumar, P.; Shanmugam, S.; Pugazhendhi, A. Chitosan nanopolymers: An overview of drug delivery against cancer. *Int. J. Biol. Macromol.* **2019**, *130*, 727–736. [CrossRef]
21. Elmehbad, N.Y.; Mohamed, N.A.; Abd El-Ghany, N.A.; Abdel-Aziz, M.M. Green synthesis of nano-silver /sodium alginate/carboxymethyl xanthan gum hydrogel and evaluation of its anti-inflammatory and anti-helicobacter pylori activity. *Cellul. Chem. Technol.* **2022**, *56*, 983–995. [CrossRef]
22. Tian, B.; Hua, S.; Liu, J. Multi-functional chitosan-based nanoparticles for drug delivery: Recent advanced insight into cancer therapy. *Carbohydr. Polym.* **2023**, *315*, 120972. [CrossRef] [PubMed]
23. Naskar, S.; Kuotsu, K.; Sharma, S. Chitosan-based nanoparticles as drug delivery systems: A review on two decades of research. *J. Drug Target.* **2018**, *27*, 379–393. [CrossRef] [PubMed]
24. Herdiana, Y.; Wathoni, N.; Shamsuddin, S.; Joni, I.M.; Muchtaridi, M. Chitosan-based nanoparticles of targeted drug delivery system in breast cancer treatment. *Polymers* **2021**, *13*, 1717. [CrossRef]
25. Ding, J.; Guo, Y. Recent advances in chitosan and its derivatives in cancer treatment. *Front. Pharmacol.* **2022**, *13*, 888740. [CrossRef]
26. Fustero, S.; Sánchez-Roselló, M.; Barrio, P.; Simón-Fuentes, A. From 2000 to Mid-2010: A Fruitful Decade for the Synthesis of Pyrazoles. *Chem. Rev.* **2011**, *111*, 6984–7034. [CrossRef] [PubMed]
27. Ansari, A.; Ali, A.; Asif, M.; Shamsuzzaman, S. Review: Biologically active pyrazole derivatives. *New J. Chem.* **2017**, *41*, 16–41. [CrossRef]
28. Uslaner, J.M.; Parmentier-Batteur, S.; Flick, R.B.; Surles, N.O.; Lam, J.S.H.; McNaughton, C.H.; Jacobson, M.A.; Hutson, P.H. Dose-dependent effect of CDPPB, the mGluR5 positive allosteric modulator, on recognition memory is associated with GluR1 and CREB phosphorylation in the prefrontal cortex and hippocampus. *Neuropharmacology* **2009**, *57*, 531–538. [CrossRef]
29. Karrouchi, K.; Radi, S.; Ramli, Y.; Taoufik, J.; Mabkhot, Y.N.; Al-Aizari, F.A.; Ansar, M. Synthesis and Pharmacological Activities of Pyrazole Derivatives: A Review. *Molecules* **2018**, *23*, 134. [CrossRef]
30. Elmehbad, N.Y.; Mohamed, N.A.; Abd El-Ghany, N.A.; Abdel-Aziz, M.M. Evaluation of the in vitro anti-inflammatory and anti-Helicobacter pylori activities of chitosan-based biomaterials modified with copper oxide nanoparticles. *Int. J. Biol. Macromol.* **2023**, *253*, 127277. [CrossRef]
31. Bennani, F.E.; Doudach, L.; Cherrah, Y.; Ramli, Y.; Karrouchi, K.; Ansar, M.; Faouzi, M.E.A. Overview of recent developments of pyrazole derivatives as an anticancer agent in different cell line. *Bioorganic Chem.* **2020**, *97*, 103470. [CrossRef]
32. Alharbi, R.A.; Alminderej, F.M.; Al-Harby, N.F.; Elmehbad, N.Y.; Mohamed, N.A. Design, synthesis, and characterization of novel bis-uracil chitosan hydrogels modified with zinc oxide nanoparticles for boosting their antimicrobial activity. *Polymers* **2023**, *15*, 980. [CrossRef]
33. Mohamed Asik, R.; Gowdhami, B.; Mohamed Jaabir, M.S.; Archunan, G.; Suganthy, N. Anticancer potential of zinc oxide nanoparticles against cervical carcinoma cells synthesized via biogenic route using aqueous extract of *Gracilaria edulis*. *Mater. Sci. Eng. C* **2019**, *103*, 109840. [CrossRef]
34. Elmehbad, N.Y.; Mohamed, N.A.; El-Ghany, N.A.A.; Abdel-Aziz, M.M. Reinforcement of the antimicrobial activity and biofilm inhibition of novel chitosan-based hydrogels utilizing zinc oxide nanoparticles. *Int. J. Biol. Macromol.* **2023**, *246*, 125582. [CrossRef]

35. Aljohar, A.Y.; Muteeb, G.; Zia, Q.; Siddiqui, S.; Aatif, M.; Farhan, M.; Khan, M.F.; Alsultan, A.; Jamal, A.; Alshoaibi, A.; et al. Anticancer effect of zinc oxide nanoparticles prepared by varying entry time of ion carriers against A431 skin cancer cells in vitro. *Front. Chem.* **2022**, *10*, 1069450. [CrossRef] [PubMed]
36. Xie, J.; Li, H.; Zhang, T.; Song, B.; Wang, X.; Gu, Z. Recent advances in ZnO nanomaterial-mediated biological applications and action mechanisms. *Nanomaterials* **2023**, *13*, 1500. [CrossRef] [PubMed]
37. Zhang, H.; Deng, Y. The synergistic effect and mechanism of doxorubicin-ZnO nanocomplexes as a multimodal agent integrating diverse anticancer therapeutics. *Int. J. Nanomed.* **2013**, *8*, 1835–1841. [CrossRef]
38. Mishra, P.K.; Mishra, H.; Ekielski, A.; Talegaonkar, S.; Vaidya, B. Zinc oxide nanoparticles: A promising nanomaterial for biomedical applications. *Drug Discov. Today* **2017**, *22*, 1825. [CrossRef]
39. Alarfaji, S.S.; Ali, I.H.; Bani-Fwaz, M.Z.; Bedair, M.A. Synthesis and Assessment of Two Malonyl Dihydrazide Derivatives as Corrosion Inhibitors for Carbon Steel in Acidic Media: Experimental and Theoretical Studies. *Molecules* **2021**, *26*, 3183. [CrossRef]
40. Kumar Verma, A.; Singh, A.K.; Manauwarul Islam, M. Synthesis, Characterization and Evaluation of Pyridopyrimidine Carboxylate Derivatives as Potential Antimicrobial and Anticancer Agents. *Int. J. Pharm. Pharm. Sci.* **2014**, *6*, 341.
41. Abdel-Motaal, M.; Aldakhili, D.A.; Farag, A.B.; Elmaaty, A.A.; Sharaky, M.; Mohamed, N.A.; Shaaban, S.; Alzahrani, A.Y.A.; Al-Karmalawy, A.A. Design and synthesis of novel multi-target tetrabromophthalimides as CBS and topo-II inhibitors and DNA intercalators. *RSC Med. Chem.* **2024**, *15*, 3800. [CrossRef]
42. Al-Harby, N.F.; Albahly, E.F.; Mohamed, N.A. Synthesis and characterization of novel uracil-modified chitosan as a promising adsorbent for efficient removal of Congo red dye. *Polymers* **2022**, *14*, 271. [CrossRef] [PubMed]
43. Al-Adiwish, W.M.; Tahir, M.I.M.; Yaacob, W.A. Synthesis of Some Novel α-Cyanoketene-N,S-acetals Derived from Secondary Aliphatic Amines and Their Use in Pyrazole Synthesis. *Synth. Commun.* **2013**, *43*, 3203–3216. [CrossRef]
44. Mohamed, N.A.; Al-Harby, N.F.; Almarshed, M.S. Enhancement of adsorption of Congo Red dye onto novel antimicrobial trimellitic anhydride isothiocyanate-cross linked chitosan hydrogels. *Polym. Bull.* **2020**, *77*, 6135–6160. [CrossRef]
45. Ahmad Yusof, N.A.; Mat Zain, N.; Pauzi, N. Synthesis of Chitosan/Zinc Oxide Nanoparticles Stabilized by Chitosan via Microwave Heating. *Bull. Chem. React. Eng. Catal.* **2019**, *14*, 450–458. [CrossRef]
46. Alharbi, R.A.; Alminderej, F.M.; Al-Harby, N.F.; Elmehbad, N.Y.; Mohamed, N.A. Preparation and characterization of a new bis-uracil chitosan-based hydrogel as efficient adsorbent for removal of anionic Congo red dye. *Polymers* **2023**, *15*, 1529. [CrossRef] [PubMed]
47. Li, J.; Sathasivam, S.; Taylor, A.; Carmalt, C.J.; Parkin, I.P. Single step route to highly transparent, conductive and hazy aluminium doped zinc oxide films. *RSC Adv.* **2018**, *8*, 42300–42307. [CrossRef]
48. Younes, I.; Rinaudo, M. Chitin and chitosan preparation from marine sources. structure, properties and applications. *Mar. Drugs* **2015**, *13*, 1133–1174. [CrossRef]
49. Zhang, Y.; Wu, C.; Zhang, N.; Fan, R.; Ye, Y.; Xu, J. Recent advances in the development of pyrazole derivatives as anticancer agents. *Int. J. Mol. Sci.* **2023**, *24*, 12724. [CrossRef]
50. Chandrasekaran, S.; Anusuya, S.; Anbazhagan, V. Anticancer, anti-diabetic, antimicrobial activity of zinc oxide nanoparticles: A comparative analysis. *J. Mol. Struct.* **2022**, *1263*, 133139. [CrossRef]
51. Singh, T.A.; Das, J.; Sil, P.C. Zinc oxide nanoparticles: A comprehensive review on its synthesis, anticancer and drug delivery applications as well as health risks. *Adv. Colloid Interface Sci.* **2020**, *286*, 102317. [CrossRef]
52. Babayevska, N.; Przysiecka, Ł.; Iatsunskyi, I.; Nowaczyk, G.; Jarek, M.; Janiszewska, E.; Jurga, S. ZnO size and shape effect on antibacterial activity and cytotoxicity profile. *Sci. Rep.* **2022**, *12*, 8148. [CrossRef] [PubMed]

polymers

MDPI

Article

Dissipative Particle Dynamics: Simulation of Chitosan–Citral Microcapsules

Wensheng Wu [1], Zhiwei Li [1,*], Dachun Feng [2], Qing Tang [1], Shuijiao Liu [1] and Wenjing Lin [3,*]

[1] Guangdong Provincial Key Laboratory of Environmental Health and Land Resource, School of Environmental and Chemical Engineering, Zhaoqing University, Zhaoqing 526061, China; 2004010015@zqu.edu.cn (W.W.); tangqing2013@126.com (Q.T.); liushuijiao2016@126.com (S.L.)
[2] College of Information Science and Technology, Zhongkai University of Agriculture and Engineering, Guangzhou 510225, China; fdczhku@sina.com
[3] School of Chemical Engineering and Light Industry, Guangdong University of Technology, Guangzhou 510006, China
* Correspondence: lizhiwei@zqu.edu.cn (Z.L.); wenjing.lin@gdut.edu.cn (W.L.)

Abstract: In this paper, the dissipative particle dynamics (DPD) method is used to simulate the self-assembly process, appearance, mesoscopic structure, and wrapping properties of microcapsules formed with citral as the core material and chitosan and sodium alginate as the single-wall materials, and with citral as the core material and chitosan-sodium alginate, chitosan–methylcellulose, sodium alginate–chitosan, and sodium alginate–methylcellulose as the double-wall materials. The effects of chitosan content and wall material composition on the structure, morphology, encapsulation performance, and stability of microcapsules are compared and analyzed. In addition, the microcapsules are deeply analyzed by using the mesoscopic structure, radial distribution function, and diffusion coefficient. This study provides a new idea and method for the preparation of citral microcapsules, and is of great significance for the design and development of new composite wall microcapsules.

Keywords: dissipative particle dynamics (DPD) simulation; microcapsules; chitosan

1. Introduction

Microencapsulation is an important technology for making functional materials, which are used for packaging and storing trace substances. Microcapsule technology plays an important role in drug encapsulation [1–7], pesticide encapsulation [8–11], food encapsulation [12], and environmental pollution treatment [13,14]. In microcapsule technology, wall materials that can be used as shell materials are generally natural, semisynthetic or fully synthetic polymer materials. Coated materials are called core materials, which can be compounds or mixtures in powder, liquid, gas, or solid states. During the preparation of microcapsules, the types of wall and core materials, the volume fraction of each component and the pH value of the system all affect the synthesis, storage, and release properties of microcapsules. Especially with regard to drug loading and release, it is necessary to consider the influence of wall materials and components on the synthesis, storage, and release properties of microcapsules when the required drugs are packaged and transported to the diseased site for treatment.

Chitosan (Cts) [15], sodium alginate (Alg) [16], and methyl cellulose (MC) [17,18] are excellent wall materials in microcapsule technology due to their advantages of wide sources, good film-forming properties, stable performance, safety, nontoxicity, green environmental protection, and biodegradability.

Essential oil has a wide range of uses. For example, it can help skin moisturize and replenish water, promote blood circulation, relieve body discomfort, aid anti-inflammation and sterilization, and help restore hair luster. Citral is an essential oil with strong physiological activities, such as anticancer [19], antimutagenic [20], anti-inflammatory [21], antioxidant [22], antiviral [23], and antibacterial activities [24]. Essential oils are highly volatile and need to be sealed for storage. Microencapsulation is a common method for addressing this issue. To optimize encapsulation performance, it is critical to understand the self-assembly dynamics between core and wall materials at the mesoscopic scale. Dissipative particle dynamic (DPD) simulation provides a powerful tool to visualize such processes [25,26]. Therefore, essential oil is often used as the core material in microencapsulation technology.

Dissipative particle dynamic (DPD) simulation [25,26] is a commonly used research method in the field of chemistry that can not only solve the problem that some experimental techniques cannot, but can also greatly reduce experimental time and cost. This simulation technique can provide a convenient and intuitive way to observe the morphological changes in microcapsules and study their properties. DPD simulation can be used to analyze the aggregation process of the core material wrapped by the wall material and to study the encapsulation performance of the microcapsules. In recent years, DPD simulation has attracted great attention in microencapsulation applications, such as nanofluids [27], drug delivery [28], water cluster structures [29], polymer self-assembly [30], surface chemistry [31–33], ionic liquid microemulsion [34], phase diagram [35,36], and environmental protection [37].

In this paper, a DPD simulation strategy is used to study the formation process of Cts–Citral microcapsules, with Cts as the main wall material, Alg and MC as the auxiliary wall materials, and citral as the core material. First, the aggregation formation process and influencing factors of single-walled material microcapsules formed by Cts-coated citral are studied. Then, based on the single-wall material microcapsule, Alg and MC are added to replace part of the Cts. The effects of Alg and MC on the formation process of Cts–Citral microcapsules are studied by adding auxiliary wall materials and comparing the encapsulation performance of Cts–Citral microcapsules with single-wall materials. This study can provide a theoretical reference for the preparation of Cts–Citral microcapsules and the optimization of encapsulation performance.

2. Method

In this study, the DPD method is employed to simulate the self-assembly process of microcapsules. The DPD parameters and interaction forces are set according to the established protocols, as described in previous studies [25,26].

2.1. Coarse-Grained Model

The simulated wall materials are Cts, Alg, and MC, and the core material is citral. The simulated environment is a neutral environment rich in water molecules. The coarsening treatment of molecules divides the molecules into different beads, and each bead represents one or more atomic clusters [38]. Using the coarse-grained beads method, the molecular structure formula of each substance and the corresponding coarse-grained model are shown in Figure 1. According to the molecular structure and the size of the beads, the molecules are divided as follows: one Alg molecule is composed of two N3 beads, one S bead, and one O3 bead; the MC molecule is composed of two N2 beads, one O1 bead, and one O2 bead; the Cts molecule is composed of two N1 beads and one O4 bead; the citral molecule is composed of one C1 bead and one C2 bead; and the three water molecules are composed of one W bead. In this simulation experiment, the molecular polymerization degree is set as 20.

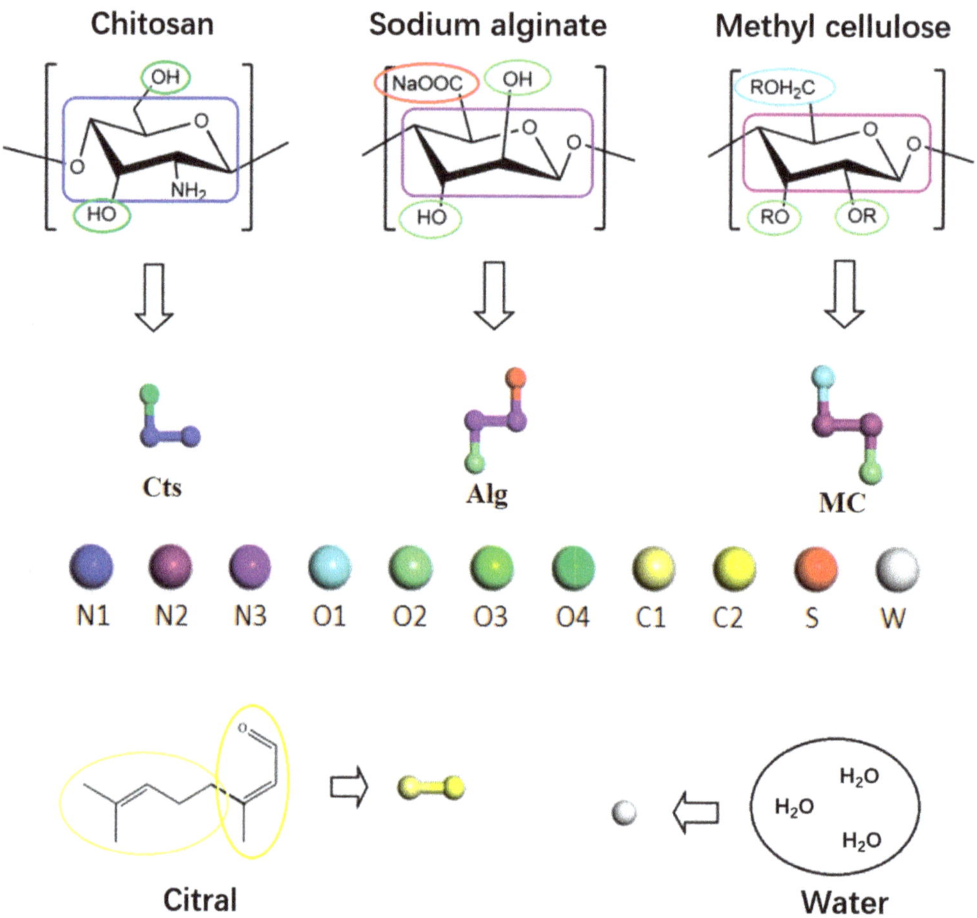

Figure 1. Molecular structures and coarse-grained models.

2.2. Interaction Parameters Between Beads

After the coarse-grained model is determined, the interaction parameters between each pair of beads can be calculated by introducing the bead repulsion parameter a_{ii} or a_{ij} and Flory Huggins parameter x_{ij} [39].

The mutual exclusion parameter is obtained using the following formula:

$$a_{ii} = 75k_BT/\rho \quad (1)$$

where a_{ii} is the repulsion parameter between the same types of particles. The compressibility of pure fluid can be set as ρ = 3 (close to that of water), and the value of k_BT = 1 can be used in the simulations. The values of the repulsion parameters between different types of particles (a_{ij}) are linearly related to the Flory–Huggins parameters (x_{ij}) according to the equation [38]:

$$a_{ij} = a_{ii} + 3.27x_{ij} \quad (2)$$

For the two different components i and j, the Flory Huggins parameter x_{ij} is calculated via the following formula:

$$x_{ij} = \frac{\Delta E^{mix} V_r}{RT\varphi_i\varphi_j V} \quad (3)$$

where R is the ideal gas constant, T is the absolute temperature, V is the total volume of the mixed component system, V_r is the relative volume, and φ_i and φ_j are the volume fractions of components i and j, respectively. ΔE^{mix} is the mixing energy, which can be calculated using the following formula:

$$\Delta E^{mix} = E_{ij} - (E_i + E_j) \quad (4)$$

where E_{ij} is the total potential energy of components i and j and E_i and E_j are the potential energy of pure components i and j, respectively. The box size of the simulation system is $200 \times 200 \times 200$ Å^3. The COMPASS force field is used in the calculation of the interaction parameters, which is carried out in the Amorphous Cell and Forcite modules of Materials Studio 7.0 software [40]. The interaction parameters between the beads are shown in Table 1.

Table 1. Interaction parameters among beads.

a_{ij}	N1	N2	O1	O2	C1	C2	S	W
N1	25							
N2	25.92	25						
O1	27.47	30.81	25					
O2	44.23	51.46	31.06	25				
C1	29.05	26.08	36.27	61.69	25			
C2	25.49	25.01	29.65	46.44	25.16	25		
S	31.77	36.46	25.88	27	43.43	34.17	25	
W	95.99	106.18	57.89	35.74	121.98	87.94	44.73	25

Note: All parameters in Table 1 are expressed in DPD reduced units, where $r_c = 1$, $k_BT = 1$, and $\rho = 3$.

It should be noted that N2 and N3 are the same beads, and O2, O3, and O4 are the same beads. In the structure of MC, there are two -OR groups in the branched chain, where R can be hydrogen-H or methyl-CH_3. To make the simulation more straightforward, two -OR groups are set as two -OH groups. Therefore, O2 beads are two -OH groups in MC, O3 beads are two -OH groups in Alg, and O4 beads are two -OH groups in Cts. Although the three beads are the same, to distinguish them in different wall materials, different numbers are set. The interaction parameters of the three beads with other beads are the same.

All simulations start from the random distribution of each component. The simulation is carried out in a cube box with periodic boundary conditions, and its side length is $200 \times 200 \times 200$ Å^3. After calculations, the average volume of the beads is 111.03 Å^3, the mass is 54 amu, and the radius is 2.98 Å. In the simulation system, the number density ρr_c^3 of the beads is set to 3, and there are three beads in each grid. That is, the grid size is 333.09 Å^3, the side length is 6.93 Å, and the cut-off radius r_c between the beads is calculated to be 6.93 Å. The polymerization degree is set to 20. According to the pre-simulation experimental results, there is almost no difference between the sections obtained by 250,000 steps and those obtained by 200,000 steps, indicating that the system has reached the microphase separation equilibrium when the simulation time is 200,000 steps. In order to ensure the accuracy of the results, the simulation time of all systems is 250,000 steps. The time integration step is 0.05 ns. All simulations are carried out in the Mesosite module of Materials Studio 7.0 software [40]. To clearly observe the shape of the beads in the box, all the results hide the water beads.

2.3. Simulation System and Scheme

The simulated system consists of citral, Cts, Alg, MC, and H_2O. First, Cts–Citral single-wall microcapsules are prepared by adding different volume fractions of Cts. Then, in the case of fixed Cts, different volume fractions of Alg or MC are added to replace part of the Cts to prepare double-wall microcapsules. By observing the formation process of the microcapsule and analyzing the simulation results, the influence of the content of each component on the formation of Cts microcapsules is found, which provides guidance for the preparation of Cts microcapsules with better performance.

It has been reported [41,42] that under the experimental conditions of 1.0 g/30 mL H_2O, 6.0 g/180 mL H_2O, and 1.6 g/50 mL H_2O for Cts, Alg, and MC, the content of citral is the largest, and the encapsulation performance and stability of microcapsules are better. The following simulation scheme is designed. In each simulation system, the volume of water is 260 mL, and the volume of citral is 1.5 mL. The amount of the remaining components and the volume fraction of all components are shown in Tables 2–5.

Table 2. Amounts and volume fractions of each component of Cts single-wall microcapsules.

	Number	H_2O (mL)	*m* (Cts) (g)	φ (H_2O)	φ (Cts)	Citral (mL)	φ (Citral)
Citral free	Cts (1)	260.00	5.2	98.87	1.13		
	Cts (2)	260.00	6.93	98.50	1.50		
	Cts (3)	260.00	8.67	98.13	1.87		
	Cts (4)	260.00	10.4	97.77	2.23		
	Cts (5)	260.00	12.13	97.40	2.60		
Containing Citral	Cts–Citral (1)	260.00	5.2	98.31	1.12	1.5	0.57
	Cts–Citral (2)	260.00	6.93	97.94	1.49	1.5	0.57
	Cts–Citral (3)	260.00	8.67	97.58	1.86	1.5	0.56
	Cts–Citral (4)	260.00	10.4	97.22	2.22	1.5	0.56
	Cts–Citral (5)	260.00	12.13	96.86	2.58	1.5	0.56

Table 3. Amounts and volume fractions of each component of Alg single-wall microcapsules.

	Number	H_2O (mL)	Alg (g)	φ (H_2O)	φ (Alg)	Citral (mL)	φ (Citral)
Citral free	Alg (1)	260.00	7.22	98.28	1.72		
	Alg (2)	260.00	7.94	98.12	1.88		
	Alg (3)	260.00	8.67	97.95	2.05		
	Alg (4)	260.00	9.39	97.78	2.22		
	Alg (5)	260.00	10.11	97.61	2.39		
Containing Citral	Alg–Citral (1)	260.00	7.22	97.73	1.71	1.5	0.56
	Alg–Citral (2)	260.00	7.94	97.56	1.87	1.5	0.56
	Alg–Citral (3)	260.00	8.67	97.40	2.04	1.5	0.56
	Alg–Citral (4)	260.00	9.39	97.23	2.21	1.5	0.56
	Alg–Citral (5)	260.00	10.11	97.07	2.37	1.5	0.56

Table 4. Amounts and volume fractions of each component of the Alg–Cts double-wall microcapsule (the amount of Alg is fixed at 14.00 g).

	Number	H_2O (mL)	Alg (g)	φ (H_2O)	Cts (g)	φ (Cts)	φ (Alg)	Citral (mL)	φ (Citral)
Citral free	Alg–Cts (1)	260.00	14.00	96.60	0.60	0.13	3.27		
	Alg–Cts (2)	260.00	14.00	96.56	0.80	0.17	3.27		
	Alg–Cts (3)	260.00	14.00	96.52	1.00	0.21	3.27		
	Alg–Cts (4)	260.00	14.00	96.48	1.20	0.25	3.27		
	Alg–Cts (5)	260.00	14.00	96.44	1.40	0.30	3.27		
Containing Citral	Alg–Cts–Citral (1)	260.00	14.00	96.07	0.60	0.13	3.25	1.50	0.55
	Alg–Cts–Citral (2)	260.00	14.00	96.03	0.80	0.17	3.25	1.50	0.55
	Alg–Cts–Citral (3)	260.00	14.00	95.98	1.00	0.21	3.25	1.50	0.55
	Alg–Cts–Citral (4)	260.00	14.00	95.94	1.20	0.25	3.25	1.50	0.55
	Alg–Cts–Citral (5)	260.00	14.00	95.90	1.40	0.30	3.25	1.50	0.55

Table 5. Simulation of Alg–MC double-wall microcapsules (Alg unchanged).

	Number	H_2O (mL)	Alg (g)	MC (g)	φ (H_2O)	φ (MC)	φ (Alg)	Citral (mL)	φ (Citral)
Citral free	Alg–MC (1)	260.00	7.00	0.40	98.22	0.12	1.66		
	Alg–MC (2)	260.00	7.00	0.60	98.16	0.17	1.66		
	Alg–MC (3)	260.00	7.00	0.80	98.11	0.23	1.66		
	Alg–MC (4)	260.00	7.00	1.00	98.05	0.29	1.66		
	Alg–MC (5)	260.00	7.00	1.20	97.99	0.35	1.66		
Containing Citral	Alg–MC–Citral (1)	260.00	7.00	0.40	97.67	0.12	1.65	1.50	0.56
	Alg–MC–Citral (2)	260.00	7.00	0.60	97.61	0.17	1.65	1.50	0.56
	Alg–MC–Citral (3)	260.00	7.00	0.80	97.55	0.23	1.65	1.50	0.56
	Alg–MC–Citral (4)	260.00	7.00	1.00	97.50	0.29	1.65	1.50	0.56
	Alg–MC–Citral (5)	260.00	7.00	1.20	97.44	0.35	1.65	1.50	0.56

3. Results and Discussion

3.1. Self-Assembly Process of Microcapsules

To ensure the balance of the system, the DPD simulation steps are set to 250,000 steps. Every 5000 steps is a frame, so the experiment has a total of 51 frames. The formation process of microcapsules can be seen from the simulation results, as shown in Figures 2–4. To facilitate observation, the water beads are hidden here. The stability of the microcapsules is supported by their consistent spherical shape and uniform size distribution, as observed in previous studies [38,43]. The radius of gyration and shape parameters were found to be consistent with those reported in similar systems.

3.1.1. Cts–Alg Double-Wall Microcapsule

From the microcapsule formation process shown in Figure 2, it can be seen that at the beginning, the molecules are randomly dispersed in the box. At 5000 steps, some of the molecules combine with the same kind of molecule and curl up with each other. Most of the molecules are close to each other, forming aggregates of different sizes, and some of the aggregates contain citral molecules. At 20,000 steps, the former small agglomerations further merge and become larger and gradually become spherical. By 150,000 steps, all the components have gathered together to form a spherical aggregate, namely, a microcapsule. However, the aggregation degree of each component cannot be clearly judged from the appearance, so the aggregates at 150,000, 200,000, and 250,000 steps are cut open to observe the cross-section, as shown in Figure 3.

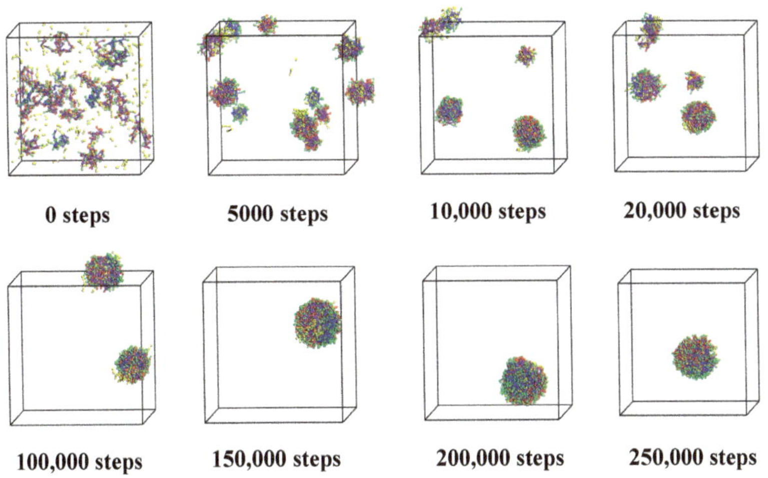

Figure 2. Formation of microcapsules.

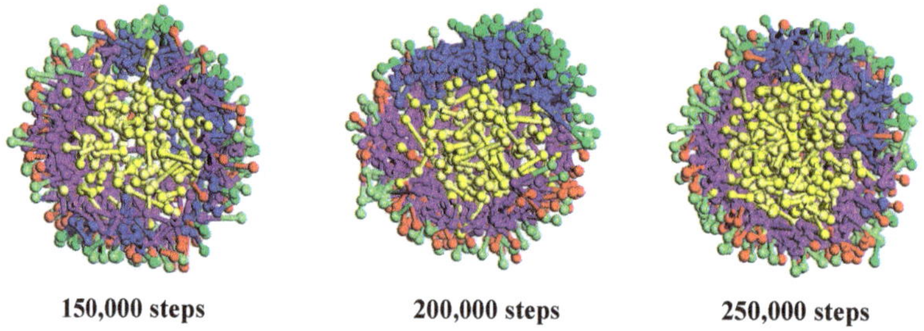

Figure 3. Cross section views of 150,000, 200,000, and 250,000 steps.

To quantitatively evaluate the structural integrity and defects of the microcapsules, we analyzed the radial distribution function (RDF) [44] and concentration profiles at different simulation steps. While radial density profiles provide valuable information about the distribution of species from the center-of-mass, the RDF was chosen to highlight the relative distances between different components, which is crucial for understanding the encapsulation mechanism. The results show that at 150,000 steps, the structure of the microcapsules is not yet fully stable, with significant fluctuations in the RDF curves, indicating that intermolecular interactions have not reached equilibrium. However, at 200,000 and 250,000 steps, the RDF curves stabilize, indicating that the system has reached microphase separation equilibrium.

3.1.2. Alg–Cts Double-Wall Material Microcapsule

As shown in Figure 4, the simulation results show that the Alg, Cts, and citral in the initial state (0 step) disperse irregularly in the water, and the polymer exhibits a random stretching state. With the development of the simulation, the citral in the aqueous phase formed small aggregates under the influence of hydrophobicity and gradually approached the polymer. With increasing simulation time, the aggregates continue to collide further and diffuse into the polymer at the same time. In 195,000 steps, single and relatively uniform spherical nanoparticles are formed. It can be clearly and directly seen that Alg and Cts can be used as composite wall materials of citral microcapsules, which provides a research basis for practical applications.

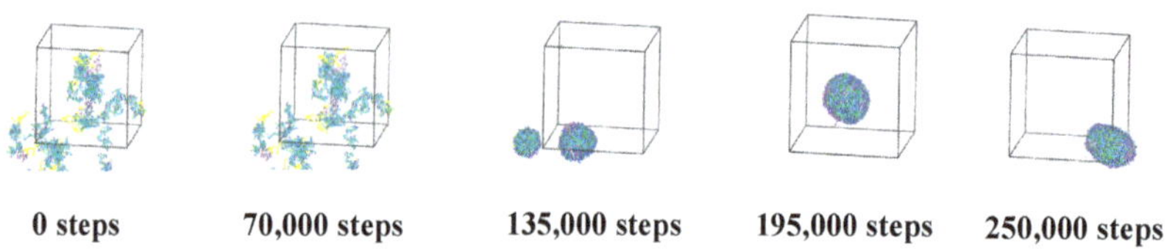

Figure 4. Self-assembly of microcapsules made of Alg and Cts double-wall materials. (Periodic boundary conditions have been removed for clarity. The images represent the microcapsule structure at specific simulation time steps).

3.1.3. Alg–MC Double-Wall Microcapsule

The number of simulation steps is 250,000, and there are 51 frames per 5000 steps. The following is taken from frames 1, 14, 27, 39, and 51 of the track file (xtd file) obtained from the fourth group of Alg–MC oil, and the results are shown in Figure 5. To observe the experimental results, the water beads are hidden. The simulation results show that Alg, MC, and citral in the initial state (0 step) disperse irregularly in water, and that the polymer exhibits a random stretching state. With the progress of the simulation, the polymer in the aqueous phase forms small aggregates under the influence of hydrophobicity and gradually approaches other polymers. As the simulation time continues to increase, the aggregates continue to collide further and diffuse into other polymers at the same time. By frame 39, complete particles have been formed. At frame 39, there is a multiparticle in a small box. When increasing the size of the box, it can be observed that 1.5 boxes can form two complete microspheres, as shown in frame 51. It is clear that Alg and MC can be used as composite wall materials of citral microcapsules, which provides a research basis for the practical application of composite wall materials.

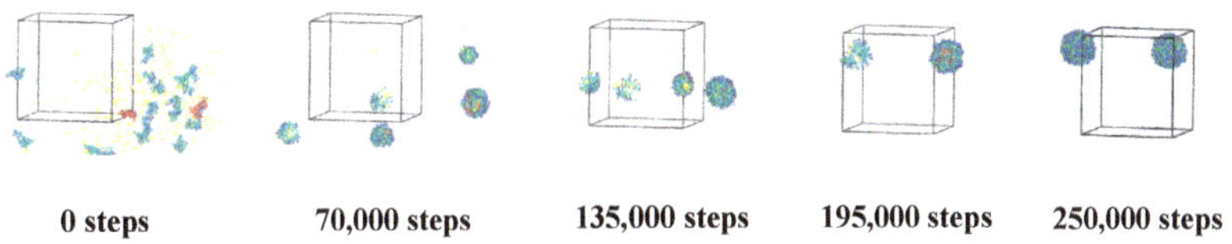

Figure 5. Self-assembly of microcapsules made of Alg and MC double-wall material. (Periodic boundary conditions have been removed for clarity. The images represent the microcapsule structure at specific simulation time steps).

3.2. Structure Analysis of Microcapsules

The ideal microcapsule structure should have good stability and intelligent aggregation to improve its sealing, which is very important to obtain a high small-molecule encapsulation ability. Therefore, first, the morphology and cross-section of microcapsules are analyzed to evaluate the distribution and aggregation of beads in microcapsules. Then, the encapsulation and bead distribution of microcapsules with different wall materials are studied more specifically and deeply from the RDF, and the internal structure and morphology of microcapsules are analyzed by comparing RDF curves.

3.2.1. Single-Wall Material

Cts–Citral Single-Wall Microcapsule

As shown in Figure 6, Cts microspheres are divided into three layers, namely O4 beads, N1 beads, and O4 beads, from the outside to the inside. The reason for the formation of such a structure is that O4 is a hydrophilic hydroxyl group, so it is distributed in the outermost layer adjacent to water molecules. The main chain of Cts is a hydrophobic carbon chain, which should be distributed far from water beads, and all of them should be in the interior. However, because Cts is a high-molecular-weight polymer, the space volume of the main chain is relatively large, so it cannot be fully squeezed in the middle of the microsphere. Therefore, the main chain is distributed in the middle layer. At the same time, a small part of the branch chain, that is, O4 beads, extends into the inner core, and the innermost layer is a small part of the branch chain, that is, O4 beads extend into it. Therefore, a three-layer structure with a hydrophilic core and shell layers and a hydrophobic medium layer is formed.

When the core material citral was added, a Cts–Citral microcapsule was formed. Morphologically, the Cts–Citral microcapsule is also a three-layer structure. From the outside to the inside of the microcapsule, O4 beads of Cts, N1 beads of Cts, and C1 and C2 beads of citral are in order. Citral has a great repulsive force with water molecules, so it is distributed in the innermost layer. The N1 bead in the main chain of Cts is a six-membered ring structure with a lipophilic carbon chain, so the N1 bead is adjacent to citral and is in the middle layer of the microcapsule. The O4 bead of the Cts branch chain is a hydrophilic group, and there is also a large repulsive force between the O4 bead and C1 and C2 beads, so O4 beads will not be adjacent to the C1 and C2 layers, but all extend to the outside. From Figure 4, when citral is added, the coating layer of the microcapsule becomes thinner. The reason is that the volume of the inner layer becomes larger, which leads to more compact binding between the main chains of Cts.

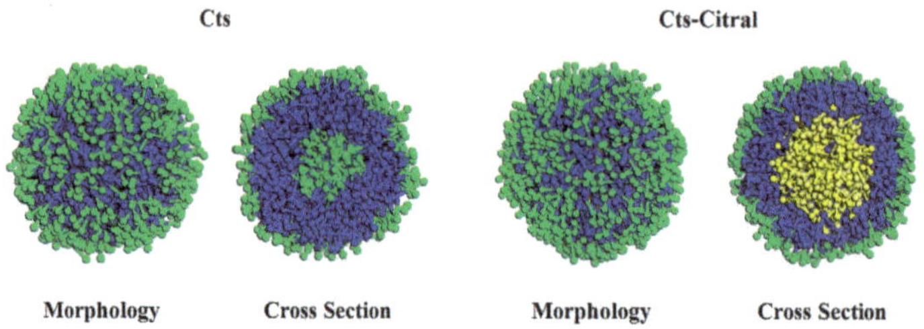

Figure 6. Morphology and cross-section of Cts microspheres and microcapsules.

It can also be seen from the concentration distribution curve in Figure 7 (The concentration profiles were calculated by radially averaging the bead density from the center-of-mass of the microcapsule, with a bin width of 1 Å) that when there is only the wall material Cts, the concentration distribution curve of O4 beads has a peak at 80 Å in the middle, indicating that the concentration of O4 beads is the highest in the middle, which confirms that the O4 beads are distributed in the innermost layer. Because of the symmetry of the microsphere, the curves of N1 and O4 are approximately symmetrical. The middle part of the O4 bead peak is clearly depressed, which indicates that most of the innermost O4 beads are extruded due to the entry of citral; therefore, the concentration of O4 beads is greatly reduced. In addition, the peaks of the N1 and O4 beads are wider than those of the C1 and C2 beads, which further explains the microcapsule structure.

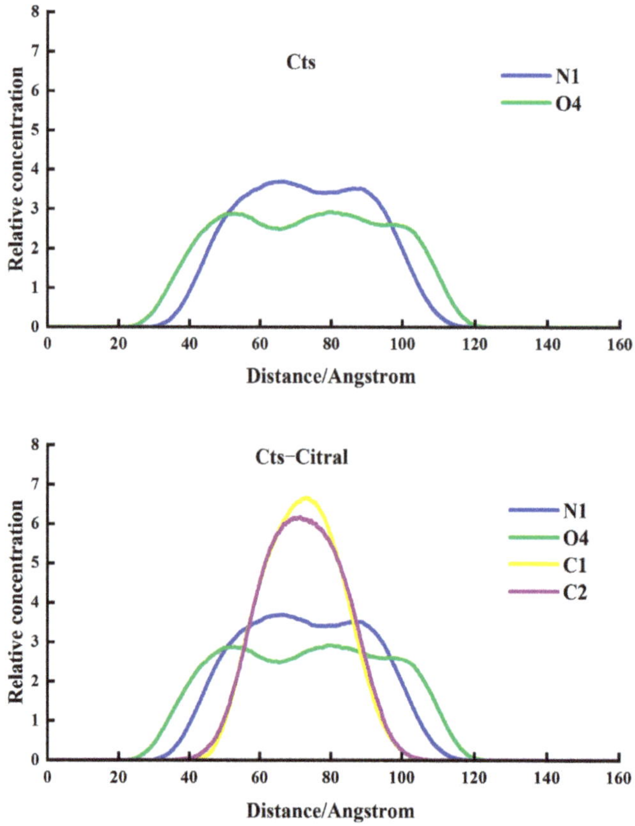

Figure 7. Concentration distribution curve of Cts microcapsules and microspheres.

(a) Effect of Cts content on the appearance and cross-section of Cts–Citral microcapsules

By changing the content of Cts, the cross-section and RDF of the microcapsule are compared to analyze the effect of Cts content on the properties of the microspheres. As shown in Figure 6, the water volume is set as 260 mL, the citral volume is 1.5 mL, and the concentration of Cts is set as 0.6 g/30 mL H_2O, 0.8 g/30 mL H_2O, 1.0 g/30 mL H_2O, 1.2 g/30 mL H_2O, and 1.4 g/30 mL H_2O. The simulation results are shown in Figure 8. The volume fraction is calculated, and the corresponding simulations are carried out.

Figure 8 shows that when the Cts content is 0.6 g/30 mL H_2O, the microspheres have obvious defects in appearance. The microsphere is not wrapped well enough to even see the citral molecules inside. The cross-section shows that some citral molecules are embedded in the main chain of Cts. Therefore, the microcapsules under this condition are not complete, the thickness of the cladding is small, and the arrangement of O4 beads is not neat enough. These disadvantages will seriously reduce the encapsulation performance of Cts microcapsules, which makes disintegration easy during storage.

When the content of Cts is increased to 0.8 g/30 mL H_2O and 1.0 g/30 mL H_2O, the cracks of the shell tended to close, and few citral molecules are exposed. The cross-section shows that the citral molecule shrinks inward, and that the citral molecule embedded in the main chain layer of the Cts molecule gradually detaches. However, from the appearance and cross-sectional point of view, there are still some defects in the shape of the microspheres, which is not a more regular sphere.

When the content of Cts is increased to 1.2 g/30 mL H_2O, few citral molecules are embedded in the main chain of Cts, and the encapsulation is further improved. The order of O4 beads is greatly improved, and the thickness of the whole microcapsule is more uniform than before. When Cts reaches the maximum content of 1.4 g/30 mL H_2O in the simulation experiment, the citral molecules in Cts are almost invisible. The whole sphere is round, and there is no citral molecule embedded in the main chain of Cts. The Cts microcapsule layer is thicker than the core material. The arrangement of O4 beads is also very neat, forming a three-layer structure of core material–N1–O4. Clearly, the Cts microcapsule with this concentration has the best encapsulation performance.

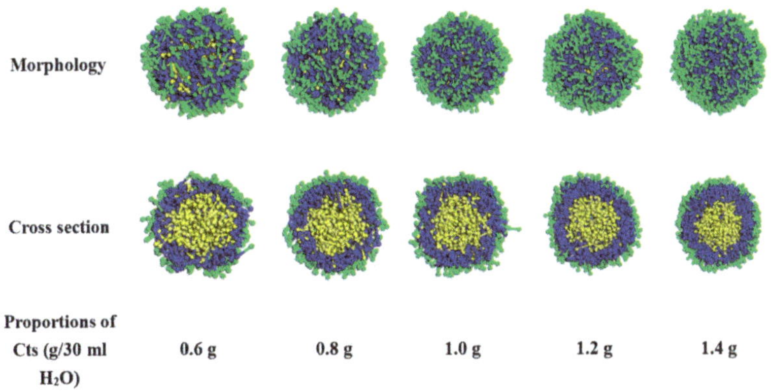

Figure 8. Appearance and cross-section of DPD simulations with different proportions of Cts to H_2O.

(b) Analysis of RDF

The RDF refers to the distribution probability of other particles relative to a given particle in space, so the RDF can reflect the distance relationship between particles. In this study, C1 beads in citral molecules are selected as reference beads, and then the RDFs of the N1 and O4 beads in Cts relative to C1 beads are studied and compared, as shown in

Figure 9. From the RDF curves of C1–N1 and C1–O4, the peak of the C1–N1 curve is higher and closer to the left end than that of C1–O4, indicating that the distance between the N1 bead and C1 bead is closer. That is, the affinity between the C1 and N1 beads is better than that between the C1 and O4 beads, and the interaction force is smaller. This conclusion is consistent with the size of the interaction parameter a_{ij} calculated above, which indicates that the RDF curve can reflect the size of the interaction parameter of each component from the side.

From the RDF curve of C1–N1, it can be seen that with increasing Cts content, the peak of the C1–N1 RDF curve gradually decreased and shifted to the right. The change trend of the RDF curve showed that the higher the content of Cts is, the greater the N1 beads spread to the outer layer; that is, the distance between the C1 and N1 beads became larger. The reason for this is that with increasing Cts molecules, the affinity between the N1 and C1 beads decreases, and the repulsive force increases. However, the space volume of the N1 bead is so large that it cannot diffuse to the inside, only to the outside. The RDF curve of C1–O4 is very similar to that of C1–N1. With increasing Cts content, the peak of the curve gradually decreased and shifted to the right. The reason is that the N1 beads diffused to the outer layer under the gradually strengthened repulsive force, so the O4 bead is forced to move to the outer layer as a whole. This analysis can also explain why the C1 beads gradually separate from the N1 beads with increasing Cts content in the cross-section. Because the affinity between the C1 and N1 beads decreased and the repulsive force increased, the C1 beads could not be embedded into the N1 beads.

By comparing the above groups of simulation experiments, it can be found that when the content of citral is fixed, the higher the content of Cts is, and the better the encapsulation performance of citral. However, considering the cost and other factors, the content of Cts should be controlled within a certain range, so it is not necessary to increase it too much, as long as it can be completely wrapped with citral.

In this study, the RDFs were averaged over the last 50,000 simulation steps (500 frames sampled every 100 steps) to ensure statistical reliability.

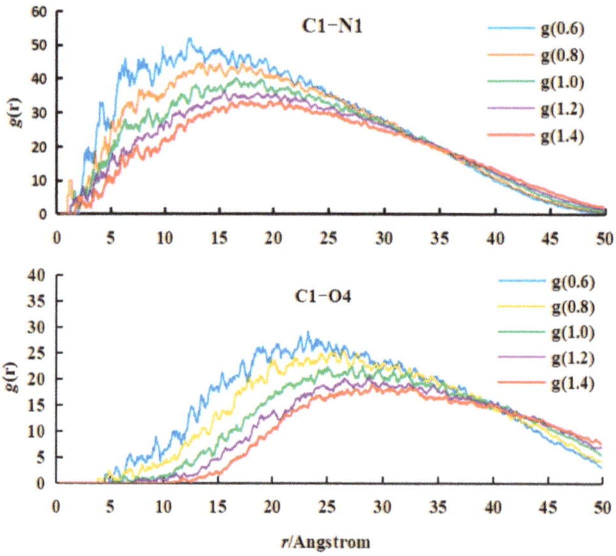

Figure 9. RDF of N1 and O4 beads with different proportions of Cts.

Alg–Citral Single-Wall Microcapsule

Figure 10 shows that under neutral simulation conditions, Alg can form a spherical core–shell structure and tightly embed the citral in it. The hydrophilic O3 beads (dark blue) are located in the outermost layer of the microcapsule, forming a hydrophilic protective shell, and N3 is mainly in the hydrophobic inner layer and tightly envelops the citral.

Compared with the profile of microcapsules, the RDF curve can more directly reflect the distribution characteristics of citral. For the same content of citral, the citral is preferentially distributed in the inner layer (hydrophobic layer) of the capsule under the condition of a lower content of wall material. Because the affinity of N3 hydrophobic beads to citral (C1 beads) is greater than that of O3, the citral is more distributed in the inner layer. With an increase in wall material content until saturation, the extra N3 beads diffuse to the outer layer. Therefore, it can be seen that with increasing wall material content, the peak of the N3 bead curve is closer to that of the O3 bead curve.

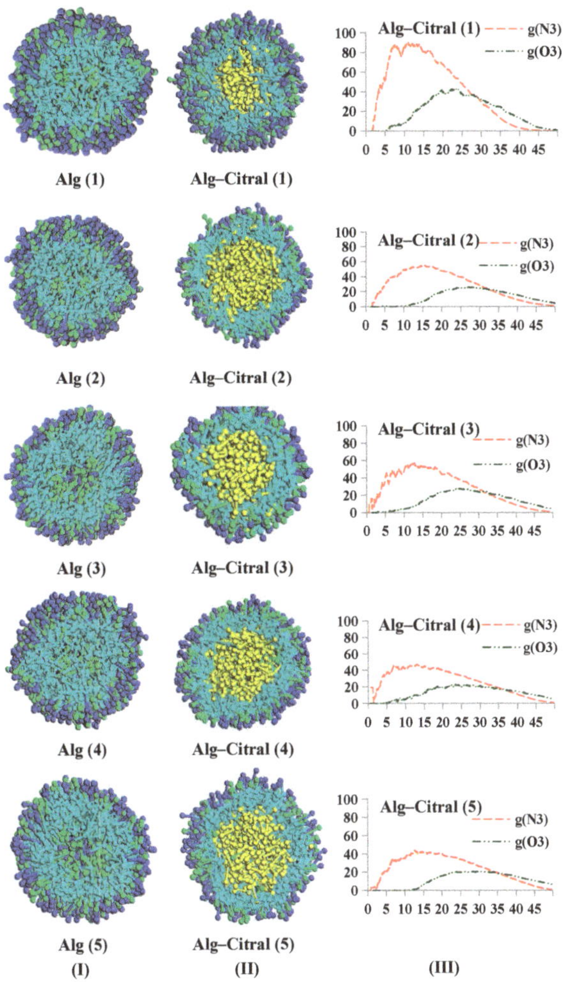

Figure 10. Alg single-wall microcapsule ((**I**) citral free; (**II**) containing citral; and (**III**) RDF curve).

3.2.2. Double-Wall Material Microcapsule

Cts–Alg–Citral and Cts–MC–Citral Microcapsules

The effects of Alg and MC on Cts–Citral microcapsules are also simulated. DPD simulation with a Cts content of 1.0 g/30 mL H_2O is selected for comparison, and the total amount of wall materials in the three groups is equal.

(a) Appearance and cross-section of microcapsules

Figure 11 shows that for the microcapsules doped with Alg or MC, both Alg and MC have obvious boundaries with Cts molecules and are not mixed with Cts molecules. The formation of this phenomenon can be explained by the interaction between the molecules of each component. Clearly, the interaction between Cts molecules is weaker than that between Cts and Alg or MC, and the corresponding repulsive forces are stronger.

Compared with the pure Cts microcapsule, the microcapsule made of Cts doped with Alg has no obvious difference in appearance and shape, and a small amount of citral molecules are embedded in the wall layer. However, when Cts doped with MC is used as the wall material, the microcapsule is quite different. First, citral molecules are more exposed in the microspheres formed by MC. Second, from a cross-sectional view, the microspheres formed by MC are obviously thinner, and the citral molecules are embedded more, with some even embedded in the most peripheral position of the microcapsule. Moreover, a thin Cts microsphere layer covered the outside of the MC microcapsule layer. This phenomenon indicates that the encapsulation performance of MC to citral is slightly worse than that of Cts, and it cannot be well coated with citral.

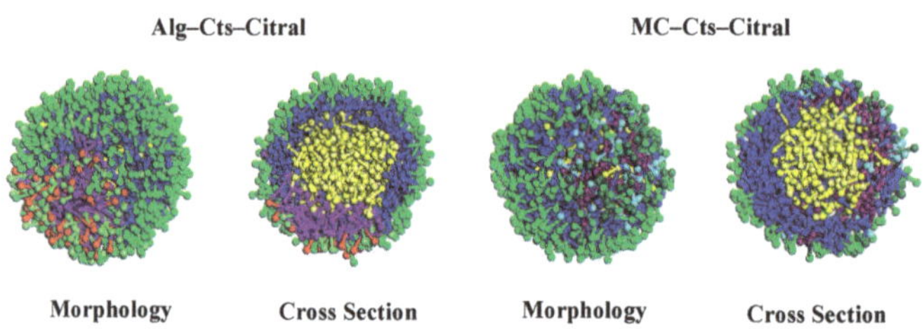

Figure 11. Morphology and cross-section of Alg–Cts–Citral and MC–Cts–Citral microcapsules.

(b) Analysis of the RDF

The results are shown in Figure 12, and the RDF curves in the figure are all based on C1 beads. Curve g(N1) represents the RDF of C1–N1 and curve g(O1) represents the RDF of C1–O1, and so on.

By comparing the three figures, the RDFs of the C1–N1 beads and C1–O4 beads are basically the same, and there is no obvious fluctuation. That is, adding Alg or MC to replace part of Cts has little effect on the molecular distribution of Cts in the microcapsule wall.

On the other hand, the peak of the C1–N2 curve is much higher than those of the C1–N1 and C1–N3 curves, which indicates that the affinity of N2 beads in the main chain of MC with C1 beads of citral is stronger than that with N1 beads of Cts or N3 beads of Alg. Therefore, compared with Cts and Alg, there are more citral molecules embedded in the molecular layer of MC.

In addition, the peak of the C1–O1 curve is much higher than those of C1–O2, C1–O3, and C1–O4, and the whole C1–O1 curve is closer to the left than the other three curves. The curve characteristics showed that the affinity between the O1 and C1 beads is stronger than that between the O2, O3, and O4 beads.

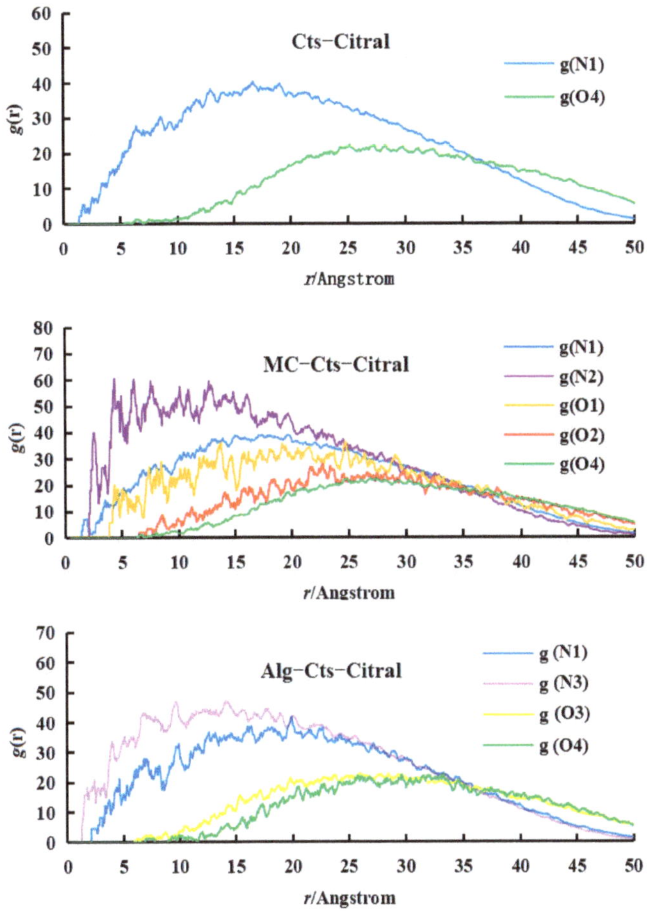

Figure 12. RDF curves of Cts–Citral, Alg–Cts–Citral, and MC–Cts–Citral microcapsules.

(c) Analysis of the concentration distribution curve of citral

As shown in Figure 13, in the system doped with Alg, the peak of the concentration distribution curve of citral is higher than that of the system without Alg, but the upper part of the peak is narrower than that of the pure Cts wall material. The area enclosed by the concentration distribution curve of the two is almost the same, which indicates that the content of the coated citral is very close. However, in the system doped with MC, the peak of the concentration distribution curve of citral is higher than those of the other two, but the width is almost the same, so the area surrounded by the concentration distribution curve is larger, which indicates that the Cts wall material doped with MC has more citral content. However, the width of the bottom of the three curves is almost the same, which indicates that the volume of microcapsules formed by the three groups of experiments is similar, and that there is little difference. Under the condition of almost no change in volume, the content of citral in the system doped with MC increases, which may be because the addition of MC weakens the interaction between the molecules and makes the molecules more closely arranged, so the volume of the system is almost unchanged.

It can be seen from the above discussion that the doping of Alg has no obvious effect on the preparation of Cts–Citral microcapsules because the properties of the microcapsules have not changed much. It is concluded that Alg can replace Cts in the preparation of Cts–Citral microcapsules. On the other hand, the doping of MC has a significant effect on the preparation of Cts–Citral. Because MC is amphiphilic and has a strong affinity for citral, more citral can be encapsulated in it. However, the encapsulation performance of MC is not very good, which easily leads to the exposure of citral molecules. Therefore, there are advantages and disadvantages for the strategy of doping MC.

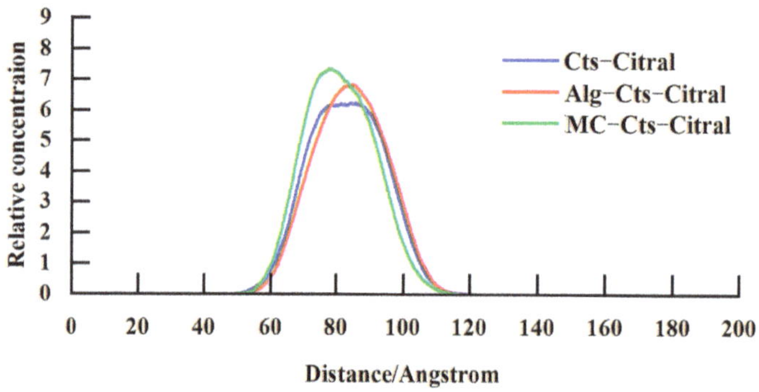

Figure 13. Concentration distribution curve of citral molecules.

Alg–Cts–Citral Double-Wall Material Microcapsule of Alg and Cts

Figure 14 shows that under neutral simulation conditions, Alg and Cts can form a spherical core–shell structure, which can tightly embed the citral in it. The N1 beads and O4 beads of Cts are mainly distributed in the outer and middle layers. The bead distribution of Alg is similar to that of single-wall Alg. Hydrophilic O3 beads (dark blue) are distributed in the outermost layer of the microcapsule, forming a hydrophilic protective layer. S beads are distributed in the middle and outer layers of the microcapsules, while N3 beads are next to the citral.

The RDF curve can reflect the distribution characteristics of citral in wall materials more directly. Under the condition of keeping the content of citral and Alg unchanged, when the content of Cts is low, N1 and N3 beads are preferentially distributed in the inner layer close to citral. The reason is that the affinity of citral to hydrophobic beads (N1 and N3 beads) is greater than that to hydrophilic beads (O3 and O4 beads). With increasing Cts content in the wall material, the inner layer gradually becomes saturated. At the same time, the space resistance of hydrophobic beads to diffuse to the outer layer of the capsule may be less than the attraction of affinity, so hydrophobic beads gradually distribute to the outer layer, which leads to the coincidence of the peaks of hydrophobic beads and hydrophilic beads. As shown in Figure 14III, the peak coincidence degree of all curves in group (4) is the highest, indicating that Cts and Alg are closely bound. The reason for this may be that there is a strong electrostatic interaction between Cts and Alg, so the compactness and stability of the composite wall material are better. Therefore, the best experimental results are obtained for the feeding ratio of group (4). It is concluded that when the contents of citral, Alg, and Cts are 0.55%, 3.25%, and 0.25%, that is, after dehydration the microcapsules have better performance when the contents of citral, Alg, and Cts are 13.6%, 80.2%, and 6.2%, respectively.

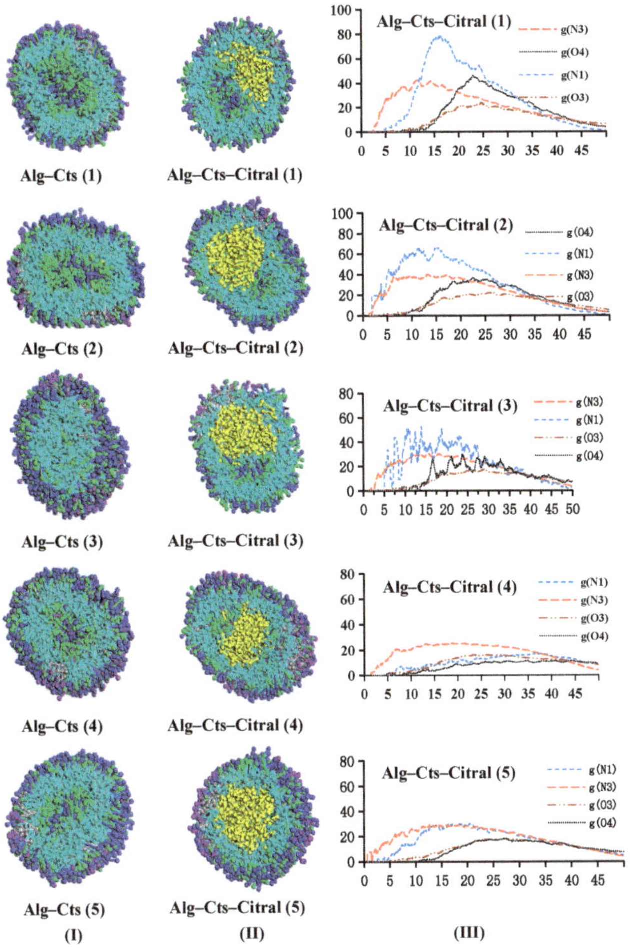

Figure 14. Alg and Cts double-wall material microcapsules ((**I**) citral free; (**II**) containing citral; and (**III**) RDF curve).

Alg–MC–Citral Double-Wall Material Microcapsule

Figure 15 shows that in aqueous solution, Alg and MC can also form a spherical core–shell structure, which can tightly embed the citral. The N2 beads and O2 beads of MC are distributed in the outer and middle layers. The bead distribution of Alg in aqueous solution without citral is similar to that of single wall Alg. Hydrophilic O3 beads (dark blue) are used as the outermost layer of the microcapsule to form a hydrophilic protective layer and S beads are distributed in the middle and outer layers of the microcapsules. In the model containing citral, the N3 bead layer is close to the citral.

Compared with the profile of microcapsules, the RDF curve can more directly reflect the distribution characteristics of citral and wall material. Under the condition of keeping the content of citral and Alg unchanged, the content of MC is gradually increased to investigate the change in bead distribution. When the content of MC is low, N2 and N3 beads are preferentially distributed in the inner layer near citral, because citral has more affinity for hydrophobic N2 and N3 beads than for hydrophilic O2 and O3 beads. When the content of MC increases, the total content of the wall material also increases, and the hydrophobic layer tends to be saturated so that the probability of hydrophobic beads distributed in the hydrophobic layer and hydrophilic layer gradually tends to be equal. Clearly, the peak value of hydrophobic beads is close to that of hydrophilic beads, but there is still a certain gap. This may be due to the amphiphilic property of MC, which has an obvious solubilization effect on citral. The hydrophobic beads of Alg (N3 beads) and the hydrophobic group of MC (N2 beads) are more likely to be distributed in the inner layer, while the hydrophilic beads of Alg (O3 beads) and the hydrophilic group of MC (O2 beads) are more likely to be distributed in the outer layer. In group (5), the peak value of the hydrophobic bead curve is higher, but the difference in the peak value of the hydrophilic layer is larger, which indicates that the hydrophilic effect between the two wall materials is not very strong. In contrast, the two peaks of group (4) are very close, which indicates that the interaction between the two wall materials is the strongest, and that the combination degree and encapsulation effect are the best. Therefore, group (4) exhibits the best result. Therefore, when the contents of citral, Alg, and MC are 0.56%, 1.65%, and 0.29%, respectively, that is, after dehydration the microcapsules demonstrate better performance when the contents of citral, Alg, and MC are 22.4%, 66.0%, and 11.6%, respectively.

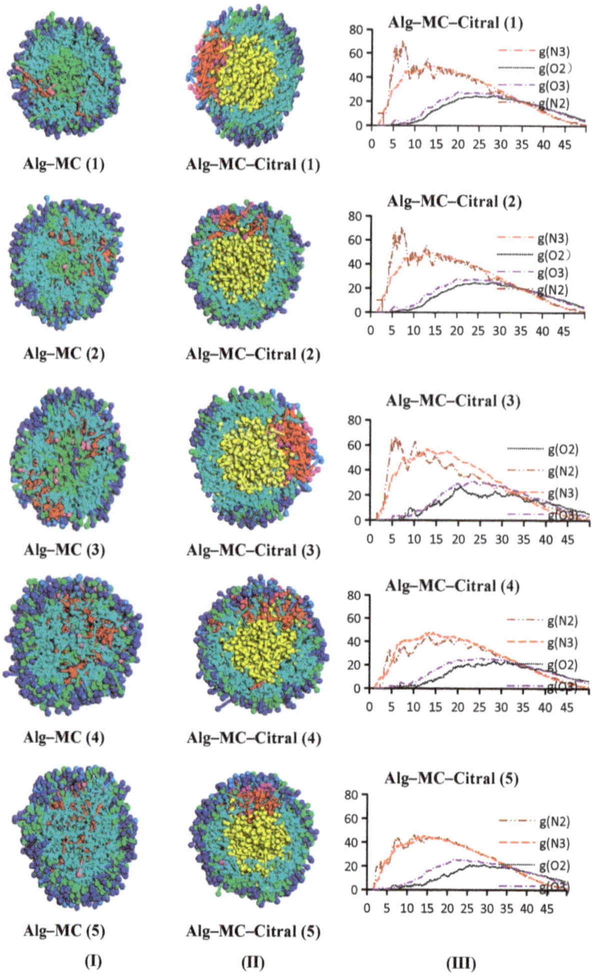

Figure 15. Alg and MC double-wall material microcapsules ((**I**) citral free; (**II**) containing citral; and (**III**) RDF curve).

Diffusion Coefficient of Citral in the Simulation Process

MSD is the distance the beads move from their original position to the second moment of their distribution in a defined time span, defined as $MSD = \frac{1}{N}\sum_{i=1}^{N}|\mathbf{r}_i(t) - \mathbf{r}_i(0)|^2$, which is related to the diffusion coefficient ($F_D = \frac{1}{6N}\lim_{t\to\infty}\frac{d}{dt}\sum_{i=1}^{N}|\mathbf{r}_i(t) - \mathbf{r}_i(0)|^2$), where r_i denotes the position vector of ith bead and N is the number of statistical beads [45]. The MSD curves are also computed in the Mesocite module of Materials Studio 7.0 software. The slope of the mean square displacement diagram can represent the diffusion coefficient. A large slope indicates a fast diffusion speed [38,43]. Figure 16 is a trend line diagram of the mean square displacement. The diffusion coefficient can be directly seen from the inclination angle of the slash. Clearly, the diffusion coefficient of the fourth group is small and close to that of the Alg single-wall material system. The slopes of the other groups hardly changed, which may be because the fourth group of Cts cations reacted completely with Alg anions. However, when the amount of Cts is too small or too large, the reaction between the Cts cation and Alg anion is not complete. Therefore, the system has excess anions or cations, resulting in electrification of the simulated system, and the formed microspheres are also unstable. Therefore, the diffusion coefficient of other groups becomes higher, and the diffusion speed of citral becomes faster.

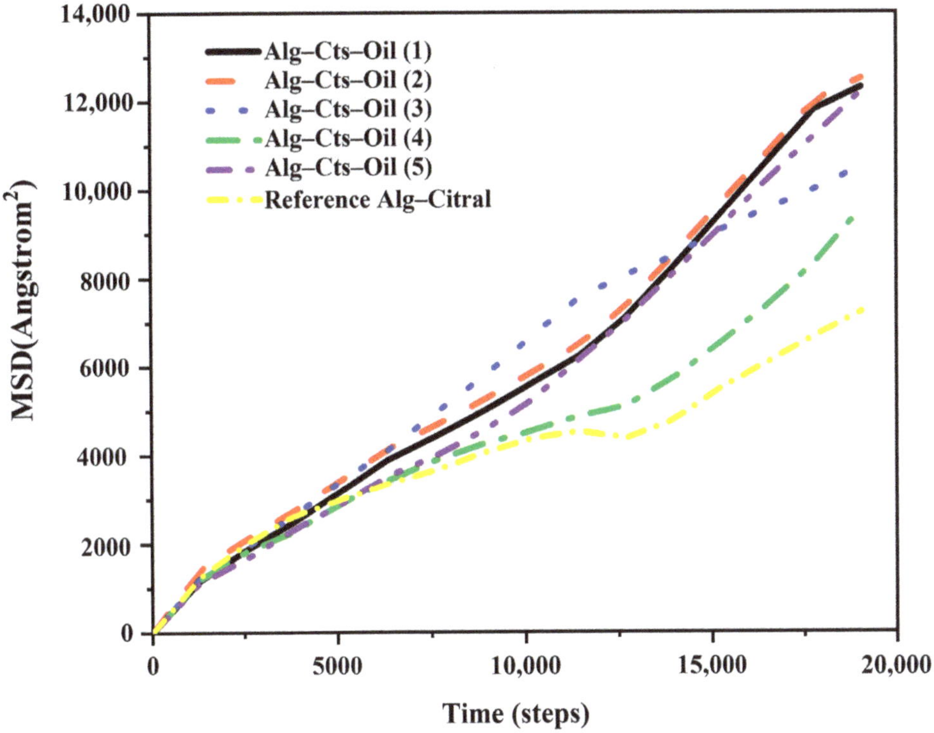

Figure 16. Mean square displacement function trend diagram of Alg and Cts double-wall materials.

As shown in Figure 17, in the simulation results of the Alg and MC composite wall material, the diffusion coefficient of the fourth group is close to that of the Alg single-wall material system. However, the diffusion coefficients of the other groups are large and almost the same. This may be because MC is amphiphilic and has an obvious solubilization effect on citral [46]. When the content of MC is high, the diffusion coefficient becomes higher, and the diffusion speed of citral becomes faster. When the content of MC is relatively small, according to the RDF in this paper, the wall material may be loose, resulting in a higher diffusion coefficient. The MSD curves exhibit different regimes, reflecting the initial rapid diffusion of components followed by a slower, more stable phase. The decreasing trend observed in Alg–MC–Citral (4) is likely due to the strong electrostatic interactions between the components, which temporarily restrict their motion before reaching equilibrium.

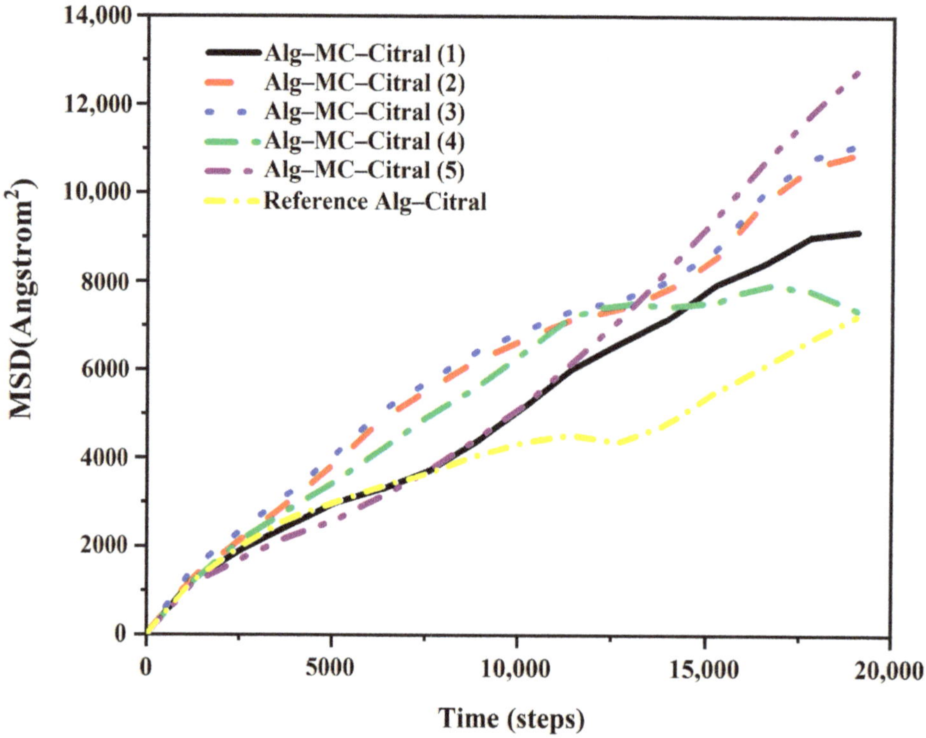

Figure 17. Mean square displacement function trend diagram of Alg and MC double-wall materials.

4. Conclusions

In this study, the DPD method is used to simulate the self-assembly process, appearance, mesoscopic structure, and wrapping properties of microcapsules formed with citral as the core material and Cts and Alg as single-wall materials, and with citral as the core material and Cts–Alg, Cts–MC, Alg–Cts, and Alg–MC as double-wall materials. The effects of Cts content and wall material composition on the structure, morphology, encapsulation performance, and stability of microcapsules are compared and explored. In addition, the microcapsules are deeply analyzed by using the mesoscopic structure, RDF, and diffusion coefficient.

The conclusions are as follows: (1) The results show that the higher the Cts content, the better the coating performance of the wall material on citral. However, considering the cost of the actual production process and other factors, the dosage of Cts should not be too high, as long as it can completely wrap citral. (2) When Alg–Cts double-wall microcapsules are prepared with water as the solvent, the microcapsules have better performance when the contents of citral, Alg, and Cts are 0.55%, 3.25%, and 0.25%, respectively. (3) When Alg–MC double wall microcapsules are prepared with water as the solvent, the microcapsules demonstrate better performance when the contents of citral, Alg, and MC are 0.56%, 1.65%, and 0.29%, respectively.

This study provides a new idea and method for the preparation of citral microcapsules and is of great significance for the design and development of new composite wall microcapsules.

Author Contributions: W.W.: funding acquisition, supervision, investigation, project administration, and writing—review. Z.L.: simulation, data analysis, writing—original draft, writing—review and editing. D.F.: writing—review. Q.T.: writing—original draft. S.L.: simulating. W.L.: data analysis and writing—review. All authors have read and agreed to the published version of the manuscript.

Funding: This work was financially supported by the Guangdong Provincial Department of Education Ordinary Universities Characteristic Innovation Project (No. 2020KTSCX158), the National College Students' Innovative Entrepreneurial Training Plan Program (No. 202010580015), Guangdong Provincial Key Laboratory of Environmental Health and Land Resource (No. 2020B121201014), and the Project of Education Department of Guangdong Province (No. 2018KQNCX288).

Institutional Review Board Statement: Not applicable.

Informed Consent Statement: Not applicable.

Data Availability Statement: The original contributions presented in the study are included in the article, and further inquiries can be directed to the corresponding authors.

Conflicts of Interest: The authors declare that they have no known competing financial interests or personal relationships that could have appeared to influence the work reported in this paper.

References

1. Ye, W.; Wang, N.; Hu, K.; Zhang, L.; Liu, A.; Pan, C.J.; Gong, T.; Liu, T.; Ding, H.Y. Bio-inspired microcapsule for targeted antithrombotic drug delivery. *RSC Adv.* **2018**, *8*, 27253–27259. [CrossRef] [PubMed]
2. Huang, D.; Sun, M.; Bu, Y.Z.; Luo, F.; Lin, C.Y.; Lin, Z.Y.; Weng, Z.Q.; Yang, F.; Wu, D.C. Microcapsule-embedded hydrogel patches for ultrasound responsive and enhanced transdermal delivery of diclofenac sodium. *J. Mater. Chem. B* **2019**, *7*, 2330–2337. [CrossRef] [PubMed]
3. Popov, A.L.; Popova, N.; Gould, D.J.; Shcherbakov, A.B.; Sukhorukov, G.B.; Ivanov, V.K. Ceria nanoparticles-decorated microcapsules as a smart drug delivery/protective system: Protection of encapsulated P. pyralis luciferase. *ACS Appl. Mater. Interfaces* **2018**, *10*, 14367–14377. [CrossRef] [PubMed]
4. Vazquez-Gonzalez, M.; Willner, I. DNA-responsive SiO_2 nanoparticles.; metal-organic frameworks.; and microcapsules for controlled drug release. *Langmuir* **2018**, *34*, 14692–14710. [CrossRef]
5. Huang, L.Y.; Wu, K.; Zhang, R.; Ji, H.B. Fabrication of multicore milli- and microcapsules for controlling hydrophobic drugs release using a facile approach. *Ind. Eng. Chem. Res.* **2019**, *58*, 17017–17026. [CrossRef]
6. Wei, L.Y.; Lu, Z.Q.; Ji, X.; Jiang, Y.K.; Ma, L. Self-assembly of hollow graphene oxide microcapsules directed by cavitation for loading hydrophobic drugs. *ACS Appl. Mater. Interfaces* **2021**, *13*, 2988–2996. [CrossRef]
7. Wang, Y.F.; Cheng, Q.Q.; Liu, J.; Tariq, Z.Z.; Zheng, Z.; Li, G.; Kaplan, D.L.; Wang, X. Tuning microcapsule shell thickness and structure with silk fibroin and nanoparticles for sustained release. *ACS Biomater. Sci. Eng.* **2020**, *6*, 4583–4594. [CrossRef]
8. Guo, X.Y.; Zhao, R.; Zhang, J.; Du, Y.J.; Yang, L.G.; Chen, L.Y.; Pang, S.; Xu, Y.; Zhang, Z.H.; Wu, X.M. A microcapsule oil dispersion for the controlled release of 1-methylcyclopropene in an open environment. *RSC Adv.* **2019**, *9*, 23465–23473. [CrossRef]
9. Suraphan, N.; Fan, L.F.; Liu, B.X.; Wu, D.C. Co-delivery of chlorantraniliprole and avermectin with a polylactide microcapsule formulation. *RSC Adv.* **2020**, *10*, 25418–25425. [CrossRef]
10. Zou, A.H.; Yang, Y.; Cheng, J.G.; Garamus, V.M.; Li, N. Construction and characterization of a novel sustained-release delivery system for hydrophobic pesticides using biodegradable polydopamine-based microcapsules. *J. Agric. Food. Chem.* **2018**, *66*, 6262–6268. [CrossRef]
11. Yang, J.L.; Zhou, Z.Y.; Liang, Y.; Tang, J.Y.; Gao, Y.H.; Niu, J.F.; Dong, H.Q.; Tang, R.; Tang, G.; Cao, Y.S. Sustainable preparation of microcapsules with desirable stability and bioactivity using phosphonium ionic liquid as a functional additive. *ACS Sustain. Chem. Eng.* **2020**, *8*, 13440–13448. [CrossRef]
12. Jia, C.H.; Huang, S.J.; Liu, R.; You, J.; Xiong, S.B.; Zhang, B.J.; Rong, J.H. Storage stability and in-vitro release behavior of microcapsules incorporating fish oil by spray drying. *Colloids Surf. A Physicochem. Eng. Asp.* **2021**, *628*, 127234. [CrossRef]

13. Fotovvat, B.; Behzadnasab, M.; Mirabedini, S.M.; EivazMohammadloo, H. Anti-corrosion performance and mechanical properties of epoxy coatings containing microcapsules filled with linseed oil and modified ceria nanoparticles. *Colloids Surf. A Physicochem. Eng. Asp.* **2022**, *648*, 129157. [CrossRef]
14. Sun, J.Y.; Li, W.; Zhan, Y.C.; Tian, L.M.; Tian, H.L. Two preparation processes for anti-corrosion and self-healing epoxy coatings containing the poly (calcium alginate) microcapsules loaded with tung oil. *Colloids Surf. A Physicochem. Eng. Asp.* **2022**, *641*, 128600. [CrossRef]
15. Prabaharan, M. Chitosan derivatives as promising materials for controlled drug delivery. *J. Biomater. Appl.* **2008**, *23*, 5–36. [CrossRef]
16. Calvo, P.; Remunan-Lopez, C.; Vila-Jato, J.L.; Alonso, M. Novel hydrophilic chitosan-polyethylene oxide nanoparticles as protein carriers. *J. Appl. Polym. Sci.* **1997**, *63*, 125–132. [CrossRef]
17. Zhang, L.; Wang, Y.F.; Liu, H.S.; Yu, L.; Liu, X.X.; Chen, L.; Zhang, N.Z. Developing hydroxypropyl methylcellulose/hydroxypropyl starch blends for use as capsule materials. *Carbohyd. Polym.* **2013**, *98*, 73–79. [CrossRef]
18. Zhang, L.; Lu, Y.Q.; Yue, L.N.; Li, Q.; Xiao, L.X.; Ding, X.L.; Guan, C.R. Microstructures, physical and sustained antioxidant properties of hydroxypropyl methylcellulose based microporous photophobic films. *Int. J. Biol. Macromol.* **2020**, *152*, 1002–1009. [CrossRef]
19. Dudai, N.; Weinstein, Y.; Krup, M.; Rabinski, T.; Ofir, R. Citral is a new inducer of caspase-3 in tumor cell lines. *Planta Medica* **2005**, *71*, 484–488. [CrossRef]
20. Rabbani, S.I.; Devi, K.; Shivananda, T.N. Studies on antimutagenic effects of citral in mice. *J. Food Agric. Environ.* **2004**, *2*, 62–64. [CrossRef]
21. Shen, Y.B.; Sun, Z.F.; Guo, X.T. Citral inhibits lipopolysaccharide-induced acute lung injury by activating PPAR-γ. *Eur. J. Pharmacol.* **2015**, *747*, 45–51. [CrossRef] [PubMed]
22. Chueca, B.; Pagán, R.; García-Gonzalo, D. Oxygenated monoterpenes citral and carvacrol cause oxidative damage in Escherichia coli without the involvement of tricarboxylic acid cycle and Fenton reaction. *Int. J. Food Microbiol.* **2014**, *187*, 189126–189131. [CrossRef] [PubMed]
23. Gilling, D.H.; Kitajima, M.; Torrey, J.R.; Bright, K.R. Mechanisms of antiviral action of plant antimicrobials against murine norovirus. *Appl. Environ. Microb.* **2014**, *80*, 4898–4910. [CrossRef] [PubMed]
24. Bergonzelli, G.E.; Donnicola, D.; Porta, N.; Corthésy-Theulaz, I.E. Essential oils as components of a diet-based approach to management of helicobacter infection. *Antimicrob. Agents Chemother.* **2003**, *47*, 3240–3246. [CrossRef]
25. Hoogerbrugge, P.J.; Koelman, J.M.V.A. Simulating microscopic hydrodynamic phenomena with dissipative particle dynamics. *EPL* **1992**, *19*, 155–160. [CrossRef]
26. Espanol, P.; Warren, P. Statistical mechanics of dissipative particle dynamics. *Europhys. Lett.* **1995**, *30*, 191–196. [CrossRef]
27. Abu-Nada, E.; Pop, I.; Mahian, O. A dissipative particle dynamics two-component nanofluid heat transfer model: Application to natural convection. *Int. J. Heat Mass Transf.* **2019**, *133*, 1086–1098. [CrossRef]
28. Feng, Y.H.; Zhang, X.P.; Zhao, Z.Q.; Guo, X.D. Dissipative particle dynamics aided design of drug delivery systems: A review. *Mol. Pharm.* **2020**, *17*, 1778–1799. [CrossRef]
29. Okuwaki, K.; Mochizuki, Y.; Doi, H.; Kawada, S.; Ozawac, T.; Yasuoka, K. Theoretical analyses on water cluster structures in polymer electrolyte membrane by using dissipative particle dynamics simulations with fragment molecular orbital based effective parameters. *RSC Adv.* **2018**, *8*, 34582–34595. [CrossRef]
30. Zhang, J.; Xu, J.C.; Wen, L.Y.; Zhang, F.S.; Zhang, L.J. The self-assembly behavior of polymer brushes induced by the orientational ordering of rod backbones: A dissipative particle dynamics study. *Phys. Chem. Chem. Phys.* **2020**, *22*, 5229–5241. [CrossRef]
31. Liang, X.P.; Wu, J.Q.; Yang, X.G.; Tu, Z.B.; Wang, Y. Investigation of oil-in-water emulsion stability with relevant interfacial characteristics simulated by dissipative particle dynamics. *Colloids Surf. A Physicochem. Eng. Asp.* **2018**, *546*, 107–114. [CrossRef]
32. Zhou, P.; Hou, J.; Yan, Y.G.; Wang, J.Q.; Chen, W. Effect of aggregation and adsorption behavior on the flow resistance of surfactant fluid on smooth and rough surfaces: A many-body dissipative particle dynamics study. *Langmuir* **2019**, *35*, 8110–8120. [CrossRef] [PubMed]
33. Wang, X.Y.; Santo, K.P.; Neimark, A.V. Modeling gas-liquid interfaces by dissipative particle dynamics: Adsorption and surface tension of cetyl trimethyl ammonium bromide at the air-water interface. *Langmuir* **2020**, *36*, 14686–14698. [CrossRef]
34. Zhang, J.W.; Chen, L.; Wang, A.; Yan, Z.C. Dissipative particle dynamics simulation of ionic liquid-based microemulsion: Quantitative properties and emulsification mechanism. *Ind. Eng. Chem. Res.* **2020**, *59*, 763–773. [CrossRef]
35. Choudhary, M.; Kamil, S.M. Phase diagram study of sodium dodecyl sulfate using dissipative particle dynamics. *ACS Omega* **2020**, *5*, 22891–22900. [CrossRef]
36. Panoukidou, M.; Wand, C.R.; Regno, A.D.; Anderson, R.L.; Carbone, P. Constructing the phase diagram of sodium laurylethoxysulfate using dissipative particle dynamics. *J. Colloid Interf. Sci.* **2019**, *557*, 34–44. [CrossRef]

37. Guo, Z.Y.; Ma, L.P.; Dai, Q.X.; Ao, R.; Liu, H.P.; Yang, J. Combined application of modified corn-core powder and sludgebased biochar for sewage sludge pretreatment: Dewatering performance and dissipative particle dynamics simulation. *Environ. Pollut.* **2020**, *265*, 115095. [CrossRef]
38. Groot, R.D.; Warren, P.B. Dissipative particle dynamics: Bridging the gap between atomistic and mesoscopic simulation. *J. Chem. Phys.* **1997**, *107*, 4423–4435. [CrossRef]
39. Li, Y.; Leng, M.T.; Cai, M.T.; Huang, L.; Chen, Y.W.; Luo, X.L. pH responsive micelles based on copolymers mPEG-PCL-PDEA: The relationship between composition and properties. *Colloid Surface B* **2017**, *154*, 397–407. [CrossRef]
40. *Materials Studio 7.0*, version 7.0; Accelrys: San Diego, CA, USA, 2013.
41. Qiu, B.N.; Zhou, Y.; Yin, X.Q.; Chen, J.H.; Yin, Y.Z.; Zhu, L. Preparation of citral microcapsules through spray-dring and the stability of the micorcapsules. *Sci. Technol. Food Ind.* **2017**, *38*, 190–195. [CrossRef]
42. Zhang, S.; Chen, J.; Yin, X.Q.; Qiu, B.N.; Zhu, L. Preparation of citral microcapsules and the effects of water adsorption on the structure of micorcapsules. *Food Sci. Technol.* **2017**, *42*, 60–66. [CrossRef]
43. Groot, R.D.; Rabone, K.L. Mesoscopic simulation of cell membrane damage, morphology change, and rupture by nonionic surfactants. *Biophys. J.* **2001**, *81*, 725–736. [CrossRef] [PubMed]
44. Ramezani, M.; Shamsara, J. Application of DPD in the design of polymeric nano-micelles as drug carriers. *J. Mol. Graphics Modell.* **2016**, *66*, 1–8. [CrossRef]
45. Zwanzig, R. Memory Effects in Irreversible Thermodynamics. *Phys Rev.* **1961**, *124*, 983–992. [CrossRef]
46. Wang, W.J.; Anderson, N.A.; Travesset, A.; Vaknin, D. Regulation of the Electric Charge in Phosphatidic Acid Domains. *J. Phys. Chem. B* **2012**, *116*, 7213–7220. [CrossRef] [PubMed]

polymers

MDPI

Article

Synthesis, Characterization, and Evaluation of Silver Nanoparticle-Loaded Carboxymethyl Chitosan with Sulfobetaine Methacrylate Hydrogel Nanocomposites for Biomedical Applications

Sonaimuthu Mohandoss [1,*,†], Kuppu Sakthi Velu [1,†], Salim Manoharadas [2], Naushad Ahmad [3], Subramanian Palanisamy [4], SangGuan You [4], Muhammad Saeed Akhtar [1,†] and Yong Rok Lee [1,*]

[1] School of Chemical Engineering, Yeungnam University, Gyeongsan 38541, Republic of Korea; sakthi.velu4@gmail.com (K.S.V.); passions.malik@gmail.com (M.S.A.)
[2] Department of Botany and Microbiology, College of Science, King Saud University, Riyadh 11451, Saudi Arabia; smanoharadas@ksu.edu.sa
[3] Department of Chemistry, College of Science, King Saud University, Riyadh 11451, Saudi Arabia; anaushad@ksu.edu.sa
[4] East Coast Life Sciences Institute, Gangneung-Wonju National University, Gangneung 25457, Republic of Korea; spalanisamy33@gwnu.ac.kr (S.P.); sangguanyou@gmail.com (S.Y.)
* Correspondence: drsmohandoss@yu.ac.kr (S.M.); yrlee@yu.ac.kr (Y.R.L.)
† These authors contributed equally to this work.

Abstract: In this study, nanocomposites of AgNPs encapsulated in carboxymethyl chitosan (CMCS) with sulfobetaine methacrylate (SB) hydrogel (AgNPs/CMCS-SB) were synthesized. The UV-Vis spectra indicated the presence of AgNPs, with a broad peak at around 424 nm, while the AgNPs-loaded CMCS-SB nanocomposite exhibited absorption peaks at 445 nm. The size and dispersion of AgNPs varied with the concentration of the $AgNO_3$ solution, affecting swelling rates: $148.37 \pm 15.63\%$, $172.26 \pm 18.14\%$, and $159.17 \pm 16.59\%$ for 1.0 mM, 3.0 mM, and 5.0 mM AgNPs/CMCS-SB, respectively. Additionally, water absorption capacity increased with AgNPs content, peaking at $11.04 \pm 0.54\%$ for the 3.0 mM AgNPs/CMCS-SB nanocomposite. Silver release from the nanocomposite was influenced by $AgNO_3$ concentration, showing rapid initial release followed by a slower rate over time for the 3.0 mM AgNPs/CMCS-SB. XRD patterns affirmed the presence of AgNPs, showcasing characteristic peaks indicative of a face-centered cubic (fcc) structure. The FTIR spectra highlighted interactions between AgNPs and CMCS-SB, with noticeable shifts in characteristic bands. In addition, SEM and TEM images validated spherical AgNPs within the CMCS-SB hydrogel network, averaging approximately 70 and 30 nm in diameter, respectively. The nanocomposite exhibited significant antibacterial activity against *S. aureus* and *E. coli*, with inhibition rates of $98.9 \pm 0.21\%$ and $99.2 \pm 0.14\%$, respectively, for the 3.0 mM AgNPs/CMCS-SB nanocomposite. Moreover, cytotoxicity assays showcased the efficacy of AgNPs/CMCS-SB against human colorectal cancer cells (HCT-116 cells), with the strongest cytotoxicity ($61.7 \pm 4.3\%$) at 100 μg/mL. These results suggest the synthesized AgNPs/CMCS-SB nanocomposites possess promising attributes for various biomedical applications, including antimicrobial and anticancer activities, positioning them as compelling candidates for further advancement in biomedicine.

Keywords: AgNPs; hydrogels; nanocomposites; release; antimicrobial; anticancer

1. Introduction

The integration of silver nanoparticles (AgNPs) with hydrogel matrices has recently gained considerable attention in biomedical research owing to the unique antimicrobial properties of AgNPs and the versatile characteristics of hydrogels for drug delivery and tissue engineering applications [1]. Among these, silver nanoparticles (AgNPs) have emerged as a particularly intriguing class of nanomaterials owing to their remarkable physical, chemical, and biological properties. The distinctive characteristics of AgNPs render them promising candidates for a wide range of applications across various sectors, including biomedical, environmental, and industrial fields [2]. The AgNPs exhibit exceptional antimicrobial activity attributed to their high surface area-to-volume ratio, which enhances contact with microbial membranes, leading to disruption of cellular integrity and inhibition of microbial growth [3]. Furthermore, AgNPs have demonstrated promising anticancer activity by inducing apoptosis and inhibiting proliferation in cancer cells, thus making them potential candidates for cancer therapy [4,5]. In biomedical applications, AgNPs have shown immense potential as antimicrobial agents, drug delivery vehicles, and diagnostic tools. Their ability to efficiently penetrate bacterial cell membranes and disrupt cellular processes has led to their exploration as alternatives to conventional antibiotics, particularly considering the rising antibiotic resistance. Furthermore, the surface functionalization of AgNPs enables the targeted delivery of therapeutic agents to specific sites within the body, reducing systemic toxicity and enhancing therapeutic efficacy [6]. Despite the notable progress in understanding and exploiting the properties of AgNPs, several challenges remain, such as the nanoparticle stability, cytotoxicity, environmental impact, and scalability of the synthesis methods that facilitate the widespread adoption of AgNPs in practical applications. Moreover, the complex interactions of AgNPs with biological systems and the environment necessitate comprehensive studies to assess their safety and environmental impact.

Hydrogels have gained significant attention in biomedical research and applications owing to their unique properties, including a high water content, biocompatibility, and tunable mechanical and chemical characteristics [7]. Among the diverse range of hydrogel-forming polymers, carboxymethyl chitosan (CMCS) is a notable and promising candidate for biomedical applications owing to its biocompatibility, biodegradability, and functional groups that can be modified to tailor its properties for specific applications [8,9]. Chitosan, which is a natural polysaccharide derived from chitin, is inherently biocompatible and has antimicrobial properties, making it an attractive material for biomedical applications [10,11]. However, the poor solubility of chitosan in physiological conditions limits its utilization, particularly in aqueous environments. Carboxymethylation of chitosan introduces carboxyl groups onto the polymer backbone, enhancing its solubility and allowing for the facile manipulation of its physicochemical properties [12]. The resulting CMCS exhibits improved water solubility, biocompatibility, and mucoadhesive properties, making it suitable for various biomedical applications, including drug delivery, tissue engineering, wound healing, and regenerative medicine [13]. Furthermore, the presence of carboxyl groups in CMCS facilitates further chemical modifications, such as crosslinking and grafting, to impart the desired functionalities and enhance its performance in specific applications. Hydrogels based on CMCS offer several advantages for biomedical applications. Owing to their hydrophilic nature and porous structure, CMCS hydrogels can absorb and retain large amounts of water and bioactive molecules, making them ideal for drug delivery, antibacterial, and anticancer systems [14]. Moreover, the mechanical properties and degradation kinetics of CMCS hydrogels can be tailored to match the requirements of different tissue engineering and regenerative medicine applications. Incorporating bioactive agents, such as growth factors, drugs, and nanoparticles, into CMCS hydrogels further expands their functionality and therapeutic potential [11]. The controlled release of bioactive molecules from CMCS hydrogels can be achieved via diffusion, degradation, or stimuli-responsive mechanisms, enabling spatiotemporal control over therapeutic delivery as well as promoting tissue regeneration and wound healing.

Sulfobetaine methacrylate (SB) is a zwitterionic polymer known for its exceptional antifouling properties and low immunogenicity; thus, it is ideal for enhancing the biocompatibility and non-fouling behavior of hydrogel matrices [15,16]. Incorporating CMCS and SB into hydrogel matrices offers numerous advantages for biomedical applications [17]. First, the synergistic interaction between CMCS and SB enhances the mechanical properties and stability of hydrogels, making them suitable for load-bearing applications in biomedical engineering [18]. Second, the non-fouling properties of SB prevent protein adsorption and cell adhesion on hydrogel surfaces, reducing the risk of inflammatory responses and improving biocompatibility [19]. Third, the stimuli-responsive behavior of CMCS-SB hydrogels enables the controlled release of encapsulated drugs or bioactive molecules in response to external triggers, facilitating spatiotemporal control over therapeutic delivery [20]. Combining CMCS and SB in hydrogel matrices demonstrates significant potential for various biomedical applications [11,18]. In drug delivery systems, CMCS-SB hydrogels offer improved drug-loading capacity, prolonged circulation times, and targeted delivery, leading to enhanced therapeutic efficacy and reduced side effects [18,19,21,22]. In tissue engineering, CMCS-SB hydrogels provide a conducive microenvironment for cell growth, proliferation, and differentiation, promoting tissue regeneration and integration. When combined with CMCS and AgNPs in hydrogel matrices, SB further improves stability, reduces protein adsorption, and enhances responsiveness to environmental stimuli, such as pH, temperature, or ionic strength [23,24]. The synergistic effects of AgNPs, CMCS, and SB hydrogels can lead to enhanced antibacterial and anticancer activities, offering new opportunities for the development of therapeutic agents with improved efficacy and reduced side effects. In this study, we aim to synthesize and characterize silver nanoparticle-loaded carboxymethyl chitosan with sulfobetaine methacrylate hydrogel (AgNPs/CMCS-SB) nanocomposites and evaluate their potential in antibacterial and anticancer applications.

2. Materials and Methods

2.1. Materials

Carboxymethyl chitosan (Carboxymethylation >/= 80%; CMCS), poly (ethylene glycol)-block-Poly (sulfobetaine methacrylate; SB) (PEG average M_n 5000, PSBMA M_n 13,000), silver nitrate ($AgNO_3$; ≥99.0%), sodium borohydride ($NaBH_4$; ≥98.0%), polyvinylpyrrolidone (MW 40,000; PVP), and ammonium persulfate ($(NH_4)_2S_2O_8$; ≥98.0%) were all purchased from Sigma-Aldrich, Seoul, Republic of Korea. The water used in the experiment underwent ultra-purification and was generated by the Milli-Q system (Milpitas, CA, USA). All additional reagents were of analytical grade and did not necessitate further purification prior to usage.

2.2. Synthesis of AgNPs

The synthesis of AgNPs via chemical reduction involves several steps [25]. Initially, a solution of $AgNO_3$ is prepared by dissolving a specified quantity of $AgNO_3$ in 100 mL of distilled water to attain a concentration of 1 mM. Concurrently, a solution of 100 mM $NaBH_4$ is prepared by dissolving it in distilled water (1.0 mL). The $AgNO_3$ solution is then heated to a predetermined temperature of 80 °C in a clean reaction vessel, with constant stirring to ensure even heating. Gradually, the 100 mM $NaBH_4$ solution is added drop by drop to the heated $AgNO_3$ solution while stirring continuously, instigating the reduction in silver ions to atoms, which subsequently aggregate to form AgNPs, leading to a color transition of the solution from colorless to yellowish-brown. To stabilize the formed AgNPs and prevent aggregation, 0.10 g of PVP is introduced into the reaction mixture. Upon completion of the reaction, the synthesized AgNPs can be characterized using various analytical techniques.

2.3. Synthesis of AgNPs-Loaded CMCS-SB Hydrogel

The synthesis of the CMCS-SB copolymer comprises several sequential steps aimed at grafting sulfobetaine methacrylate (SB) onto carboxymethyl chitosan (CMCS) [26]. Initially, 2.0 g of CMCS is dissolved in 100 mL of a 1 vol% acetic acid aqueous solution. The mixture is then heated to 65 °C while maintaining constant stirring at 500 rpm throughout the reaction. Following this, 0.2 g of $(NH_4)_2S_2O_8$ is added to the CMCS solution, and N_2 gas is bubbled through it for 30 min, creating an oxygen-free environment crucial for the polymerization reaction. Meanwhile, SB is dissolved in deionized water, and this solution is gradually incorporated into the reaction system. The reaction continues for 6 h at a consistent temperature and stirring rate. Upon completion, the solution is cooled, followed by a 72-h dialysis process using a dialysis bag with a molecular cutoff weight ranging from 8000 to 14,000, aiming to eliminate small molecules, salts, and unreacted monomers. Finally, the solution undergoes freeze-drying to yield the CMCS-SB hydrogels in a dry form ready for collection and subsequent utilization in various applications.

To prepare the AgNPs-loaded CMCS-SB hydrogels, the synthesized AgNPs are incorporated into the CMCS-SB hydrogel matrix during the gelation process [27]. Initially, different concentrations of $AgNO_3$ (1.0, 3.0, and 5.0 mM), synthesized through chemical reduction, are dispersed in the CMCS-SB solution before gelation. This dispersion is achieved by thoroughly mixing in a water bath at 90 °C to ensure the even distribution of AgNPs within the CMCS-SB hydrogel matrix solution. Following this, the gelation process is initiated by adjusting the pH of the CMCS-SB solution to 6, depending on the specific gelation mechanism employed. Upon completion of gelation, the AgNPs-loaded CMCS-SB hydrogel forms, ready for further characterization and utilization across various applications.

2.4. Characterizations

The UV-vis spectra of the AgNPs, CMCS-SB hydrogel, and AgNPs/CMCS-SB nanocomposites with varying irradiation times were obtained by a UV-vis spectrophotometer (UV 3220-Optizen, Daejeon, Republic of Korea) in the wavelength range of 200–800 nm. The XRD patterns of the AgNPs, CMCS-SB hydrogel, and AgNPs/CMCS-SB nanocomposites were obtained using a Cu Kα radiation source (k = 1.54064 Å) generated by an (PANalytic X'Pert Philips, MRD model, Tokyo, Japan) X-ray diffractometer operating at a voltage and current of 40 kV and 40 mA, respectively. To assess the chemical composition of the AgNPs, CMCS-SB hydrogel, and AgNPs/CMCS-SB nanocomposites, FTIR analysis was conducted by employing a spectrometer (Perkin Elmer, Shelton, CT, USA) with a resolution of 4 cm^{-1}, utilizing an attenuated total reflection (ATR) in the wavenumber range of 4000–400 cm^{-1}. The surface morphology of the materials was examined via SEM (JEOL JSM-6490LA, Tokyo, Japan), whereas the internal structural morphologies of the AgNPs and AgNPs/CMCS-SB nanocomposites were characterized via TEM (200 kV, JEOLJEM-2100, Tokyo, Japan). For the SEM analysis, the samples were prepared by casting 0.1 mg/mL of water suspensions onto a silicon substrate and allowing them to dry prior to examination.

2.5. Swelling and Water Absorption

The swelling, water solubility, and water absorption of the CMCS-SB hydrogels and AgNPs-loaded CMCS-SB hydrogels were assessed via a gravimetric approach. Initially, both types of hydrogels were cut into 3 cm × 3 cm. These segments were then subjected to drying in an oven at 45 °C until a constant weight (Wd) was achieved. Subsequently, the first set of segments was immersed in 30 mL of PBS buffer solution with a pH of 7.4 for 24 h. After removal from the solution, any excess surface liquid was carefully removed using filter paper, followed by reweighing (W_{ts}) the segments. The swelling of both types of hydrogels was determined using Equation (1) as follows:

$$\text{Swelling degree (\%)} = (W_{ts} - W_{td})/W_{td} \times 100 \tag{1}$$

Subsequently, the swollen CMCS-SB hydrogels and AgNPs-loaded CMCS-SB hydrogels were subjected to drying in an oven at 45 °C until a constant weight (W_{t1}) was achieved. The water solubility of both types of hydrogels was then determined using Equation (2) as follows:

$$\text{Water solubility (\%)} = (W_{td} - W_{t1})/W_{td} \times 100 \tag{2}$$

The remaining groups of the CMCS-SB hydrogels and AgNPs-loaded CMCS-SB hydrogels were exposed to 25 °C and a relative humidity of 50% for 24 h, after which they were weighed (W_{t2}). The water absorption of the films was then determined using Equation (3) as follows:

$$\text{Water absorption (\%)} = (W_{t2} - W_{td})/W_{td} \times 100 \quad (3)$$

2.6. Release of Silver from AgNPs/CMCS-SB Nanocomposites

The release profiles of silver from the AgNPs/CMCS-SB nanocomposites were obtained by monitoring the optical density (O.D.) at various time intervals. Specifically, approximately 1.0 g of the AgNPs/CMCS-SB nanocomposites were placed in a flask containing 10 mL of water at 37 °C. The flask was then placed on a rotary shaker and oscillated at a frequency of 60 rpm. The amount of silver released was quantified by measuring the O.D. at its peak wavelength (420 nm) using a UV-vis spectrophotometer (UV 3220-Optizen, Daejeon, Republic of Korea).

2.7. Antibacterial Study

The antibacterial efficacy of the prepared nanocomposites (CMCS-SB hydrogel, AgNPs, and AgNPs/CMCS-SB nanocomposite) against *S. aureus* and *E. coli* was evaluated using methods adapted from Yalei Liu et al., 2022 [27]. Initially, 0.5 g of each material was prepared in sterile test tubes. Then, 20 μL of the bacterial suspension (OD_{600} = 0.1) was added to the surface of each material in separate test tubes. After inoculating at 37 °C for 2 h, 980 μL of sterile medium was added to suspend any remaining bacteria. A control group with the same bacterial suspension but without contact with the materials was also prepared. Subsequently, 100 μL of the bacterial suspension from each sample was spread onto agar plates using a sterile coating rod. After incubating at 37 °C for 24 h, bacterial colonies on the plates were identified and counted using ImageJ software. The counting process was repeated six times, and the average was calculated. The antimicrobial performance was quantified using the following Equation (4):

$$\text{Mortality (\%)} = \text{blank control group} - \text{sample group}/\text{blank control group} \times 100\% \quad (4)$$

2.8. In-Vitro Cytotoxicity Study

Initially, HCT-116 cells were seeded in a 96-well microplate containing the cell culture medium (DMEM with 1% penicillin-streptomycin and 10% FBS) at a density of 1×10^6 cells/mL. The plates were incubated at 37 °C under 5% CO_2 for 24 h. Subsequently, the cells were treated with different concentrations of the CMCS-SB hydrogel, AgNPs, and AgNPs/CMCS-SB nanocomposite for 24 h. Before treatment, the samples were sterilized via UV irradiation and then extracted at a concentration of 10 mg/mL at 37 °C for 24 h. The resulting extract solution was sterilized by filtration and stored for subsequent use. Cells without treatment served as the control. The experiment was performed in triplicate. During the incubation period, the freshly prepared WST-1 solution (20 μL) was added to the cells, followed by an additional 45 min of incubation. The absorbance was then read at 450 nm. The cell viability was assessed using Equation (5) as follows:

$$\text{Cell viability (\%)} = \text{Absorbance of test}/\text{Absorbance of control} \times 100 \quad (5)$$

To gain a comprehensive understanding, morphological changes in the HCT-116 cells were examined using fluorescence microscopy. The cells were seeded in a six-well plate at a density of 1×10^5 cells/well and incubated for 24 h. The cells were then treated with the samples for 24 h at 37 °C, after which the cells were washed with 1× PBS and fixed with a methanol: acetic acid solution (3:1, *v*/*v*). Subsequently, a staining procedure using 4′,6-diamidino-2-phenylindole/Propidium iodide (DAPI/PI) was performed to distinguish live and dead cells. The fluorescent dyes (DAPI/PI) enabled visualization using a Nikon Research Inverted Microscope, the ECLIPSE TS2R-C-AL (Tokyo, Japan), allowing for detailed assessments of the cellular morphology.

3. Results and Discussion

3.1. Optical Properties of AgNPs/CMCS-SB Nanocomposite

The UV-Vis spectra of the AgNPs exhibit a broad peak with a maximum absorption at approximately 424 nm (Figure 1a), corresponding to the Plasmon absorbance characteristic of AgNPs [28]. Conversely, the CMCS-SB hydrogel lacks such a peak owing to its amorphous nature. The interaction between the Ag^+ ions and functional groups within the polymer network, such as the –OH, $-NH_2$, and –C=O groups, facilitates the reduction of Ag^+ ions under UV irradiation, leading to the formation of AgNPs within the cross-linked polymer matrix [29]. The hydrogel matrix serves as an excellent host for the incorporation of nanoparticles, with the polymer structure controlling the nucleation and growth of AgNPs. Consequently, the presence of AgNPs enhances the overall performance of the AgNPs/CMCS-SB nanocomposite [30]. The apparent color change from violet to dark brown in the AgNPs-loaded CMCS-SB hydrogels coincides with the absorption peaks shifting to approximately 445 nm in the UV-Vis spectra [31]. This shift in the absorption peaks is attributed to the surface plasmon resonance (SPR) phenomenon, which is influenced by the diameter of the larger AgNPs, which exhibit a red-shifted SPR peak at longer wavelengths, whereas nanoparticle aggregates further extend this shift. The formation of AgNPs within the CMCS-SB hydrogels was confirmed via SPR alignment with the typical λ_{max} values of the AgNPs within the visible range of 350–550 nm. The prepared AgNPs/CMCS-SB nanocomposite involves swelling the cross-linked CMCS-SB hydrogels in $AgNO_3$ solutions of varying concentrations (1.0, 3.0, and 5.0 mM), as illustrated in Figure 1b. Higher degrees of hydrolysis result in a greater absorption of the Ag^+ solution by the hydrogel, leading to the formation of larger AgNPs and a broader peak in the UV-Vis spectral region of 350–450 nm [32]. However, as the concentration of the $AgNO_3$ solution increases from 3.0 to 5.0 mM, the intensity of the plasmon absorbance decreases, indicating the presence of larger AgNPs and the formation of multi-nanoparticle aggregates. Conversely, in the non-hydrolyzed AgNPs/CMCS-SB nanocomposite system, the shift in peaks is more pronounced, indicating the prevalence of smaller nanoparticles within the matrix [33]. This highlights the influence of hydrolysis on the size distribution and aggregation of AgNPs within the CMCS-SB nanocomposite system.

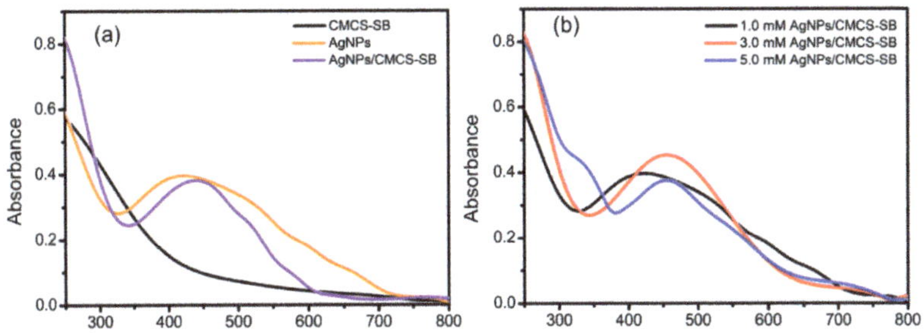

Figure 1. UV-Vis absorption spectra of the (**a**) CMCS-SB hydrogel, AgNPs (1.0 mM), and AgNPs/CMCS-SB nanocomposite, and (**b**) CMCS-SB hydrogel with AgNPs at different concentrations (1.0, 3.0, and 5.0 mM).

3.2. Swelling and Water Absorption of AgNPs/CMCS-SB Nanocomposite

The swelling, water absorption, and water solubility characteristics of both the prepared CMCS-SB hydrogel and AgNPs/CMCS-SB nanocomposite were precisely evaluated, as summarized in Table 1. Notably, 1.0, 3.0, and 5.0 mM AgNPs/CMCS-SB nanocomposite demonstrated remarkable swelling rates of 148.37 ± 15.63%, 172.26 ± 18.14%, and 159.17 ± 16.59%, respectively [34]. These rates significantly surpassed those of the CMCS-SB hydrogel alone, which exhibited a swelling rate of 138.19 ± 14.83%. The prominent improvement is primarily owing to the graft copolymerization of CMCS and SB, which introduces numerous –COOH, -SO_3, and quaternary –NH_2 groups onto the molecular chains [35]. This modification resulted in a substantial increase in the hydration capacity, enhanced the hydrophilicity, and improved the water absorption capabilities. Moreover, the 3.0 mM AgNPs/CMCS-SB nanocomposite exhibited notable improvements in the water absorption and swelling properties compared to both the CMCS-SB hydrogel and other nanocomposite modifications [36]. It can be attributed to the disruption of the -SH bonds between the AgNPs and CMCS-SB, consequently reducing the crystallinity and facilitating the ingress of water molecules. Consequently, the 3.0 mM AgNPs/CMCS-SB nanocomposite demonstrated notably superior water absorption and swelling properties compared to its other counterparts. Additionally, the water solubility of the 3.0 mM AgNPs/CMCS-SB nanocomposite proportionally decreased as the amount of AgNPs added to the CMCS-SB hydrogel increased [37]. Although the CMCS-SB hydrogel exhibited partial dissolution in water at room temperature, incorporating AgNPs improved the hydration ability of the nanocomposite without compromising its water solubility. Consequently, the 3.0 mM AgNPs/CMCS-SB nanocomposite, when employed as a food packaging film, displayed favorable water retention and shape preservation properties, demonstrating its promising potential for such applications.

Table 1. Swelling, water absorption, and water solubility of the CMCS-SB hydrogel and AgNPs/CMCS-SB nanocomposite.

Sample	Swelling Rate %	Water Absorption %	Water Solubility %
CMCS-SB	138.19 ± 14.83	9.38 ± 0.72	7.16 ± 0.48
1.0 mM AgNPs/CMCS-SB	148.37 ± 15.63	9.48 ± 0.66	8.99 ± 0.41
3.0 mM AgNPs/CMCS-SB	172.26 ± 18.14	11.04 ± 0.54	10.06 ± 0.37
5.0 mM AgNPs/CMCS-SB	159.17 ± 16.59	10.57 ± 0.33	9.67 ± 0.24

3.3. Release of Silver from Hydrogels

As illustrated in Figure 2a, the 3.0 mM AgNPs/CMCS-SB nanocomposite displayed a release pattern of silver ions characterized by an initial rapid phase followed by a nearly constant release rate [38]. This behavior can be attributed to the heightened hydrophilicity of the matrix associated with an increased $AgNO_3$ concentration. Conversely, variations in $AgNO_3$ concentration (1.0 and 5.0 mM) resulted in corresponding changes in the initial rate of silver ion release, with a significant increase in release rate observed only after an extended incubation period of approximately 18 h. This prolonged release duration may be attributed to the slower swelling nature of the matrix, observed with both low and high $AgNO_3$ contents [39]. Moreover, the linear relationship plotted between various time intervals and AgNPs release from AgNPs/CMCS-SB nanocomposite exhibited a strong linear response within the time range of up to 24 h, with correlation coefficients (R^2 = 0.9126, 0.9902, and 0.9510) as depicted in Figure 2b–d. Based on these findings, the 3.0 mM AgNPs/CMCS-SB nanocomposite was chosen for further characterization due to its optimal release kinetics. This AgNPs/CMCS-SB nanocomposite strikes a balance between rapid initial release and sustained long-term efficacy, rendering it a promising candidate for subsequent investigations and potential applications in antimicrobial and anticancer formulations.

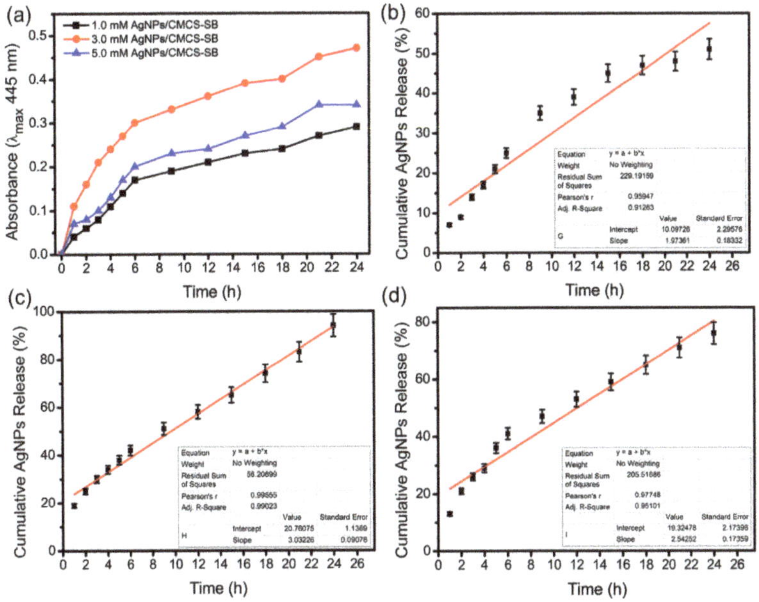

Figure 2. (**a**) Release of silver ions from the AgNPs/CMCS-SB nanocomposite with $AgNO_3$ at different concentrations (1.0, 3.0, and 5.0 mM) and Linearity of (**b**) 1.0 mM, (**c**) 3.0 mM, and (**d**) 5.0 mM of AgNPs from the AgNPs/CMCS-SB nanocomposite.

3.4. XDR Pattern of AgNPs/CMCS-SB Nanocomposite

The crystal structures of the CMCS-SB hydrogel, AgNPs, and AgNPs/CMCS-SB nanocomposite were elucidated via XRD, as illustrated in Figure 3a. The XRD pattern of the CMCS-SB hydrogel displays broad peaks, indicative of its amorphous nature [40]. The XRD pattern of the synthesized AgNPs was analyzed to confirm their identity. Several Bragg reflections with 2θ values of 38.14°, 44.87°, 61.31°, and 77.19° corresponding to the (111), (200), (220), and (311) lattice planes, respectively, were observed [41]. These peaks are characteristic of a face-centered cubic (fcc) structure, commonly associated with metallic silver nanoparticles. Notably, the intensity of the peak corresponding to the (111) plane was higher than that of the other planes. The broadening of these peaks is typical for nano-sized particles, indicating the presence of AgNPs. The XRD pattern of the AgNPs/CMCS-SB nanocomposite also displays diffraction signals at 2θ values of 37.82°, 45.57°, 60.38°, and 77.29°, corresponding to the (111), (200), (220), and (311) diffraction planes of fcc AgNPs [42]. This further confirms the incorporation of AgNPs into the nanocomposite structure. Additionally, the presence of Ag^+ ions within the network of the nanocomposite was inferred, which can be explained by the incomplete reduction of Ag^+ ions into Ag^0 by the borohydride ions. The cross-linked polymer network may have impeded the diffusion of borohydride ions into the CMCS-SB hydrogel network, thereby resulting in this observation.

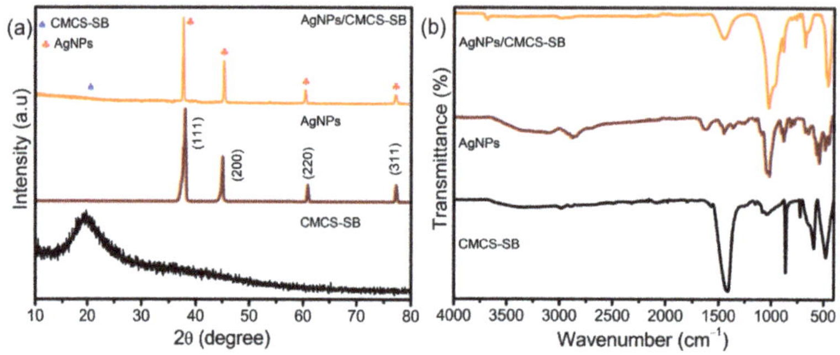

Figure 3. (**a**) XRD pattern and (**b**) FTIR spectra of the CMCS-SB hydrogel, AgNPs, and AgNPs/CMCS-SB nanocomposite.

3.5. FTIR Analysis of AgNPs/CMCS-SB Nanocomposite

The FTIR transmission spectra for the prepared CMCS-SB hydrogel, AgNPs, and AgNPs/CMCS-SB nanocomposite are presented in Figure 3b. The FTIR analysis of the CMCS-SB hydrogel reveals distinct absorption peaks corresponding to various functional groups within the molecules. In the FTIR spectrum of SB, the peaks at 1792 and 1599 cm^{-1} correspond to the –C=O group and $-CH_3$ stretching vibrations, respectively, within the methyl acrylate structure [43]. Additionally, the peaks at 1154 and 596 cm^{-1} are attributed to the stretching vibrations of the $-SO_3$ group and the C-S bond, respectively. Furthermore, a prominent absorption peak at 1427 cm^{-1} was observed, corresponding to the $-CH_2$ absorption peak of the quaternary $-NH_2$ group in the betaine molecule. A comparison of the infrared spectrum of the graft copolymer CS-SBMA with that of the pure CMCS and SB reveals absorption peaks at 1792, 1599, 1427, 1154, and 596 cm^{-1} in the CS-SBMA hydrogel spectrum. These peaks are associated with the respective functional groups (–C=O, $-SO_3$, $-NH_2$ groups) present in SB. In the FTIR spectrum of the AgNPs, characteristic peaks were observed at 3365, 2918, 1622, and 1012 cm^{-1}, which were attributed to the cyclic –OH, $-CH_2$, –C=O, and C–N functional groups, respectively [44]. Changes were observed in the FTIR spectrum of the AgNPs stabilized in the CMCS-SB hydrogel, with the disappearance of absorption bands at 1657 and 1600 cm^{-1} representing the CMCS $-CONH_2$ and $-NH_2$ groups, and the emergence of a new band at 1677 cm^{-1} indicating the attachment of silver to the nitrogen atom. Additionally, variations in the shape and peak positions of the $-NH_2$ and –OH bands at 3647 cm^{-1} occurred owing to their contribution to the reduction and stabilization processes [32,45]. Although there was no strong interaction of the AgNPs with C=O groups in this spectral region, a shift in the band corresponding to the N–H groups of the polymer matrix suggests a weak interaction between the $-NH_2$ groups of the polymer chains and AgNPs, stabilizing the nanosystem.

3.6. SEM Analysis of AgNPs/CMCS-SB Nanocomposite

The SEM images depict a smooth surface morphology of the CMCS-SB hydrogel with interconnected pores (Figure 4a). The hydrogel exhibits a porous structure with large, open, and channel-like structures, which indicate regions of water permeation and interaction sites for external stimuli with the hydrophilic groups of the copolymers [46]. Figure 4b presents predominantly spherical AgNPs that are size-dependent [47]. Additionally, to assess the formation of AgNPs within the CMCS-SB hydrogel, SEM images at both lower and higher magnifications are presented (Figure 4c,d). The presence of AgNPs within the CMCS-SB hydrogel networks suggests that when Ag^+ ions on the surface of the swollen hydrogel are reduced by $NaBH_4$, the resulting AgNPs increase the gel porosity, providing a pathway for the reducing agents to enter the bulk of the hydrogel and produce the AgNPs/CMCS-SB nanocomposite [48,49]. Furthermore, incorporating AgNPs into the CMCS-SB hydrogel apparently affects the average pore size, which is likely owing to the interaction of the functional groups with the metallic particles, displacing the water molecules and modifying the structural characteristics of the hydrogel.

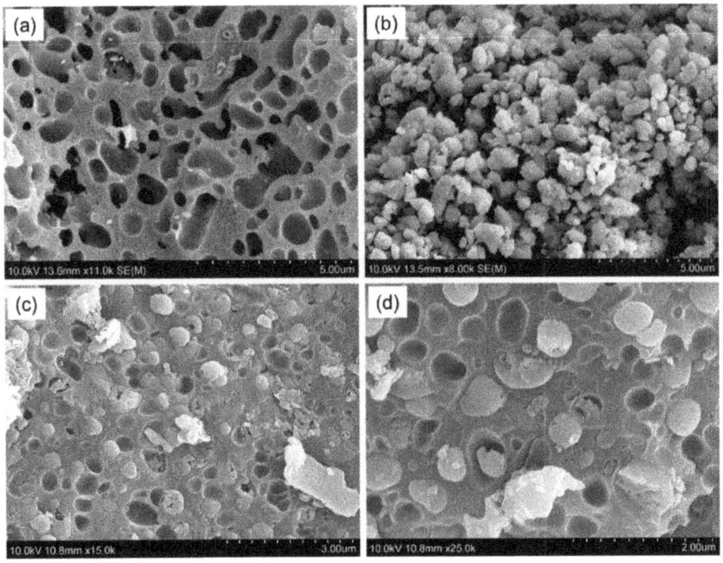

Figure 4. SEM images of the (**a**) CMCS-SB hydrogel, (**b**) AgNPs, and (**c**,**d**) AgNPs/CMCS-SB nanocomposite at different magnifications.

3.7. TEM Analysis of AgNPs/CMCS-SB Nanocomposite

The TEM was utilized to examine the morphology of the prepared AgNPs and AgNPs/CMCS-SB nanocomposite, as depicted in Figure 5. Figure 5a presents the TEM micrograph of AgNPs, revealing spherical nanoparticles with various sizes [50,51]. Figure 5b highlights AgNPs with diameters of approximately 70 nm, which is consistent with the surface plasmon resonance peak observed at 450 nm, indicative of their size distribution. Additionally, TEM images were obtained to observe the presence of AgNPs within the CMCS-SB hydrogel network (Figure 5c) [52,53], for which AgNPs/CMCS-SB nanocomposites were finely ground, equilibrated in distilled water for three days, and then sonicated to facilitate the release of AgNPs from the swollen hydrogel network. This phenomenon can be attributed to factors such as the increased mesh size of the network, the relaxation of polymeric chains entangled around the AgNPs, and decreased binding between the stabilized AgNPs and electron-rich nitrogen atoms of the macromolecular chain. Figure 5d illustrates the AgNPs/CMCS-SB nanocomposites with an average diameter of 30 nm, clearly demonstrating the in-situ formation of AgNPs within the CMCS-SB hydrogel network. Overall, the TEM images confirm the spherical morphology of AgNPs and demonstrate their successful incorporation within the CMCS-SB hydrogel network, validating the effectiveness of the nanocomposite synthesis process.

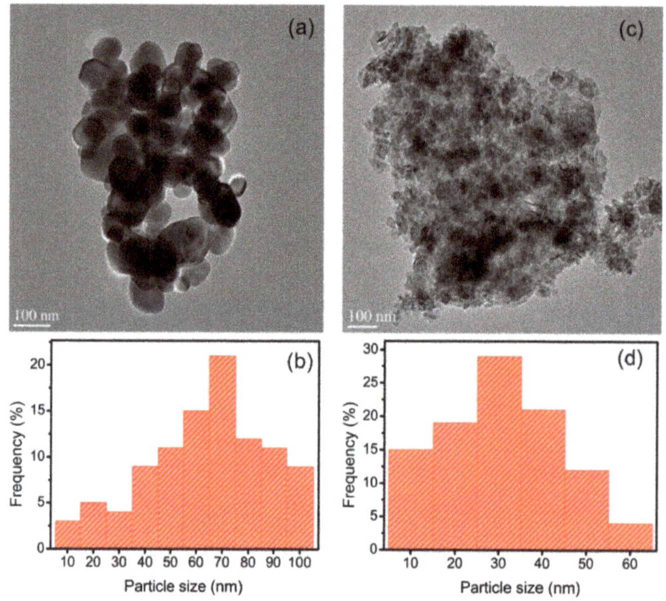

Figure 5. (**a**,**c**) TEM images and (**b**,**d**) histogram of the particle size distribution of the AgNPs and AgNPs/CMCS-SB nanocomposite.

3.8. Antibacterial Properties of AgNPs/CMCS-SB Nanocomposite

The antibacterial properties of the control, CMCS-SB hydrogel, AgNPs, and AgNPs/CMCS-SB nanocomposite were evaluated using the plate counting method against *S. aureus* and *E. coli* bacteria (Figure 6). The bacterial growth on agar plates served as an indicator of the antibacterial activity. The initial antibacterial effect of CMCS likely stems from the presence of amino and carboxyl groups, which can interact with various components on the bacterial surface [8,54], which disrupts cell membrane function, leading to damage or destruction. Although the mechanism of AgNPs is not fully understood, AgNPs are known to release silver ions upon contact with water [55]. These silver ions bind to negatively charged thiol groups (-SH) within the bacterial proteins, thus increasing the membrane permeability and denaturation of the cellular proteins. This process ultimately inhibits bacterial growth and achieves sterilization. Furthermore, the antimicrobial effect of silver ions is long-lasting because they can be reduced back into silver atoms after killing the bacteria. A comparative analysis revealed that although both the AgNPs and CMCS-SB hydrogel exhibited certain antimicrobial properties, incorporating AgNPs into the CMCS-SB hydrogel significantly enhanced its efficacy [36,56]. The superior antibacterial activity of the AgNPs/CMCS-SB nanocomposite, primarily attributed to the presence of AgNPs, was confirmed via experiments on *S. aureus* and *E. coli* bacteria (Figure 6). The AgNPs/CMCS-SB nanocomposite demonstrated significantly improved efficacy against both *S. aureus* (98.9 $\pm$ 0.21%) and *E. coli* (99.2 $\pm$ 0.14%) compared to the control group, CMCS-SB hydrogel (55.2 $\pm$ 0.11% and 57.2 $\pm$ 0.17%), and AgNPs (72.8 $\pm$ 0.14% and 75.9 $\pm$ 0.13%). These findings highlight the superior antibacterial activity of the AgNPs/CMCS-SB nanocomposite, suggesting its potential for use in various biomedical applications.

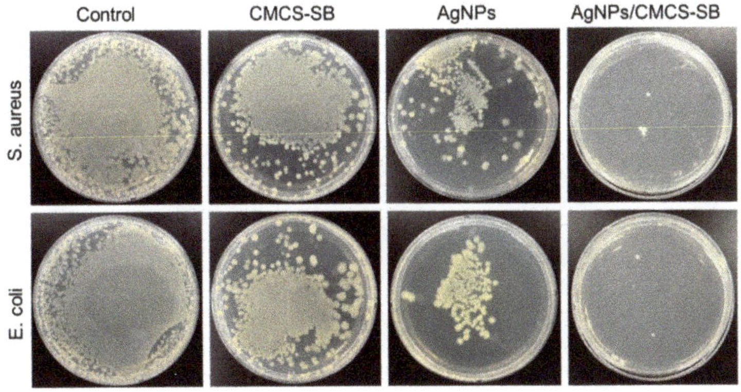

Figure 6. Evaluation of the antibacterial performance of the control, CMCS-SB hydrogel, AgNPs, and AgNPs/CMCS-SB nanocomposite against *E. coli* and *S. aureus*.

3.9. In-Vitro Cytotoxicity and Imaging of AgNPs/CMCS-SB Nanocomposite

In-vitro cytotoxicity tests were conducted to further assess the anticancer activity of the CMCS-SB hydrogel, AgNPs, and AgNPs/CMCS-SB nanocomposite. As shown in Figure 7, HCT-116 cells were co-cultured with the leaching solution of the CMCS-SB hydrogel, AgNPs, and AgNPs/CMCS-SB nanocomposite at various concentrations (5, 25, 50, 75, and 100 μg/mL) for 24 h. The cell survival was assessed using the WST-1 method, revealing a concentration-dependent cytotoxic effect for all the tested samples. Notably, the AgNPs/CMCS-SB nanocomposite exhibited the strongest cytotoxicity (61.7 ± 4.3%) against the HCT-116 cells at 100 μg/mL, suggesting a possible synergistic effect between the AgNPs and CMCS-SB hydrogel [57,58].

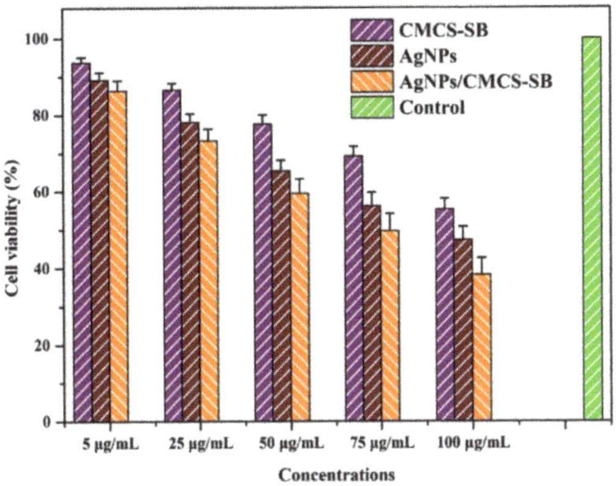

Figure 7. Cell viability of the CMCS-SB hydrogel, AgNPs, and AgNPs/CMCS-SB nanocomposite with HCT-116 cells for 24 h.

This observation supports the potential application of the AgNPs/CMCS-SB nanocomposite in cancer prevention. Fluorescence microscopy revealed that the CMCS-SB hydrogel and AgNPs alone did not significantly increase the cellular uptake of the material. The combination of CMCS-SB hydrogel and AgNPs significantly increased the cellular uptake, suggesting the effective delivery of the AgNPs/CMCS-SB nanocomposite to cancer cells [59,60]. To better understand this mechanism, HCT-116 cancer cells were incubated with a constant concentration of the CMCS-SB hydrogel, AgNPs, and AgNPs/CMCS-SB nanocomposite (80 μg/mL) for 24 h, followed by DAPI/PI staining to assess the differentiation in cell morphology (Figure 8). The increased presence of PI-stained cells in the AgNPs/CMCS-SB nanocomposite group suggests apoptosis induction. Overall, the results indicated that the AgNPs/CMCS-SB nanocomposite exhibited stronger anticancer activities than those of the CMCS-SB hydrogel and AgNPs alone, suggesting that the CMCS-SB hydrogel in the presence of AgNPs may enhance the cellular uptake efficiency and contribute to enhanced anticancer effects.

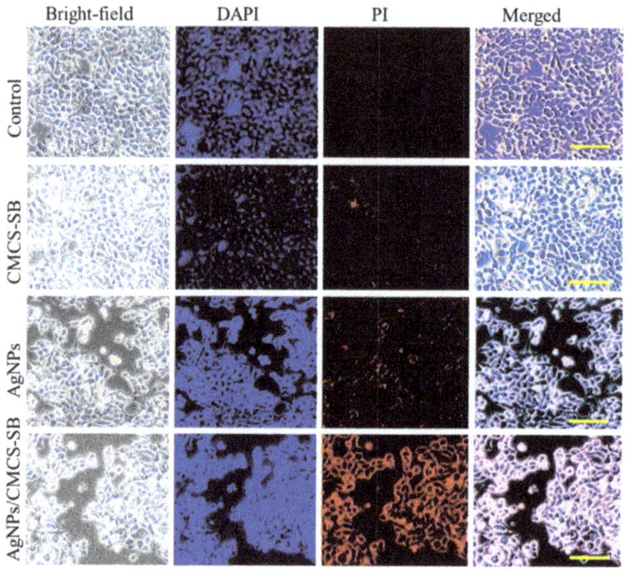

Figure 8. Cell culture effect of the control, CMCS-SB hydrogel, AgNPs, and AgNPs/CMCS-SB nanocomposite with the HCT-116 cells for 24 h on the morphological changes after treatment at 80 μg/mL for 24 h using DAPI/PI with HCT-116 cells. Scale bar~100 μm.

Based on these results, the novelty statements for the integration of AgNPs with CMCS-SB hydrogel matrices offer a novel approach in biomedical research combining the unique antimicrobial properties of AgNPs with the versatile characteristics of CMCS-SB hydrogels for drug delivery and tissue engineering applications. The synergistic interaction between CMCS and SB in CMCS-SB hydrogel matrices presents a novel strategy to enhance mechanical properties, stability, and controlled drug release capabilities for biomedical applications. The surface morphology imaging confirms the successful incorporation and distribution of AgNPs within the CMCS-SB hydrogel network, providing novel insights into the synthesis process and nanoparticle morphology, which are vital for understanding the AgNPs/CMCS-SB nanocomposite structure-property relationships. The incorporation of SB into CMCS-SB hydrogel matrices to improve stability, reduce protein adsorption, and enhance responsiveness to environmental stimuli represents a novel advancement in enhancing the biocompatibility and functionality of hydrogel-based systems. The superior antibacterial efficacy of the AgNPs/CMCS-SB nanocomposite against common pathogens such as *S. aureus* and *E. coli*, surpassing that of the hydrogel and AgNPs alone, underscores its novelty as a promising candidate for biomedical applications requiring enhanced antimicrobial properties. The concentration-dependent cytotoxicity of the nanocomposite against HCT-116 cancer cells, along with its ability to induce apoptosis and enhance cellular uptake, presents a novel approach to cancer therapy utilizing hydrogel-based nanocomposites, paving the way for innovative biomedical applications.

4. Conclusions

In this study, we successfully synthesized and characterized silver nanoparticle-loaded carboxymethyl chitosan with sulfobetaine methacrylate hydrogel (AgNPs/CMCS-SB) nanocomposites for biomedical applications. The synthesis process resulted in UV-Vis spectroscopy indicating the presence of AgNPs, characterized by a broad peak around 424 nm. Upon integration into the AgNPs/CMCS-SB nanocomposite, AgNPs exhibited absorption peaks at 445 nm in UV-Vis spectra. The size and dispersion of AgNPs varied depending on the concentration of the $AgNO_3$ solution, impacting absorbance intensity. Additionally, nanocomposites showed increased swelling rates of 148.37 $\pm$ 15.63%, 172.26 $\pm$ 18.14%, and 159.17 $\pm$ 16.59% for AgNPs/CMCS-SB concentrations of 1.0, 3.0, and 5.0 mM, respectively. The water absorption capacity increased with AgNPs content peaking at 11.04 $\pm$ 0.54% for the 3.0 mM AgNPs/CMCS-SB nanocomposite. Conversely, water solubility decreased with increasing AgNPs concentration, with the 3.0 mM AgNPs/CMCS-SB nanocomposite showing the lowest solubility at 10.06 $\pm$ 0.37%. Silver release from the nanocomposite depended on $AgNO_3$ concentration, notably with the 3.0 mM AgNPs/CMCS-SB exhibiting rapid initial release followed by a slower rate. The XRD patterns confirmed AgNPs in the nanocomposite, displaying characteristic peaks of a face-centered cubic (fcc) structure. In addition, the FTIR spectra suggested interactions between AgNPs and CMCS-SB hydrogel functional groups, with peak shifts indicating silver attachment to nitrogen atoms. Moreover, the SEM and TEM images confirmed the presence of spherical AgNPs within the CMCS-SB hydrogel network with average diameters of approximately 70 and 30 nm, respectively. The AgNPs/CMCS-SB nanocomposite demonstrated potent antibacterial activity against *S. aureus* and *E. coli*, with inhibition rates of 98.9 $\pm$ 0.21% and 99.2 $\pm$ 0.14%, respectively, for the 3.0 mM AgNPs/CMCS-SB nanocomposite. Notably, AgNPs/CMCS-SB nanocomposite exhibited strong cytotoxicity against HCT-116 cells with fluorescence microscopy, indicating increased intracellular accumulation and suggesting potential for cancer treatment. Compared to conventional polysaccharide-based hydrogel systems, the AgNPs/CMCS-SB nanocomposites exhibited superior swelling rates and water absorption capacities, attributed to the enhanced hydrophilicity and hydration ability conferred by the graft copolymerization of CMCS and SB. The potent antibacterial and anticancer activities demonstrated by the AgNPs/CMCS-SB nanocomposites surpass those of existing polysaccharide hydrogel formulations, underscoring their efficacy in combating microbial infections and cancerous cell proliferation. Further research is warranted to explore their efficacy in vivo and to optimize their formulation for specific therapeutic applications. The multifunctional nature of these nanocomposites is significantly promising for addressing various biomedical challenges and improving patient outcomes in the future.

Author Contributions: Conceptualization, S.M. (Sonaimuthu Mohandoss), K.S.V., S.P. and M.S.A.; methodology, S.M. (Sonaimuthu Mohandoss), K.S.V., M.S.A. and N.A.; software, S.M. (Sonaimuthu Mohandoss), K.S.V., S.M. (Salim Manoharadas) and N.A.; validation, S.M. (Sonaimuthu Mohandoss), K.S.V., S.P. and M.S.A.; formal analysis, S.M. (Salim Manoharadas), K.S.V., S.P. and M.S.A.; investigation, S.M. (Salim Manoharadas), K.S.V., S.P. and M.S.A.; resources, S.M. (Sonaimuthu Mohandoss), K.S.V., S.P. and M.S.A.; data curation, S.M. (Sonaimuthu Mohandoss), K.S.V., S.M. (Salim Manoharadas), S.P. and M.S.A.; writing—original draft preparation, S.M. (Sonaimuthu Mohandoss), K.S.V. and M.S.A.; writing—review and editing, S.M. (Sonaimuthu Mohandoss), K.S.V. and M.S.A.; visualization, S.Y. and Y.R.L.; supervision, S.Y. and Y.R.L.; project administration, S.M. (Sonaimuthu Mohandoss), S.M. (Salim Manoharadas) and N.A.; funding acquisition, S.M. (Sonaimuthu Mohandoss), S.M. (Salim Manoharadas) and N.A. All authors have read and agreed to the published version of the manuscript.

Funding: This work was supported by an NRF grant funded by the Korean government (2021R1F1A-1061566). The authors thank the Researchers Supporting Project for funding this work through Researchers Supporting Project number (RSPD2024R708), King Saud University, Riyadh, Saudi Arabia.

Institutional Review Board Statement: Not applicable.

Data Availability Statement: Data can be requested from the corresponding author due to privacy.

Acknowledgments: The authors would like to acknowledge the financial support provided by the NRF grant by the Korean government & Researchers Supporting Project number (RSPD2024R708), King Saud University, Riyadh, Saudi Arabia to conduct this research.

Conflicts of Interest: The authors declare no conflicts of interest.

References

1. Burdușel, A.C.; Gherasim, O.; Grumezescu, A.M.; Mogoantă, L.; Ficai, A.; Andronescu, E. Biomedical Applications of Silver Nanoparticles: An up-to-Date Overview. *Nanomaterials* **2018**, *8*, 681. [CrossRef] [PubMed]
2. Xu, L.; Wang, Y.Y.; Huang, J.; Chen, C.Y.; Wang, Z.X.; Xie, H. Silver Nanoparticles: Synthesis, Medical Applications and Biosafety. *Theranostics* **2020**, *10*, 8996–9031. [CrossRef]
3. Bruna, T.; Maldonado-Bravo, F.; Jara, P.; Caro, N. Silver Nanoparticles and Their Antibacterial Applications. *Int. J. Mol. Sci.* **2021**, *22*, 7202. [CrossRef] [PubMed]
4. Duan, H.; Wang, D.; Li, Y. Green Chemistry for Nanoparticle Synthesis. *Chem. Soc. Rev.* **2015**, *44*, 5778–5792. [CrossRef] [PubMed]
5. Takáč, P.; Michalková, R.; Čižmáriková, M.; Bedlovičová, Z.; Balážová, Ľ.; Takáčová, G. The Role of Silver Nanoparticles in the Diagnosis and Treatment of Cancer: Are There Any Perspectives for the Future? *Life* **2023**, *13*, 466. [CrossRef] [PubMed]
6. Gomes, H.I.O.; Martins, C.S.M.; Prior, J.A.V. Silver Nanoparticles as Carriers of Anticancer Drugs for Efficient Target Treatment of Cancer Cells. *Nanomaterials* **2021**, *11*, 964. [CrossRef] [PubMed]
7. Ho, T.C.; Chang, C.C.; Chan, H.P.; Chung, T.W.; Shu, C.W.; Chuang, K.P.; Duh, T.H.; Yang, M.H.; Tyan, Y.C. Hydrogels: Properties and Applications in Biomedicine. *Molecules* **2022**, *27*, 2902. [CrossRef] [PubMed]
8. Shariatinia, Z. Carboxymethyl Chitosan: Properties and Biomedical Applications. *Int. J. Biol. Macromol.* **2018**, *120*, 1406–1419. [CrossRef] [PubMed]
9. Chen, Z.; Zhao, J.; Wu, H.; Wang, H.; Lu, X.; Shahbazi, M.A.; Wang, S. A Triple-Network Carboxymethyl Chitosan-Based Hydrogel for Hemostasis of Incompressible Bleeding on Wet Wound Surfaces. *Carbohydr. Polym.* **2023**, *303*, 120434. [CrossRef]
10. Yadav, M.; Kaushik, B.; Rao, G.K.; Srivastava, C.M.; Vaya, D. Advances and Challenges in the Use of Chitosan and Its Derivatives in Biomedical Fields: A Review. *Carbohydr. Polym. Technol. Appl.* **2023**, *5*, 100323. [CrossRef]
11. Taokaew, S.; Kaewkong, W.; Kriangkrai, W. Recent Development of Functional Chitosan-Based Hydrogels for Pharmaceutical and Biomedical Applications. *Gels* **2023**, *9*, 277. [CrossRef]
12. Jimtaisong, A.; Saewan, N. Utilization of Carboxymethyl Chitosan in Cosmetics. *Int. J. Cosmet. Sci.* **2014**, *36*, 12–21. [CrossRef] [PubMed]
13. Geng, Y.; Xue, H.; Zhang, Z.; Panayi, A.C.; Knoedler, S.; Zhou, W.; Mi, B.; Liu, G. Recent Advances in Carboxymethyl Chitosan-Based Materials for Biomedical Applications. *Carbohydr. Polym.* **2023**, *305*, 120555. [CrossRef] [PubMed]
14. Singha, I.; Basu, A. Chitosan Based Injectable Hydrogels for Smart Drug Delivery Applications. *Sens. Int.* **2022**, *3*, 100168. [CrossRef]
15. Sun, H.; Chang, M.Y.Z.; Cheng, W.I.; Wang, Q.; Commisso, A.; Capeling, M.; Wu, Y.; Cheng, C. Biodegradable Zwitterionic Sulfobetaine Polymer and Its Conjugate with Paclitaxel for Sustained Drug Delivery. *Acta Biomater.* **2017**, *64*, 290–300. [CrossRef]
16. Sun, H.; Yan, L.; Zhang, R.; Lovell, J.F.; Wu, Y.; Cheng, C. A Sulfobetaine Zwitterionic Polymer-Drug Conjugate for Multivalent Paclitaxel and Gemcitabine Co-Delivery. *Biomater. Sci.* **2021**, *9*, 5000–5010. [CrossRef]
17. SchAnemann, E.; Koc, J.; Karthaüser, J.F.; Ozcan, O.; Schanzenbach, D.; Schardt, L.; Rosenhahn, A.; Laschewsky, A. Sulfobetaine Methacrylate Polymers of Unconventional Polyzwitterion Architecture and Their Antifouling Properties. *Biomacromolecules* **2021**, *22*, 1494–1508. [CrossRef]
18. Lee, S.Y.; Lee, Y.; Le Thi, P.; Oh, D.H.; Park, K.D. Sulfobetaine Methacrylate Hydrogel-Coated Anti-Fouling Surfaces for Implantable Biomedical Devices. *Biomater. Res.* **2018**, *22*, 3–9. [CrossRef] [PubMed]
19. Zhang, J.; Qian, S.; Chen, L.; Chen, L.; Zhao, L.; Feng, J. Highly Antifouling Double Network Hydrogel Based on Poly(Sulfobetaine Methacrylate) and Sodium Alginate with Great Toughness. *J. Mater. Sci. Technol.* **2021**, *85*, 235–244. [CrossRef]
20. Fathi, M.; Zangabad, P.S.; Majidi, S.; Barar, J.; Erfan-Niya, H.; Omidi, Y. Stimuli-Responsive Chitosan-Based Nanocarriers for Cancer Therapy. *BioImpacts* **2017**, *7*, 269–277. [CrossRef]
21. Chen, Z.; Yao, J.; Zhao, J.; Wang, S. Injectable Wound Dressing Based on Carboxymethyl Chitosan Triple-Network Hydrogel for Effective Wound Antibacterial and Hemostasis. *Int. J. Biol. Macromol.* **2023**, *225*, 1235–1245. [CrossRef]
22. Sahiner, M.; Yilmaz, A.S.; Ayyala, R.S.; Sahiner, N. Carboxymethyl Chitosan Microgels for Sustained Delivery of Vancomycin and Long-Lasting Antibacterial Effects. *Gels* **2023**, *9*, 708. [CrossRef]
23. Xiang, J.; Bai, Y.; Huang, Y.; Lang, S.; Li, J.; Ji, Y.; Peng, B.; Liu, G. A Zwitterionic Silver Nanoparticle-Incorporating Injectable Hydrogel with a Durable and Efficient Antibacterial Effect for Accelerated Wound Healing. *J. Mater. Chem. B* **2022**, *10*, 7979–7994. [CrossRef] [PubMed]
24. Dong, P.; Feng, J.; Zhang, D.; Li, C.; Shi, Q.S.; Xie, X. In Situ Synthesis of Amply Antimicrobial Silver Nanoparticle (AgNP) by Polyzwitterionic Copolymers Bearing Hydroxyl Groups. *React. Funct. Polym.* **2020**, *153*, 104609. [CrossRef]
25. Quintero-Quiroz, C.; Acevedo, N.; Zapata-Giraldo, J.; Botero, L.E.; Quintero, J.; Zárate-Trivinõ, D.; Saldarriaga, J.; Pérez, V.Z. Optimization of Silver Nanoparticle Synthesis by Chemical Reduction and Evaluation of Its Antimicrobial and Toxic Activity. *Biomater. Res.* **2019**, *23*, 1–15. [CrossRef]

26. Li, Q.; Ai, R.; Fan, J.; Fu, X.; Zhu, L.; Zhou, Q.; Chen, L.; Ma, W.; Li, Y.; Liu, L. AgNPs-Loaded Chitosan/Sodium Alginate Hydrogel Film by in-Situ Green Reduction with Tannins for Enhancing Antibacterial Activity. *Mater. Today Commun.* **2024**, *38*, 107927. [CrossRef]
27. Liu, Y.; Mao, J.; Guo, Z.; Hu, Y.; Wang, S. Polyvinyl Alcohol/Carboxymethyl Chitosan Hydrogel Loaded with Silver Nanoparticles Exhibited Antibacterial and Self-Healing Properties. *Int. J. Biol. Macromol.* **2022**, *220*, 211–222. [CrossRef]
28. Ansari, M.; Ahmed, S.; Abbasi, A.; Khan, M.T.; Subhan, M.; Bukhari, N.A.; Hatamleh, A.A.; Abdelsalam, N.R. Plant Mediated Fabrication of Silver Nanoparticles, Process Optimization, and Impact on Tomato Plant. *Sci. Rep.* **2023**, *13*, 18048. [CrossRef]
29. Carapeto, A.P.; Ferraria, A.M.; do Rego, A.M.B. Unraveling the Reaction Mechanism of Silver Ions Reduction by Chitosan from so Far Neglected Spectroscopic Features. *Carbohydr. Polym.* **2017**, *174*, 601–609. [CrossRef] [PubMed]
30. Aldakheel, F.M.; Mohsen, D.; El Sayed, M.M.; Alawam, K.A.; Binshaya, A.K.S.; Alduraywish, S.A. Silver Nanoparticles Loaded on Chitosan-g-PVA Hydrogel for the Wound-Healing Applications. *Molecules* **2023**, *28*, 3241. [CrossRef]
31. Huang, S.; Yu, Z.; Zhang, Y.; Qi, C.; Zhang, S. In Situ Green Synthesis of Antimicrobial Carboxymethyl Chitosan-Nanosilver Hybrids with Controlled Silver Release. *Int. J. Nanomed.* **2017**, *12*, 3181–3191. [CrossRef]
32. Diniz, F.R.; Maia, R.C.A.P.; Rannier, L.; Andrade, L.N.; Chaud, M.V.; da Silva, C.F.; Corrêa, C.B.; de Albuquerque Junior, R.L.C.; da Costa, L.P.; Shin, S.R.; et al. Silver Nanoparticles-Composing Alginate/Gelatine Hydrogel Improves Wound Healing In Vivo. *Nanomaterials* **2020**, *10*, 390. [CrossRef]
33. Constantin, M.; Lupei, M.; Bucatariu, S.M.; Pelin, I.M.; Doroftei, F.; Ichim, D.L.; Daraba, O.M.; Fundueanu, G. PVA/Chitosan Thin Films Containing Silver Nanoparticles and Ibuprofen for the Treatment of Periodontal Disease. *Polymers* **2023**, *15*, 4. [CrossRef] [PubMed]
34. Vimala, K.; Mohan, Y.M.; Sivudu, K.S.; Varaprasad, K.; Ravindra, S.; Reddy, N.N.; Padma, Y.; Sreedhar, B.; MohanaRaju, K. Fabrication of Porous Chitosan Films Impregnated with Silver Nanoparticles: A Facile Approach for Superior Antibacterial Application. *Colloids Surf. B Biointerfaces* **2010**, *76*, 248–258. [CrossRef]
35. Chen, J.; Wu, J.; Raffa, P.; Picchioni, F.; Koning, C.E. Superabsorbent Polymers: From Long-Established, Microplastics Generating Systems, to Sustainable, Biodegradable and Future Proof Alternatives. *Prog. Polym. Sci.* **2022**, *125*, 101475. [CrossRef]
36. Amiri, N.; Ghaffari, S.; Hassanpour, I.; Chae, T.; Jalili, R.; Kilani, R.T.; Ko, F.; Ghahary, A.; Lange, D. Antibacterial Thermosensitive Silver–Hydrogel Nanocomposite Improves Wound Healing. *Gels* **2023**, *9*, 542. [CrossRef]
37. Salem, S.S.; Hashem, A.H.; Sallam, A.A.M.; Doghish, A.S.; Al-Askar, A.A.; Arishi, A.A.; Shehabeldine, A.M. Synthesis of Silver Nanocomposite Based on Carboxymethyl Cellulose: Antibacterial, Antifungal and Anticancer Activities. *Polymers* **2022**, *14*, 3352. [CrossRef] [PubMed]
38. Xiong, J.; Ghori, M.Z.; Henkel, B.; Strunskus, T.; Schürmann, U.; Deng, M.; Kienle, L.; Faupel, F. Tuning Silver Ion Release Properties in Reactively Sputtered Ag/TiOx Nanocomposites. *Appl. Phys. A Mater. Sci. Process.* **2017**, *123*, 470. [CrossRef]
39. Farasati Far, B.; Naimi-Jamal, M.R.; Jahanbakhshi, M.; Hadizadeh, A.; Dehghan, S.; Hadizadeh, S. Enhanced Antibacterial Activity of Porous Chitosan-Based Hydrogels Crosslinked with Gelatin and Metal Ions. *Sci. Rep.* **2024**, *14*, 7505. [CrossRef]
40. Suresh, R.; Deepa, M.; Sudha, P.N.; Gomathi, T.; Pavithra, S.; Moganavally, P. Synthesis, Characterization, Biological and Catalytic Activity of Carboxymethyl Chitosan Schiff Base Metal Complexes. *Indian J. Geo-Mar. Sci.* **2022**, *51*, 423–431. [CrossRef]
41. Garibo, D.; Borbón-Nuñez, H.A.; de León, J.N.D.; García Mendoza, E.; Estrada, I.; Toledano-Magaña, Y.; Tiznado, H.; Ovalle-Marroquin, M.; Soto-Ramos, A.G.; Blanco, A.; et al. Green Synthesis of Silver Nanoparticles Using Lysiloma Acapulcensis Exhibit High-Antimicrobial Activity. *Sci. Rep.* **2020**, *10*, 12805. [CrossRef] [PubMed]
42. Venkatesham, M.; Ayodhya, D.; Madhusudhan, A.; Veera Babu, N.; Veerabhadram, G. A Novel Green One-Step Synthesis of Silver Nanoparticles Using Chitosan: Catalytic Activity and Antimicrobial Studies. *Appl. Nanosci.* **2014**, *4*, 113–119. [CrossRef]
43. Anand Ganesh, V.; Kundukad, B.; Cheng, D.; Radhakrishnan, S.; Ramakrishna, S.; Van Vliet, K.J. Engineering Silver-Zwitterionic Composite Nanofiber Membrane for Bacterial Fouling Resistance. *J. Appl. Polym. Sci.* **2019**, *136*, 47580. [CrossRef]
44. Ahsan, A.; Farooq, M.A. Therapeutic Potential of Green Synthesized Silver Nanoparticles Loaded PVA Hydrogel Patches for Wound Healing. *J. Drug Deliv. Sci. Technol.* **2019**, *54*, 101308. [CrossRef]
45. Nadtoka, O.; Kutsevol, N.; Naumenko, A.; Virych, P. Photochemical Synthesis and Characterization of Hydrogel–Silver Nanoparticle Composites. *Res. Chem. Intermed.* **2019**, *45*, 4069–4080. [CrossRef]
46. Nadtoka, O.; Virych, P.; Bezugla, T.; Doroschuk, V.; Lelyushok, S.; Pavlenko, V.; Yeshchenko, O.; Kutsevol, N. Antibacterial Hybrid Hydrogels Loaded with Nano Silver. *Appl. Nanosci.* **2022**, *12*, 629–636. [CrossRef]
47. Hajj, F.E.; Hasan, A.; Nakhleh, J.; Osta, M.; Darwish, G.; Karam, P.; Nassereddine, M. Nanosilver loaded GelMA hydrogel for antimicrobial coating of biomedical implants. In Proceedings of the 2015 International Conference on Advances in Biomedical Engineering (ICABME), Beirut, Lebanon, 16–18 September 2015; pp. 189–192.
48. Murali Mohan, Y.; Vimala, K.; Thomas, V.; Varaprasad, K.; Sreedhar, B.; Bajpai, S.K.; Mohana Raju, K. Controlling of Silver Nanoparticles Structure by Hydrogel Networks. *J. Colloid Interface Sci.* **2010**, *342*, 73–82. [CrossRef] [PubMed]
49. Rojas, M.A.; Amalraj, J.; Santos, L.S. Biopolymer-Based Composite Hydrogels Embedding Small Silver Nanoparticles for Advanced Antimicrobial Applications: Experimental and Theoretical Insights. *Polymers* **2023**, *15*, 3370. [CrossRef] [PubMed]
50. Chou, K.S.; Huang, K.C.; Lee, H.H. Fabrication and Sintering Effect on the Morphologies and Conductivity of Nano-Ag Particle Films by the Spin Coating Method. *Nanotechnology* **2005**, *16*, 779–784. [CrossRef]

51. Mekkawy, A.I.; El-Mokhtar, M.A.; Nafady, N.A.; Yousef, N.; Hamad, M.; El-Shanawany, S.M.; Ibrahim, E.H.; Elsabahy, M. In Vitro and in Vivo Evaluation of Biologically Synthesized Silver Nanoparticles for Topical Applications: Effect of Surface Coating and Loading into Hydrogels. *Int. J. Nanomed.* **2017**, *12*, 759–777. [CrossRef]
52. Henríquez, C.M.G.; del Carmen Pizarro Guerra, G.; Vallejos, M.A.S.; de la Fuente, S.D.R.; Flores, M.T.U.; Jimenez, L.M.R. In Situ Silver Nanoparticle Formation Embedded into a Photopolymerized Hydrogel with Biocide Properties. *J. Nanostructure Chem.* **2014**, *4*, 119–132. [CrossRef]
53. Cha, H.R.; Ramesh Babu, V.; Krishna Rao, K.S.V.; Kim, Y.H.; Mei, S.; Joo, W.H.; Lee, Y.I. Fabrication of Amino Acid Based Silver Nanocomposite Hydrogels from Pva-Poly(Acrylamide-Co-Acryloyl Phenylalanine) and Their Antimicrobial Studies. *Bull. Korean Chem. Soc.* **2012**, *33*, 3191–3195. [CrossRef]
54. Rumon, M.M.H.; Akib, A.A.; Sultana, F.; Moniruzzaman, M.; Niloy, M.S.; Shakil, M.S.; Roy, C.K. Self-Healing Hydrogels: Development, Biomedical Applications, and Challenges. *Polymers* **2022**, *14*, 4539. [CrossRef] [PubMed]
55. Anees Ahmad, S.; Sachi Das, S.; Khatoon, A.; Tahir Ansari, M.; Afzal, M.; Saquib Hasnain, M.; Kumar Nayak, A. Bactericidal Activity of Silver Nanoparticles: A Mechanistic Review. *Mater. Sci. Energy Technol.* **2020**, *3*, 756–769. [CrossRef]
56. Aldakheel, F.M.; Sayed, M.M.E.; Mohsen, D.; Fagir, M.H.; El Dein, D.K. Green Synthesis of Silver Nanoparticles Loaded Hydrogel for Wound Healing; Systematic Review. *Gels* **2023**, *9*, 530. [CrossRef] [PubMed]
57. Han, X.; He, J.; Wang, Z.; Bai, Z.; Qu, P.; Song, Z.; Wang, W. Fabrication of Silver Nanoparticles/Gelatin Hydrogel System for Bone Regeneration and Fracture Treatment. *Drug Deliv.* **2021**, *28*, 319–324. [CrossRef] [PubMed]
58. Aldakheel, F.M.; Mohsen, D.; El Sayed, M.M.; Fagir, M.H.; El Dein, D.K. Employing of Curcumin–Silver Nanoparticle-Incorporated Sodium Alginate-Co-Acacia Gum Film Hydrogels for Wound Dressing. *Gels* **2023**, *9*, 780. [CrossRef] [PubMed]
59. Maher, S.; Kalil, H.; Liu, G.; Sossey-Alaoui, K.; Bayachou, M. Alginate-Based Hydrogel Platform Embedding Silver Nanoparticles and Cisplatin: Characterization of the Synergistic Effect on a Breast Cancer Cell Line. *Front. Mol. Biosci.* **2023**, *10*, 1242838. [CrossRef]
60. Dong, Q.; Zu, D.; Kong, L.; Chen, S.; Yao, J.; Lin, J.; Lu, L.; Wu, B.; Fang, B. Construction of Antibacterial Nano-Silver Embedded Bioactive Hydrogel to Repair Infectious Skin Defects. *Biomater. Res.* **2022**, *26*, 36. [CrossRef]

polymers

MDPI

Article

Chitosan Supports Boosting $NiCo_2O_4$ for Catalyzed Urea Electrochemical Removal Application

Fowzia S. Alamro [1], Mahmoud A. Hefnawy [2,*], Sherif S. Nafee [3], Nada S. Al-Kadhi [1], Rami Adel Pashameah [4], Hoda A. Ahmed [2,5] and Shymaa S. Medany [2,*]

1 Department of Chemistry, College of Science, Princess Nourah Bint Abdulrahman University, P.O. Box 84428, Riyadh 11671, Saudi Arabia
2 Chemistry Department, Faculty of Science, Cairo University, Giza 12613, Egypt
3 Physics Department, Faculty of Science, King Abdulaziz University, Jeddah 21589, Saudi Arabia
4 Department of Chemistry, Faculty of Applied Science, Umm Al-Qura University, Makkah 24230, Saudi Arabia
5 Chemistry Department, College of Sciences, Taibah University, Yanbu 30799, Saudi Arabia
* Correspondence: maadel@cu.edu.eg (M.A.H.); shymaasamir80@cu.edu.eg (S.S.M.)

Abstract: Currently, wastewater containing high urea levels poses a significant risk to human health. Else, electrocatalytic methodologies have the potential to transform urea present in urea-rich wastewater into hydrogen, thereby contributing towards environmental conservation and facilitating the production of sustainable energy. The characterization of the $NiCo_2O_4$@chitosan catalyst was performed by various analytical techniques, including scanning electron microscopy (SEM) and X-ray photoelectron spectroscopy (XPS). Furthermore, the activity of electrodes toward urea removal was investigated by several electrochemical techniques. As a function of current density, the performance of the modified $NiCo_2O_4$@chitosan surface was employed to remove urea using electrochemical oxidation. Consequently, the current density measurement was 43 mA cm^{-2} in a solution of 1.0 M urea and 1.0 M KOH. Different kinetic characteristics were investigated, including charge transfer coefficient (α), Tafel slope (29 mV dec^{-1}), diffusion coefficient (1.87×10^{-5} cm^2 s^{-1}), and surface coverage 4.29×10^{-9} mol cm^{-2}. The electrode showed high stability whereas it lost 10.4% of its initial current after 5 h of urea oxidation.

Keywords: urea removal; spinel oxide: electrochemical oxidation; nickel cobaltite

1. Introduction

The oxidation of urea, also known as UOR, presents a promising solution to address energy, environmental, and healthcare challenges. This is attributed to its eco-friendly, cost-effective, and sustainable processing methods [1–3]. In conjunction with electrocatalysts, electrochemical methods can potentially enhance molecular conversion on the electrode surface. This process can be facilitated by renewable electricity and can serve the purpose of achieving various objectives such as energy storage and conversion, environmental remediation, and electroanalysis [4–6]. Specifically, sewage containing urea has the potential to be transformed into gaseous byproducts using UOR technologies in alkaline environments and producing energy through a well-planned design [7]. Nevertheless, it was observed that urea experienced decomposition primarily into N_2 and CO_2 when subjected to an acidic environment using an applied potential exceeding 1.7 V relative to the normal hydrogen electrode (NHE) [3]. When the electrolyte maintains a neutral state, the decomposition of urea primarily yields nitrite and nitrate ions, leading to the generation of CO_2 [2].

Electrochemical systems have been employed in diverse contexts with distinct arrangements and objectives, leading to heterogeneous urea conversion patterns. Although the electrochemical treatment process is currently in its early stages of development, its exceptional characteristics present encouraging prospects for advancement in energy, environment, and healthcare [8–12]. Thus, the electrochemical oxidation of urea can be employed for both urea removal application (wastewater treatment) [13], and fuel cell (energy conversion application) [14].

The surfaces composed of nickel were subject to a significant modification to improve their electronic properties, physical characteristics, and electrochemical activity. As a result, various bimetallic electrocatalysts based on nickel have been documented in the literature as Ni-Mo [15,16], Ni-Co [17,18], Ni-Rh [19], Ni-Mn [20,21], and Ni-Cu [22–24].

AB_2O_4 spinel oxides, characterized by a basic structure, have been found to possess remarkable chemical and thermal stability, rendering them highly suitable for diverse catalytic applications. The metallic cations A and B are integral components of this class of compounds. Transition metal oxides with a spinel phase are considered the most appealing anodic materials for electrochemical applications [25]. $NiCo_2O_4$, a type of nickel-based spinel oxide, has gained significant attention in various applications such as fuel cells, oxygen evolution reactions, electrochemical sensors, Li-ion batteries, and supercapacitors [26–30].

Chitosan, a derivative of chitin, is frequently employed in diverse applications. Therefore, it is an option for the creation of new chitosan products. These advancements in fermentation technology have allowed the production of chitosan with unique physiochemical characteristics that differ from those found in waste materials. As a result, this presents a promising opportunity to develop innovative chitosan-based products. An alternative to the traditional sources, such as crab shells, is being considered. Chitosan is a frequently utilized material for immobilization purposes owing to its favorable environmental properties, high absorption capacity, notable layer-forming abilities, superior permeability, increased thermal stability, sturdy mechanical strength, biocompatibility, and ease of accessibility [31].

Chitosan has special structural and functional qualities, such as non-toxicity, hydrophilicity, excellent adhesion, biocompatibility, environmental sustainability, antibacterial and antimicrobial characteristics, and non-carcinogenicity. These characteristics make it a very adaptable and widely used chemical in numerous fields [32–36].

Herein, chitosan is employed to boost the activity of nickel cobalt spinel oxide toward urea electrooxidation. The facile synthesis of nickel cobaltite-based composite was used for electrode fabrication. Comparative studies were performed between $NiCo_2O_4$@Chitosan and unmodified $NiCo_2O_4$. The modified electrode was employed as an efficient electrode for electrochemical urea removal. Thus, different electrochemical techniques were used to judge the electrode performance. Additionally, kinetic parameters were calculated to well-understand the electrochemical oxidation process.

2. Experimental

2.1. Synthesis of $NiCo_2O_4$

The $NiCo_2O_4$ was synthesized by hydrothermal technique. A mixture comprising $CoCl_2.6H_2O$ (6 mmol), $NiCl_2.6H_2O$ (3 mmol), urea (8 mmol), NH_4F (25 mmol), and DI water (40 mL) was subjected to magnetic stirring for 30 min after its mixing in a beaker. The solution was introduced into a 50 mL stainless steel reactor with a polytetrafluoroethylene (PTFE) lining. The mixture was subjected to a consistent temperature of 130 °C for 8 h. Following the natural cooling of the reactor to ambient temperature, the sample containing precursors underwent a 30-min ultrasonic cleaning process with deionized water to eliminate any ionic impurities and loose deposition. The crystalline particles of $NiCo_2O_4$ were ultimately acquired through annealing at a temperature of 400 °C in an air environment for 2 h, with a heating rate of 2 °C per minute.

2.2. Synthesis of $NiCo_2O_4$ Supported Chitosan

The nickel cobaltite chitosan composite was synthesized by combining a chitosan solution with $NiCo_2O_4$ nanoparticles. 1.5 g of chitosan was introduced to 60 mL of absolute ethanol in a beaker. Subsequently, the mixture was subjected to gentle stirring while gradually increasing the temperature. A quantity of 1.5 g of nickel cobaltite nanoparticles was introduced into the mixture. The temperature of the solution was reduced to the ambient temperature of the surrounding environment. Introducing $NiCo_2O_4$ into the chitosan solution resulted in crosslinking and subsequent encapsulation of $NiCo_2O_4$ nanoparticles. The polyelectrolytic nature of chitosan in acidic environments is attributed to the protonation of its $-NH_2$ functional groups. The consequential equilibrium reaction delineates the ionization state.

Consequently, approximately 3 mL of 10% acetic acid was added to the mixture and agitated until the solution exhibited a thick consistency. After ten minutes, the mixture was subjected to filtration and subsequently washed with distilled water. The final composite was dried in an oven, which was maintained at a temperature of 80 °C for 3 h.

2.3. Electrode Fabrication

The working electrode was a glassy carbon electrode with a 0.0707 cm^2 surface area. A gentle emery paper polish was applied after it had been cleansed with ethanol and double-distilled water. The cast solution was then created by ultrasonically dispersing 10 mg of the catalyst powder ($NiCo_2O_4$ or $NiCo_2O_4$@Chitosan) in 0.75 mL of ethanol and 0.25 mL of 5 wt% Nafion for 1 h. The modified electrodes ($NiCo_2O_4$ or $NiCo_2O_4$@Chitosan) were created as follows: 30 μL of catalyst solution was sprayed onto the electrode's surface and left to dry for 6 h at 60 °C. The Autolab PGSTAT128N was used to conduct all electrochemical experiments. NOVA (Version 2.1, Metrohm Autolab, Utrecht, The Netherlands), an electrochemistry application, fits the impedance spectrum. The counter and reference electrodes were Pt wire and Ag/AgCl/KCl (sat.); respectively. However, NiCo and NiCo@Chit were used to represent the modified electrodes $NiCo_2O_4$ or $NiCo_2O_4$@Chitosan; respectively, and used as working electrodes for urea electrochemical elimination in alkaline medium applications.

3. Result and Discussion

3.1. Characterizations of Morphology, Microstructure, and Composition

Figure 1a displays the Ni 2p spectrum, which manifests multiple prominent peaks. These peaks are subjected to fitting procedures, which involve the identification of the $2p_{3/2}$ and $2p_{1/2}$ peaks and the satellite peaks. The spectral peaks observed at 854.3 and 873.4 eV are attributed to the Ni^{2+} component, whereas the peaks detected at 856.1 and 873.2 eV are associated with the Ni^{3+} component in NiCo, as reported by Hao et al. [37]. The spectral features observed at 863.1 and 879.4 eV are identified as satellite peaks. The spectrum of Co 2p comprises two doublets resulting from spin-orbit coupling and two satellite peaks, as depicted in Figure 1b. The distinctive doublet peaks indicate the presence of the Co^{3+} component observed at 781.2 and 795.2 eV. The characteristic doublet peaks can identify the Co^{2+} component observed at 782.1 and 797.6 eV. The satellite peak observed at 788.6 and 804.2 eV can be attributed to the Co^{3+} and Co^{2+}/Co^{3+} components, respectively, as reported by Marco et al. [38]. The spectrum of O1s (as depicted in Figure 1c) can be effectively modeled by three distinct peaks at 530.2, 531.1, and 533.18 eV; respectively. These peaks indicate metal-oxygen bonds, and oxygen defects [39–41]. The XPS spectrum of C1s (see Figure 1d), three peaks can be observed at binding energy of 287.1, 286.6, 285.2 eV attributed for C-O, C-N, and C-C; respectively [42,43].

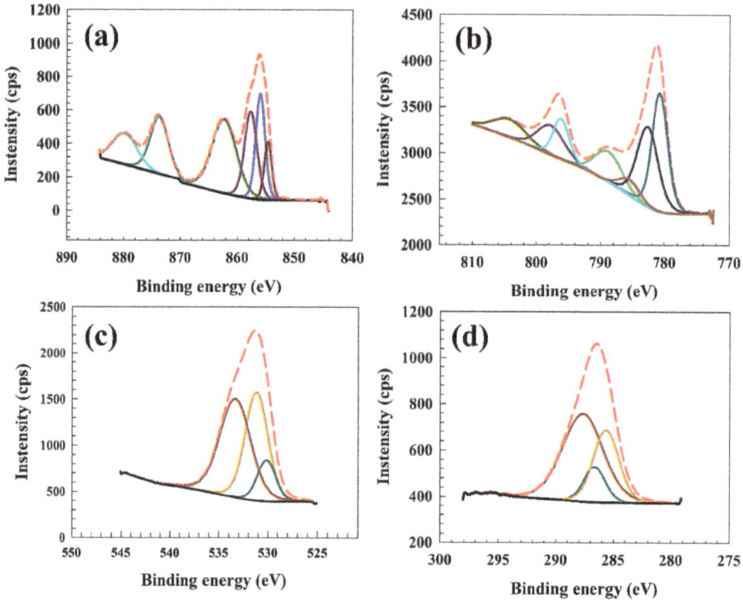

Figure 1. XPS of NiCo nanoparticles (**a**) Ni2p, (**b**) Co2p, (**c**) O1s, (**d**) C1s.

The chemical structures of as-prepared NiCo and NiCo@Chit were confirmed using powder X-Ray diffraction technique. Figure 2 shows the XRD chart of as-prepared $NiCo_2O_4$ powder. Thus, several peak observed at 2θ equaled to 31, 37, 44, 58, 65 and 76 that attributed to the reference card of (JCPDS #20-0781) [44]. For the chitosan-based sample, the intensity of the peak decreased because of embedding the nanoparticles in chitosan sheets. The interaction between chitosan and $NiCo_2O_4$ lead to change in lattice structure [45–47].

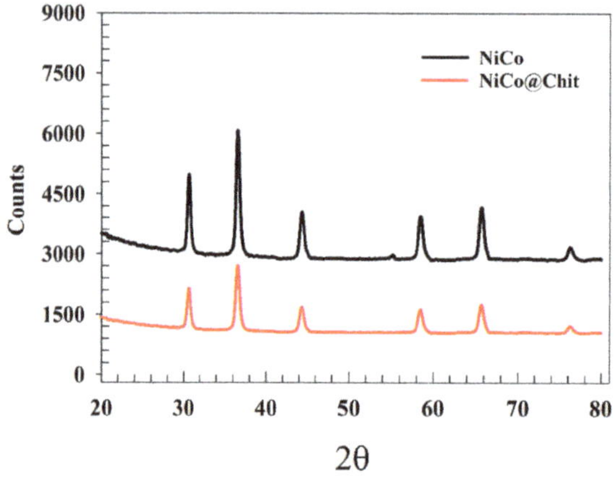

Figure 2. XRD as prepared NiCo, and NiCo@Chit.

The morphological characteristics of the NiCo nanostructures in their initial state were examined utilizing scanning electron microscopy (SEM), as illustrated in Figure 3a. The particles ranged in size 35~80 nm. The small particle size of NiCo indicates the higher activity of the prepared materials. Figure 3b shows the NiCo incorporated into the chitosan sheets. The well-distribution of the NiCo on chitosan sheets can explain the electrode's high electrochemical activity toward urea electrochemical removal. Presence of chitosan can promote urea adsorption.

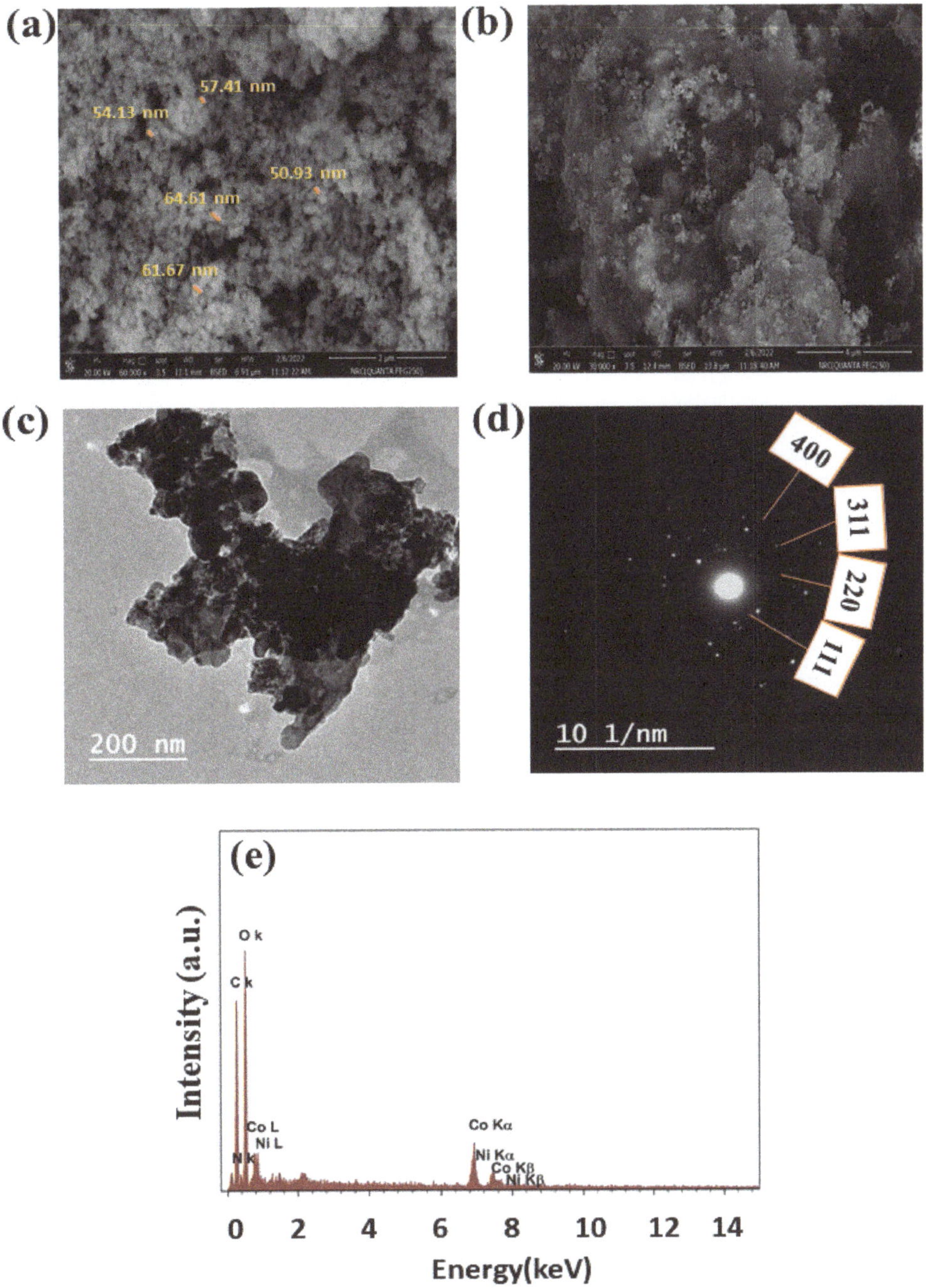

Figure 3. SEM of (**a**) NiCo and (**b**) NiCo@Chit, (**c**) TEM of NiCo@Chit, (**d**) diffraction pattern of NiCo@Chit, (**e**) EDX of NiCo sample.

The conventional method for determining the dimensions of NiCo nanoparticles was the utilization of Transmission Electron Microscopy (TEM). The average particle size of NiCo was approximately ~60 nanometers. Figure 3c shows the TEM of NiCo@Chit. Thus, the nanosphere of NiCo was observed to be attached to the chitosan sheets. The corresponding TEM diffraction patterns are used to confirm the formation of NiCo on the chitosan sheet. As represented in Figure 3d, d-spacing was used to find the Miller indices (hkl) using ImageJ software. However, the observed rings can be attributed to planes of (400), (311), (220), and (111); respectively. The elemental analysis of NiCo@Chit was estimated by EDX. As a result, EDX indicates that Ni, Co, O, C, and N are present. Figure 3e displays the elemental compositions of the NiCo@Chit sample. As a result, the elemental percentages displayed in the inset figure match the target structure of NiCo, which has a Ni/Co ratio of 1 to 2.

3.2. Urea Electrooxidation

The modified GC/NiCo and GC/NiCo@Chit activity was investigated by cyclic voltammetry in a solution of 1.0 M urea and 1.0 M KOH. Activating electrodes composed of nickel is a pivotal stage in the electrochemical oxidation of urea, therefore the electrode performance was enhanced by an activation process; firstly. The outcome of this process is the creation of a Ni-form that exhibits a high degree of electrocatalytic activity, specifically NiOOH. The activation process was executed through cyclic voltammetry (CV) with a scan rate of 100 mV s^{-1} for 150 cycles, utilizing a solution containing 1.0 M KOH(see Figure 4) [48]. The phenomenon of NiOOH formation leads to an increase in current during successive cycles. With an increase in the number of potential sweeps, there is a corresponding increase in the thickness of NiOOH layer. This can be attributed to the presence of OH^- ions, which facilitate the rate of conversion between $Ni(OH)_2$ and NiOOH according to the following Equation (1) [49–52]:

$$6\,Ni(OH)_2 + 6\,OH^- \leftrightarrow 6\,NiOOH + 6\,H_2O + 6\,e^- \quad (1)$$

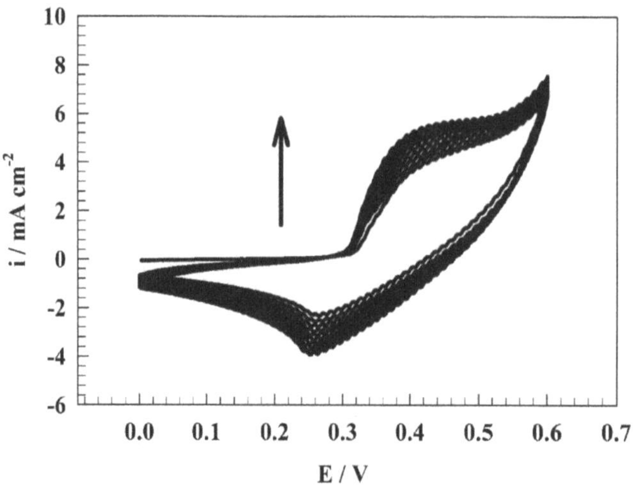

Figure 4. CVs of activation for GC/NiCo@Chit electrode.

The generated NiOOH species is mainly used for the electrochemical oxidation of urea depending on the following Equation (2):

$$6\,NiOOH + CO(NH_2)_2 + H_2O \leftrightarrow 6\,Ni(OH)_2 + N_2 + CO_2 \quad (2)$$

Figure 5a shows CVs of the modified NiCo and NiCo@Chit in 1.0 M KOH. One redox peak can be observed at a potential range of 0.3 to 0.45 V for conversion of $Ni(OH)_2$ and NiOOH. Additionally, the urea oxidation can be represented in Figure 5b. Thus, strong oxidation peaks at potential ~0.5 V are attributed to the conversion of urea. However, a sample of NiCo@Chit utilized high activity compared to the unmodified NiCo sample. The presence of chitosan could enhance the activity toward urea electrochemical oxidation in the alkaline medium. The reason for higher activity toward urea electrochemical oxidation may be explained by the ability of chitosan to adsorb urea along with the extended surface area and enhancement of mechanical and chemical stability of chitosan-based samples compared with the unmodified NiCo samples [53–55]. Comparative studies between chitosan-based and unmodified NiCo were performed using several approaches. Table 1 summarizes some of the results of the NiCo and NiCo@Chit surfaces.

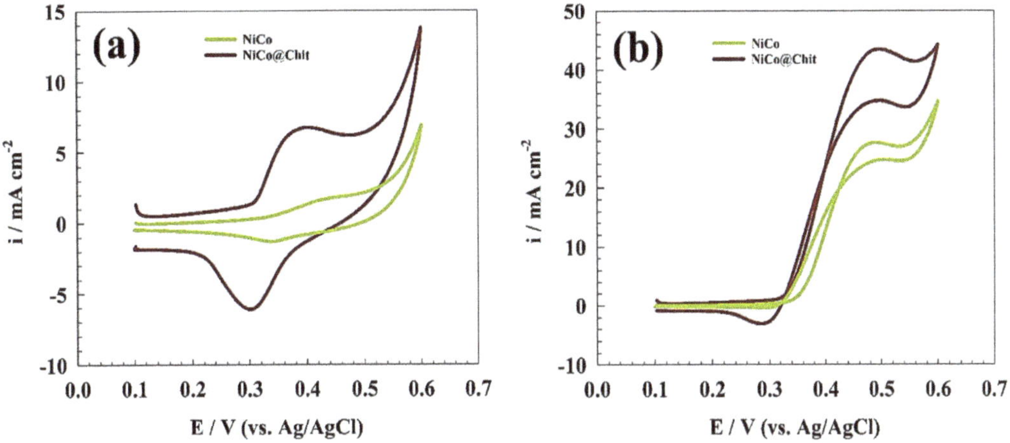

Figure 5. CVs of comparison between NiCo and NiCo@Chit (**a**) in the absence and (**b**) in the presence of 1.0 M urea.

Table 1. Electrochemical parameters for NiCo and NiCo@Chit electrodes.

Electrode	Anodic Current (mA cm^{-2})	Onset Potential (V)	E_{pa} (V)	Tafel Slope mV dec^{-1}	Diffusion Coefficient (cm^2 s^{-1})	Surface Coverage (Γ)/(mol cm^{-2})
NiCo	27	0.35	0.5	44	5.98×10^{-6}	9.34×10^{-10}
NiCo@Chit	43	0.32	0.49	29	1.87×10^{-5}	4.29×10^{-9}

Furthermore, an investigation was conducted on the electrooxidation of urea across a range of concentrations that extend from 0.05 to 1.0 M. Surface saturation was not observed within the concentration range under investigation, as depicted in Figure 6a,b. The specific anodic peak current of the electrooxidation of urea exhibits a positive correlation with the urea concentration (see Figure 6c,d. The results of this study suggest that the electrode under investigation may be suitable for use in applications involving high concentrations of urea, such as in wastewater treatment and direct urea fuel cells (DUFCs). The comparison between the modified NiCo@Chit electrode and others reported in the literature is listed in Table 2.

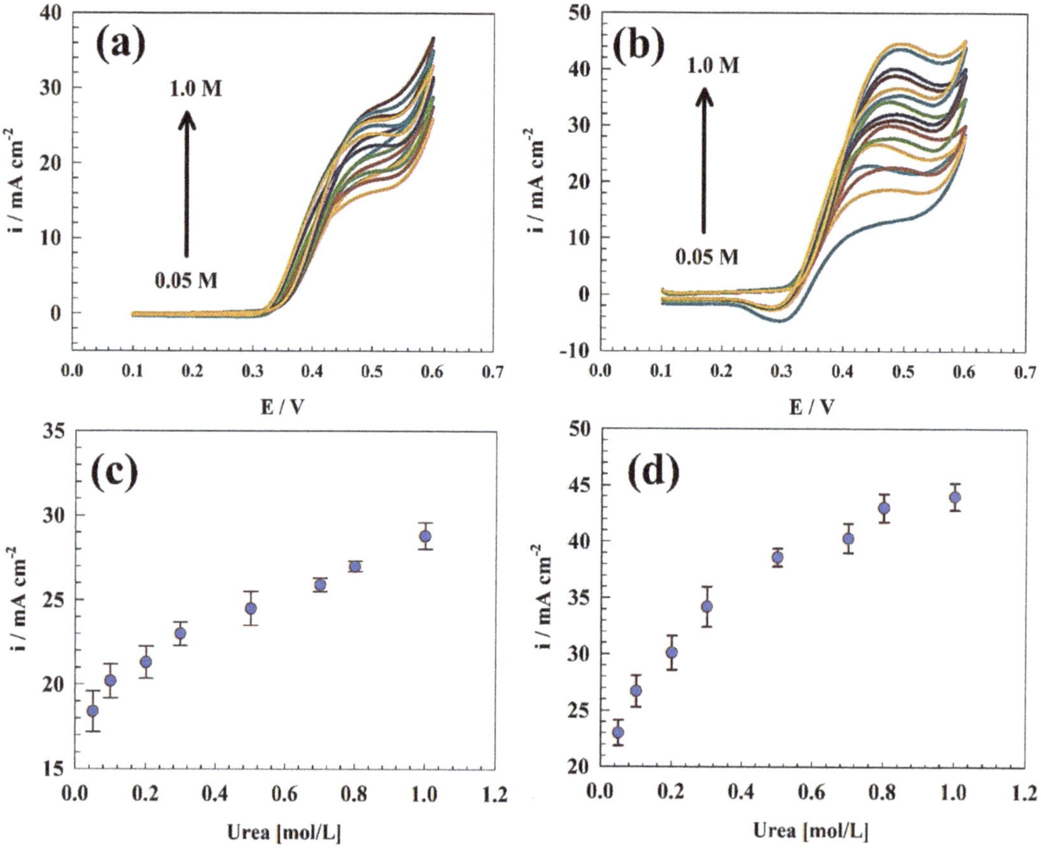

Figure 6. CVs of (**a**) GC/NiCo and (**b**) GC/NiCo@Chit electrodes in 1.0 M KOH and a wide range of urea concentrations. Relation between the urea oxidation current vs. urea concentrations using (**c**) GC/NiCo, and (**d**) GC/NiCo@Chit.

Table 2. Comparison between different surfaces for urea electrochemical oxidation in an alkaline medium.

Electrode	Fuel Concentration (M)	Electrolyte Concentration (M)	Scan Rate (mV s^{-1})	Oxidation Current (mA cm^{-2})	References
$NiCo_2O_4$@Chitosan	1.0	1.0	20	43	This work
$Ni_{0.85}Se$/rGO	0.5	1.0	50	10	[56]
$Ni_{0.9}Cu_{0.1}$	0.3	0.5	20	32	[23]
IN738 supper alloy	1.0	1.0	20	12	[57]
NiO-MnOx/Polyaniline	0.3	0.5	20	16	[58]
$Ni(OH)_2$ meshes	0.3	1.0	50	20	[59]

3.3. *Urea Oxidation Kinetics*

To achieve an in-depth understanding of urea electrochemical oxidation, kinetic parameters were estimated for the oxidation of nitrite over the modified electrodes.

Additionally, various scan rates were used with the modified electrodes NiCo and NiCo@Chit in a solution of 1.0 M KOH as represented in Figure 7a,b. The following Equation (3) was used to estimate the surface coverage:

$$i = (n^2F^2/4RT)\ A\ \nu\ \Gamma^* \quad (3)$$

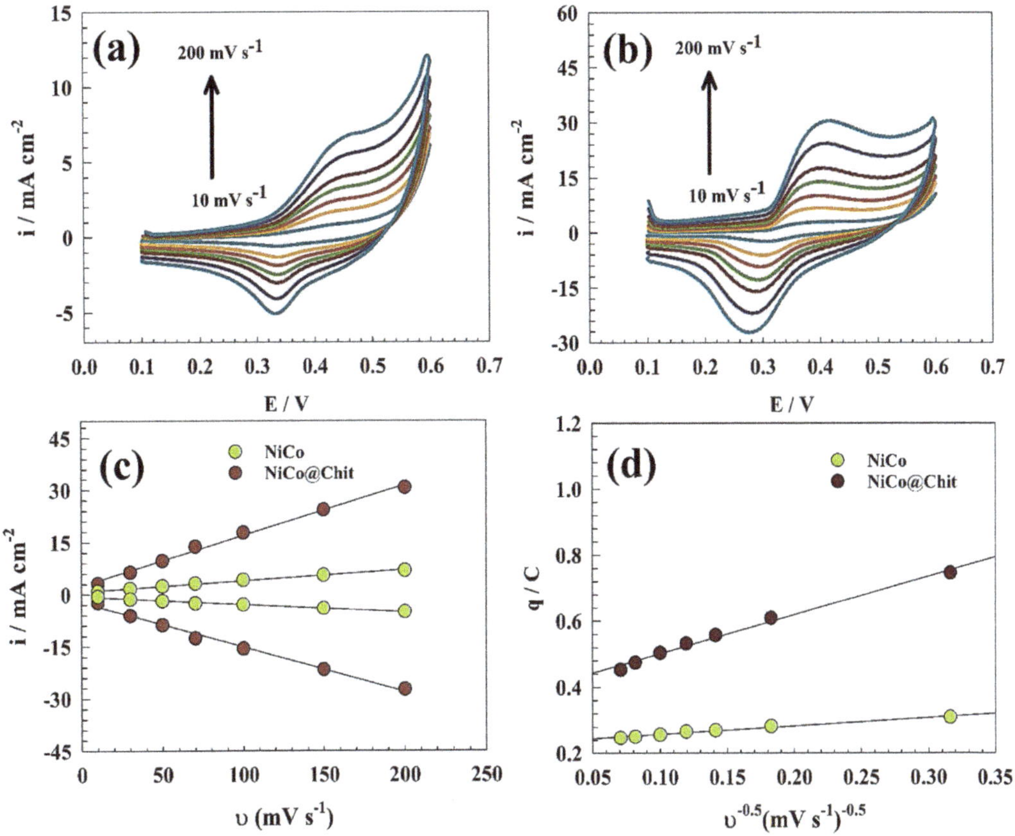

Figure 7. CVs of (**a**) NiCo and (**b**) NiCo@Chit at different scan rates (10 to 200 mV s^{-1}) in 1.0 M KOH in the absence of urea. (**c**) Linear relation between redox current versus the scan rate. (**d**) Relation between the charge versus the reciprocal of the square root of the scan rate.

Where A is surface area, ν scan rate, and Γ* surface coverage, n is the number of electrons, F is the Faraday constant, R is the universal gas constant, and T is the measurement temperature.

As shown in Figure 7c, the relationship between the scan rate and the anodic peak current will reveal the surface coverage. The surface coverage of the modified electrodes, NiCo and NiCo@Chit, was 9.34×10^{-10} mol cm^{-2} and 4.29×10^{-9} mol cm^{-2}; respectively. The larger surface coverage can be observed due to the NiCo@Chit sample's increased surface activity when urea conversion is compared to unmodified NiCo.

The following relation (Equation (4)) was utilized to confirm that the active sites are evenly dispersed on the surface of the chitosan support electrochemically [60]:

$$q = q_{\infty} + a\ \nu^{-0.5} \quad (4)$$

Where q_∞ is the maximum quantity of the charge related to the "outer" surface of active material in Coulombs, q is the charge calculated in Coulombs for various potential scan rates in CV, a is constant (slope of the relation), and ν is the potential scan rate (mV s^{-1}) (see Figure 7d).

The intercept value of NiCo@Chit exhibits a significantly higher magnitude than that of pristine NiCo. According to the results, it can be observed that the NiCo@Chit composite possesses active sites that are 1.65 times greater than those of pristine NiCo, thus implying a higher efficiency for urea electrochemical removal. The catalytic reaction is expected to experience significant acceleration on the surface of NiCo@Chit, owing to the abundant active sites of the catalyst that are uniformly distributed and highly effective.

The CVs of the NiCo and NiCo@Chit modifications were presented in Figure 8a,b; respectively. The measurements were utilized in a solution of 1.0 M urea and 1.0 M KOH, with a scan rate ranging from 5 to 400 mV s^{-1} (vs. Ag/AgCl).

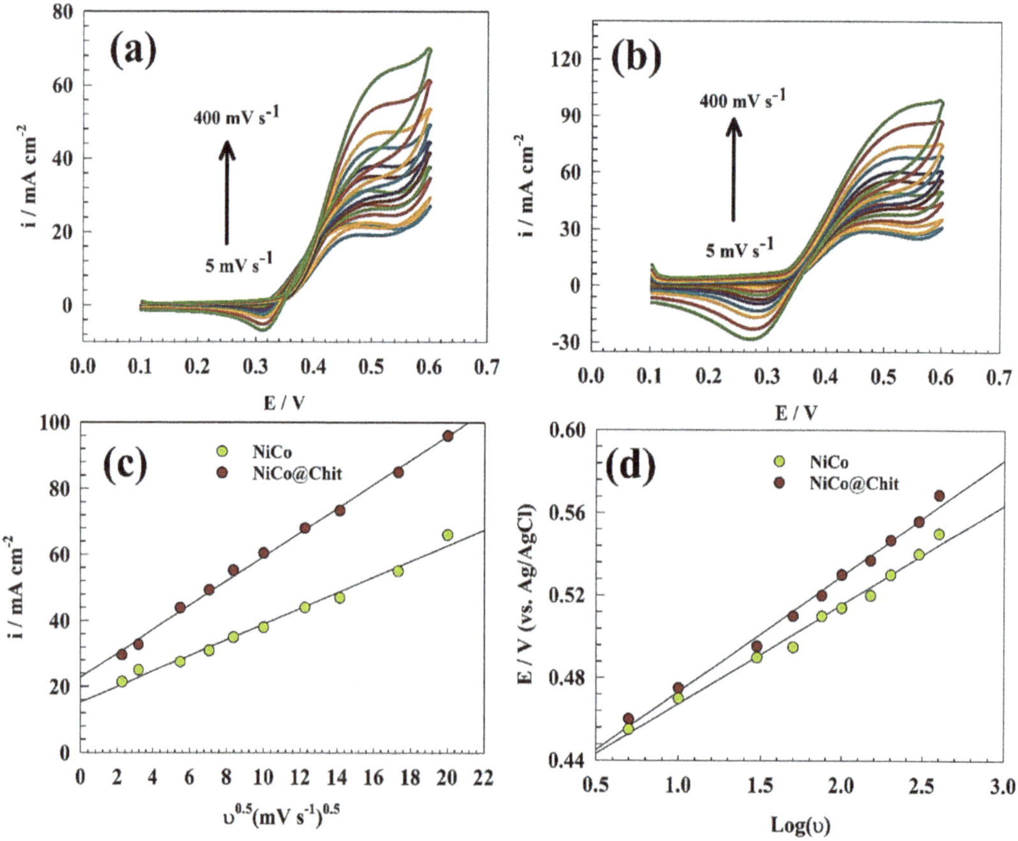

Figure 8. CVs of different modified surfaces (**a**) NiCo and (**b**) NiCo@Chit in a solution of 1.0 M urea and 1.0 M KOH at different scan rates 5 to 400 mV s^{-1}. (**c**) Linear relation between anodic current versus the square root of scan rate. (**d**) Linear relation between anodic peak potential versus the logarithmic scan rate.

Thus, Randles-Sevcik equation can be employed to calculate the diffusion coefficient (D) for irreversible processes, according to Equation (5) [20,61]:

$$I_p = 2.99 \times 10^5\ n\ A\ C_o\ [(1-\alpha)\ n_o\ D\ \nu]^{0.5} \quad (5)$$

The equation mentioned above denotes the relationship between various parameters, namely the urea oxidation current(i), the number of electrons represented by (n), the surface area of the electrode denoted by (A), the analyte diffusion coefficient represented by (D), the analyte concentration denoted by (C_o), and the scan rate represented by (ν).

The Randles-Sevick method was employed to estimate the diffusion coefficient. This was achieved by establishing a linear correlation between the current of nitrite oxidation and the square root of the scan rate, as illustrated in Figure 8c. The diffusion coefficients for NiCo and NiCo@Chit electrodes are reported as 5.98×10^{-6} and 1.87×10^{-5} $cm^2\ s^{-1}$; respectively. The enhanced diffusion coefficient observed for a surface based on chitosan can be attributed to the increased capacity of chitosan to adsorb urea.

Figure 8d illustrates a linear correlation between the peak potential and the logarithm of the scan rate across various modified surfaces. The confirmation of reversibility can be established through the positive shift of the Ep with an increase in the scan rate. The Laviron Equation (6) for irreversible reactions was utilized to observe a change in the location of the peak potential through an increase in the scan rate values [62,63]:

$$E_{pa}(V) = E^{\circ} - \frac{RT}{\propto nF} \ln \frac{RTk_s}{\propto nF} + \frac{RT}{\propto nF} \ln v \tag{6}$$

The previously mentioned variables, namely E_{pa} denoting peak potential, R representing the universal gas constant, E° signifying formal potential, T indicating temperature, n denoting the number of electrons, v representing scanning rate, and F representing the Faraday constant, are of significance in the academic context.

The transfer coefficient (α) is a kinetic parameter that indicates the propensity of a reaction to proceed in the oxidation/reduction direction. A preference for oxidation direction is observed when the value of (α) is less than 0.5. The transfer coefficients were computed for NiCo and NiCo@Chit using Laviron relation, which involved determining the linear correlation between Log (ν) and E_{pa}. The resulting transfer coefficients were 0.46 and 0.53 for NiCo and NiCo@Chit; respectively. The symmetry factor and charge transfer coefficient (α) suggest that urea oxidation on NiCo@Chit has a better reputation than NiCo. However, linear correlation indicates the adsorption of urea onto electrode surfaces.

The endurance of the electrode in the face of uninterrupted electrooxidation is the most important in the context of urea elimination. Chronoamperometry was utilized to investigate the enduring stability of the electrode for the electrooxidation of urea. Figure 9 depicts the chronoamperogram of the NiCo and NiCo@Chit-modified electrodes in a solution containing 1.0 M urea and 1.0 M KOH while maintaining a constant oxidation potential of 0.5 V (vs. Ag/AgCl). After 5 h, the electrodes' oxidation current density exhibited a decrease of 12.3 and 10.4% for NiCo and NiCo@Chit; respectively. The present reduction is attributed to the electrocatalyst surface's mechanical corrosion, incompletely oxidized urea accumulation, and metal carbonate formation due to the adsorption of the generated carbon monoxide [64]. Nevertheless, minor variations in the oxidation current indicate the enhanced durability of the electrodes to the electrochemical oxidation of urea over an extended period.

Electrochemical impedance spectroscopy was employed to ascertain the charge transfer resistance across various electrode surfaces. Figure 10a depicts Nyquist plots of various modified electrodes (GC/NiCo and GC/NiCo@Chit) in a solution containing 1.0 M urea and 1.0 M KOH at 0.5 V (vs. Ag/AgCl). The observation of the double semi-circuit suggests that the process of two-charge transfer warrants consideration. The equivalent fitting circuit corresponding to the statement has been presented in the inset of Figure 10a. The constant phase element (CPE) is employed instead of the capacitive element to account for the non-homogeneity of the electrode surfaces. The resistance values denoted by Rs, R_1, R_2, Q_1, and Q_2 pertain to the outer and inner layers' solution resistance, charge transfer resistance, and constant phase element (CPE). Table 3 presents the fitting parameters that were computed. The NiCo@Chit exhibited a charge transfer resistance of 103 $\Omega\ cm^2$ while NiCo surfaces displayed a resistance of 230 $\Omega\ cm^2$. The enhanced activity of NiCo@Chit in urea oxidation compared to unmodified NiCo can be attributed to the lower charge transfer resistance. However, the EIS data represented in Table 3 confirm the data obtained from the cyclic voltammetry that the modified chitosan composite has higher activity toward urea oxidation due to the high surface area and adsorption ability [55,65].

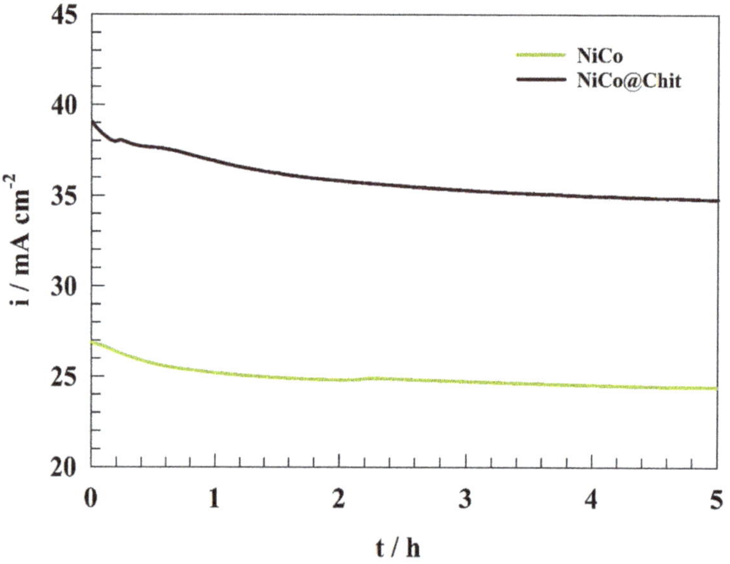

Figure 9. Chronoamperograms of NiCo and NiCo@Chit at constant 0.5 V (vs. Ag/AgCl).

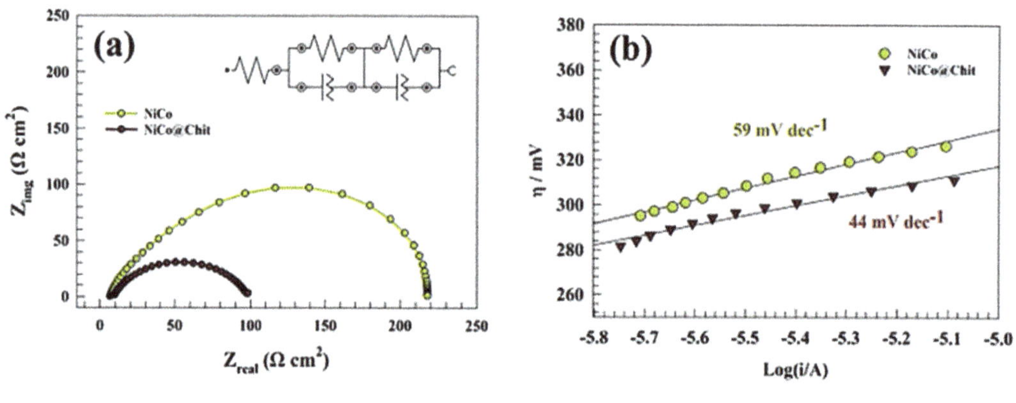

Figure 10. (**a**) Nyquist plots of NiCo and NiCo@Chit surfaces at 0.5 V potential. (**b**) Tafel plots of NiCo and NiCo@Chit.

Table 3. EIS parameters for NiCo and NiCo@Chit electrodes.

Electrode	R_s	R_1	Q_1		R_2	Q_2	
	Ω cm^2	Ω cm^2	Y_0	N	Ω cm^2	Y_0	m
NiCo	3.2	6.76	0.0005621	0.5154	230	0.002180	0.8322
NiCo@Chit	2.5	7.56	0.0013547	0.6523	103	0.003715	0.7354

Figure 10b depicts Tafel analysis employed to investigate the electrochemical kinetics of urea removal by utilizing the Tafel equation. The Tafel slopes computed for the GC/NiCo and GC/NiCo@Chit electrodes are 44 and 29 mV dec^{-1}; respectively. The NiCo@Chit sample exhibits a lower Tafel slope, suggesting that the oxidation of urea over the surfaces modified by chitosan is more favorable than the unmodified NiCo surface. The calculated Tafel slopes are comparable with other reported catalysts for urea removal like 22 mV dec^{-1}, 21.5 mV dec^{-1}, and 26.4 mV dec^{-1} for Ni-MOF, $LaNiO_3$, and FeOOH; respectively [11,50,66].

4. Conclusions

The present study reports the successful preparation of a Nickel-based spinel oxide ($NiCo_2O_4$) by hydrothermal techniques. The synthesized materials were supported on chitosan sheets to enhance the efficiency of the spinel oxide to electrochemical urea removal.

A comparative analysis was utilized between the performances of pristine $NiCo_2O_4$ versus $NiCo_2O_4$@Chit surfaces. Including functionalized carbon materials in NiCo oxides enhances structural stability, thereby mitigating surface poisoning and ensuring compatibility between the electrocatalyst and glassy carbon surface.

The extended surface area of nickel-containing chitosan enhances its efficacy in facilitating urea removal. $NiCo_2O_4$@Chitosan has been identified as a highly promising material due to its exceptional electrochemical properties. The lower Tafel slopes for chitosan-modified surface indicates the higher thermodynamic favorability.

Author Contributions: Formal analysis, M.A.H., S.S.M., N.S.A.-K., F.S.A., R.A.P., S.S.N. and H.A.A.; Funding acquisition, N.S.A.-K., F.S.A., R.A.P., M.A.H., S.S.N., S.S.M. and H.A.A.; Methodology, M.A.H., S.S.M., H.A.A. and R.A.P.; Data curation, M.A.H., S.S.N. and S.S.M.; Conceptualization, S.S.M. and M.A.H.; Validation, S.S.M. and M.A.H.; Project administration, N.S.A.-K., F.S.A., M.A.H., S.S.M. and S.S.N.; Resources and Software, S.S.M. and M.A.H.; Writing—original draft, S.S.M. and M.A.H.; Writing—review and editing, N.S.A.-K., F.S.A., R.A.P., S.S.M., H.A.A. and M.A.H. All authors have read and agreed to the published version of the manuscript.

Funding: This research was funded by "Princess Nourah Bint Abdulrahman University Researchers Supporting Project number (PNURSP2023R107), Princess Nourah Bint Abdulrahman University, Riyadh, Saudi Arabia".

Institutional Review Board Statement: Not Applicable.

Data Availability Statement: The data used for research described in this manuscript are available upon request from corresponding authors: shymaasamir80@cu.edu.eg; shymaa@sci.cu.edu.eg (S.S.M); maadel@cu.edu.eg; maahefnawy@gmail.com (M.A.H).

Acknowledgments: The authors extend their sincere appreciation to "Princess Nourah Bint Abdulrahman University Researchers Supporting Project number (PNURSP2023R107), Princess Nourah Bint Abdulrahman University, Riyadh, Saudi Arabia".

Conflicts of Interest: The authors state that the publishing of this work does not include any conflict of interest.

References

1. Zhu, B.; Liang, Z.; Zou, R. Designing advanced catalysts for energy conversion based on urea oxidation reaction. *Small* **2020**, *16*, 1906133. [CrossRef]
2. Urbańczyk, E.; Sowa, M.; Simka, W. Urea removal from aqueous solutions—A review. *J. Appl. Electrochem.* **2016**, *46*, 1011–1029. [CrossRef]
3. Wang, H.; Zheng, X.; Fang, L.; Lu, S. Urea Electrooxidation in Alkaline Environment: Fundamentals and Applications. *ChemElectroChem* **2023**, *10*, e202300138. [CrossRef]
4. Cho, K.; Hoffmann, M.R. Urea degradation by electrochemically generated reactive chlorine species: Products and reaction pathways. *Environ. Sci. Technol.* **2014**, *48*, 11504–11511. [CrossRef] [PubMed]
5. Wang, X.; Li, J.; Duan, Y.; Li, J.; Wang, H.; Yang, X.; Gong, M. Electrochemical Urea Oxidation in Different Environment: From Mechanism to Devices. *ChemCatChem* **2022**, *14*, e202101906. [CrossRef]
6. Xu, W.; Wu, Z.; Tao, S. Urea-Based Fuel Cells and Electrocatalysts for Urea Oxidation. *Energy Technol.* **2016**, *4*, 1329–1337. [CrossRef]

7. Gurrola, M.P.; Cruz, J.C.; Espinosa-Lagunes, F.I.; Martínez-Lázaro, A.; Ledesma-García, J.; Arriaga, L.G.; Escalona-Villalpando, R.A. Perspective of Use of Pd/rGO in a Direct Urea Microfluidic Fuel Cell. *Catalysts* **2023**, *13*, 788. [CrossRef]
8. Kumar, S.; Bukkitgar, S.D.; Singh, S.; Pratibha; Singh, V.; Reddy, K.R.; Shetti, N.P.; Venkata Reddy, C.; Sadhu, V.; Naveen, S. Electrochemical sensors and biosensors based on graphene functionalized with metal oxide nanostructures for healthcare applications. *ChemistrySelect* **2019**, *4*, 5322–5337. [CrossRef]
9. Shetti, N.P.; Bukkitgar, S.D.; Reddy, K.R.; Reddy, C.V.; Aminabhavi, T.M. Nanostructured titanium oxide hybrids-based electrochemical biosensors for healthcare applications. *Colloids Surf. B Biointerfaces* **2019**, *178*, 385–394. [CrossRef]
10. Hefnawy, M.A.; Fadlallah, S.A.; El-Sherif, R.M.; Medany, S.S. Competition between enzymatic and non-enzymatic electrochemical determination of cholesterol. *J. Electroanal. Chem.* **2023**, *930*, 117169. [CrossRef]
11. Eliwa, A.S.; Hefnawy, M.A.; Medany, S.S.; Deghadi, R.G.; Hosny, W.M.; Mohamed, G.G. Ultrasonic-assisted synthesis of nickel metal-organic framework for efficient urea removal and water splitting applications. *Synth. Met.* **2023**, *294*, 117309. [CrossRef]
12. Hefnawy, M.A.; Nafady, A.; Mohamed, S.K.; Medany, S.S. Facile green synthesis of Ag/carbon nanotubes composite for efficient water splitting applications. *Synth. Met.* **2023**, *294*, 117310. [CrossRef]
13. Najafinejad, M.S.; Chianese, S.; Fenti, A.; Iovino, P.; Musmarra, D. Application of Electrochemical Oxidation for Water and Wastewater Treatment: An Overview. *Molecules* **2023**, *28*, 4208. [CrossRef] [PubMed]
14. Galindo-de-la-Rosa, J.; Álvarez, A.; Gurrola, M.P.; Rodríguez-Morales, J.A.; Oza, G.; Arriaga, L.G.; Ledesma-García, J. Alcohol Dehydrogenase Immobilized on TiO2 Nanotubes for Ethanol Microfluidic Fuel Cells. *ACS Sustain. Chem. Eng.* **2020**, *8*, 10900–10910. [CrossRef]
15. Yu, Z.-Y.; Lang, C.-C.; Gao, M.-R.; Chen, Y.; Fu, Q.-Q.; Duan, Y.; Yu, S.-H. Ni–Mo–O nanorod-derived composite catalysts for efficient alkaline water-to-hydrogen conversion via urea electrolysis. *Energy Environ. Sci.* **2018**, *11*, 1890–1897. [CrossRef]
16. Wang, H.; Jiao, X.; Zeng, W.; Zhang, Y.; Jiao, Y. Electrodeposition NiMoSe ternary nanoshperes on nickel foam as bifunctional electrocatalyst for urea electrolysis and hydrogen evolution reaction. *Int. J. Hydrogen Energy* **2021**, *46*, 37792–37801. [CrossRef]
17. Wang, X.-H.; Hong, Q.-L.; Zhang, Z.-N.; Ge, Z.-X.; Zhai, Q.-G.; Jiang, Y.-C.; Chen, Y.; Li, S.-N. Two-dimensional nickel–cobalt bimetallic hydroxides towards urea electrooxidation. *Appl. Surf. Sci.* **2022**, *604*, 154484. [CrossRef]
18. Khalafallah, D.; Ouyang, C.; Zhi, M.; Hong, Z. Carbon Anchored Epitaxially Grown Nickel Cobalt-Based Carbonate Hydroxide for Urea Electrooxidation Reaction with a High Activity and Durability. *ChemCatChem* **2020**, *12*, 2283–2294. [CrossRef]
19. Mirzaei, P.; Bastide, S.; Dassy, A.; Bensimon, R.; Bourgon, J.; Aghajani, A.; Zlotea, C.; Muller-Bouvet, D.; Cachet-Vivier, C. Electrochemical oxidation of urea on nickel-rhodium nanoparticles/carbon composites. *Electrochim. Acta* **2019**, *297*, 715–724. [CrossRef]
20. Hefnawy, M.A.; Fadlallah, S.A.; El-Sherif, R.M.; Medany, S.S. Nickel-manganese double hydroxide mixed with reduced graphene oxide electrocatalyst for efficient ethylene glycol electrooxidation and hydrogen evolution reaction. *Synth. Met.* **2021**, *282*, 116959. [CrossRef]
21. Basumatary, P.; Lee, U.H.; Konwar, D.; Yoon, Y.S. An efficient tri-metallic anodic electrocatalyst for urea electro-oxidation. *Int. J. Hydrogen Energy* **2020**, *45*, 32770–32779. [CrossRef]
22. Wei, D.; Tang, W.; Wang, Y. Hairy sphere-like Ni9S8/CuS/Cu2O composites grown on nickel foam as bifunctional electrocatalysts for hydrogen evolution and urea electrooxidation. *Int. J. Hydrogen Energy* **2021**, *46*, 20950–20960. [CrossRef]
23. Hefnawy, M.A.; Fadlallah, S.A.; El-Sherif, R.M.; Medany, S.S. Synergistic effect of Cu-doped NiO for enhancing urea electrooxidation: Comparative electrochemical and DFT studies. *J. Alloys Compd.* **2022**, *896*, 162857. [CrossRef]
24. Wala, M.; Blacha-Grzechnik, A.; Stolarczyk, A.; Bajkacz, S.; Dydo, P.; Simka, W. Unexpected electrochemical oxidation of urea on a new NiCuGO composite catalyst. *Int. J. Hydrogen Energy* **2023**. [CrossRef]
25. Narayanasamy, S.; Jayaprakash, J. Carbon cloth/nickel cobaltite ($NiCo_2O_4$)/polyaniline (PANI) composite electrodes: Preparation, characterization, and application in microbial fuel cells. *Fuel* **2021**, *301*, 121016. [CrossRef]
26. Hua, B.; Zhang, W.; Wu, J.; Pu, J.; Chi, B.; Jian, L. A promising $NiCo_2O_4$ protective coating for metallic interconnects of solid oxide fuel cells. *J. Power Sources* **2010**, *195*, 7375–7379. [CrossRef]
27. Chen, R.; Wang, H.-Y.; Miao, J.; Yang, H.; Liu, B. A flexible high-performance oxygen evolution electrode with three-dimensional $NiCo_2O_4$ core-shell nanowires. *Nano Energy* **2015**, *11*, 333–340. [CrossRef]
28. Zhang, J.; Sun, Y.; Li, X.; Xu, J. Fabrication of $NiCo_2O_4$ nanobelt by a chemical co-precipitation method for non-enzymatic glucose electrochemical sensor application. *J. Alloys Compd.* **2020**, *831*, 154796. [CrossRef]
29. Li, J.; Xiong, S.; Liu, Y.; Ju, Z.; Qian, Y. High electrochemical performance of monodisperse $NiCo_2O_4$ mesoporous microspheres as an anode material for Li-ion batteries. *ACS Appl. Mater. Interfaces* **2013**, *5*, 981–988. [CrossRef]
30. Chen, H.; Jiang, J.; Zhang, L.; Qi, T.; Xia, D.; Wan, H. Facilely synthesized porous $NiCo_2O_4$ flowerlike nanostructure for high-rate supercapacitors. *J. Power Sources* **2014**, *248*, 28–36. [CrossRef]
31. Annu; Raja, A.N. Recent development in chitosan-based electrochemical sensors and its sensing application. *Int. J. Biol. Macromol.* **2020**, *164*, 4231–4244. [CrossRef]
32. Rinaudo, M. Chitin and chitosan: Properties and applications. *Prog. Polym. Sci.* **2006**, *31*, 603–632. [CrossRef]
33. Kas, H.S. Chitosan: Properties, preparations and application to microparticulate systems. *J. Microencapsul.* **1997**, *14*, 689–711. [CrossRef] [PubMed]
34. de Alvarenga, E.S. Characterization and properties of chitosan. *Biotechnol. Biopolym.* **2011**, *91*, 48–53.

35. Aranaz, I.; Alcántara, A.R.; Civera, M.C.; Arias, C.; Elorza, B.; Heras Caballero, A.; Acosta, N. Chitosan: An overview of its properties and applications. *Polymers* **2021**, *13*, 3256. [CrossRef] [PubMed]
36. Kabir, S.M.F.; Sikdar, P.P.; Haque, B.; Bhuiyan, M.A.R.; Ali, A.; Islam, M.N. Cellulose-based hydrogel materials: Chemistry, properties and their prospective applications. *Prog. Biomater.* **2018**, *7*, 153–174. [CrossRef]
37. Hao, C.; Zhou, S.; Wang, J.; Wang, X.; Gao, H.; Ge, C. Preparation of hierarchical spinel $NiCo_2O_4$ nanowires for high-performance supercapacitors. *Ind. Eng. Chem. Res.* **2018**, *57*, 2517–2525. [CrossRef]
38. Marco, J.F.; Gancedo, J.R.; Gracia, M.; Gautier, J.L.; Ríos, E.; Berry, F.J. Characterization of the nickel cobaltite, $NiCo_2O_4$, prepared by several methods: An XRD, XANES, EXAFS, and XPS study. *J. Solid State Chem.* **2000**, *153*, 74–81. [CrossRef]
39. Shim, J.-J. Three-dimensional nickel foam/graphene/$NiCo_2O_4$ as high-performance electrodes for supercapacitors. *J. Power Sources* **2015**, *273*, 110–117.
40. Adhikari, S.; Kwon, Y.; Kim, D.-H. Three-dimensional core–shell structured $NiCo_2O_4$@CoS/Ni-Foam electrocatalyst for oxygen evolution reaction and electrocatalytic oxidation of urea. *Chem. Eng. J.* **2020**, *402*, 126192. [CrossRef]
41. Kim, J.-G.; Pugmire, D.L.; Battaglia, D.; Langell, M.A. Analysis of the $NiCo_2O_4$ spinel surface with Auger and X-ray photoelectron spectroscopy. *Appl. Surf. Sci.* **2000**, *165*, 70–84. [CrossRef]
42. Lin, W.-C.; Lo, W.-C.; Li, J.-X.; Wang, Y.-K.; Tang, J.-F.; Fong, Z.-Y. In situ XPS investigation of the X-ray-triggered decomposition of perovskites in ultrahigh vacuum condition. *npj Mater. Degrad.* **2021**, *5*, 13. [CrossRef]
43. Dementjev, A.P.; de Graaf, A.; van de Sanden, M.C.M.; Maslakov, K.I.; Naumkin, A.V.; Serov, A.A. X-Ray photoelectron spectroscopy reference data for identification of the C3N4 phase in carbon–nitrogen films. *Diam. Relat. Mater.* **2000**, *9*, 1904–1907. [CrossRef]
44. Medany, S.S.; Hefnawy, M.A. Nickel–cobalt oxides decorated Chitosan electrocatalyst for ethylene glycol oxidation. *Surf. Interfaces* **2023**, *40*, 103077. [CrossRef]
45. Tiwari, S.; Rathore, G.; Patra, N.; Yadav, A.K.; Bhattacharya, D.; Jha, S.N.; Tseng, C.M.; Liu, S.W.; Biring, S.; Sen, S. Oxygen and cerium defects mediated changes in structural, optical and photoluminescence properties of Ni substituted CeO2. *J. Alloys Compd.* **2019**, *782*, 689–698. [CrossRef]
46. Ikram, M.; Shahzadi, A.; Hayat, S.; Nabgan, W.; Ul-Hamid, A.; Haider, A.; Noor, M.; Goumri-Said, S.; Kanoun, M.B.; Ali, S. Novel Ta/chitosan-doped CuO nanorods for catalytic purification of industrial wastewater and antimicrobial applications. *RSC Adv.* **2022**, *12*, 16991–17004. [CrossRef] [PubMed]
47. Vijayaprasath, G.; Murugan, R.; Palanisamy, S.; Prabhu, N.M.; Mahalingam, T.; Hayakawa, Y.; Ravi, G. Role of nickel doping on structural, optical, magnetic properties and antibacterial activity of ZnO nanoparticles. *Mater. Res. Bull.* **2016**, *76*, 48–61. [CrossRef]
48. Vedharathinam, V.; Botte, G.G. Understanding the electro-catalytic oxidation mechanism of urea on nickel electrodes in alkaline medium. *Electrochim. Acta* **2012**, *81*, 292–300. [CrossRef]
49. Hefnawy, M.A.; Medany, S.S.; El-Sherif, R.M.; Fadlallah, S.A. Green synthesis of NiO/Fe3O4@chitosan composite catalyst based on graphite for urea electro-oxidation. *Mater. Chem. Phys.* **2022**, *290*, 126603. [CrossRef]
50. Galal, A.; Atta, N.F.; Hefnawy, M.A. Lanthanum nickel oxide nano-perovskite decorated carbon nanotubes/poly (aniline) composite for effective electrochemical oxidation of urea. *J. Electroanal. Chem.* **2020**, *862*, 114009. [CrossRef]
51. Atta, N.F.; El-Sherif, R.M.A.; Hassan, H.K.; Hefnawy, M.A.; Galal, A. Conducting Polymer-Mixed Oxide Composite Electrocatalyst for Enhanced Urea Oxidation. *J. Electrochem. Soc.* **2018**, *165*, J3310–J3317. [CrossRef]
52. Vedharathinam, V.; Botte, G.G. Direct evidence of the mechanism for the electro-oxidation of urea on Ni (OH) 2 catalyst in alkaline medium. *Electrochim. Acta* **2013**, *108*, 660–665. [CrossRef]
53. Feng, Y.Q.; Liang, Z.Y.; Meng, S.X. Adsorption of urea nitrogen onto chitosan coated dialdehyde cellulose under biocatalysis of immobilized urease: Equilibrium and kinetic. *Biochem. Eng. J.* **2005**, *24*, 65–72. [CrossRef] [PubMed]
54. Puspita, A.; Prawati, G.; Fatimah, I. Chitosan-modified smectite clay and study on adsorption-desorption of urea. *Chem. Eng. Trans.* **2017**, *56*, 1645–1650.
55. Zhou, Y.; Yang, Y.; Guo, X.; Chen, G. Effect of molecular weight and degree of deacetylation of chitosan on urea adsorption properties of copper chitosan. *J. Appl. Polym. Sci.* **2003**, *89*, 1520–1523. [CrossRef]
56. Zhao, L.; Chang, Y.; Jia, M.; Jia, J.; Wen, Z. Monodisperse Ni0·85Se nanocrystals on rGO for high-performance urea electrooxidation. *J. Alloys Compd.* **2021**, *852*, 156751. [CrossRef]
57. Hefnawy, M.A.; Medany, S.S.; El-Sherif, R.M.; El-Bagoury, N.; Fadlallah, S.A. High-performance IN738 superalloy derived from turbine blade waste for efficient ethanol, ethylene glycol, and urea electrooxidation. *J. Appl. Electrochem.* **2023**, *53*, 1337–1348. [CrossRef]
58. Hefnawy, M.A.; Medany, S.S.; El-Sherif, R.M.; Fadlallah, S.A. NiO-MnOx/Polyaniline/Graphite Electrodes for Urea Electrocatalysis: Synergetic Effect between Polymorphs of MnOx and NiO. *ChemistrySelect* **2022**, *7*, e202103735. [CrossRef]
59. Ding, Y.; Li, Y.; Xue, Y.; Miao, B.; Li, S.; Jiang, Y.; Liu, X.; Chen, Y. Atomically thick Ni (OH) 2 nanomeshes for urea electrooxidation. *Nanoscale* **2019**, *11*, 1058–1064. [CrossRef]
60. Wang, Y.; Yin, Z.; Yan, G.; Wang, Z.; Li, X.; Guo, H.; Wang, J. New insight into the electrodeposition of NiCo layered double hydroxide and its capacitive evaluation. *Electrochim. Acta* **2020**, *336*, 135734. [CrossRef]
61. Galal, A.; Atta, N.F.; Hefnawy, M.A. Voltammetry study of electrocatalytic activity of lanthanum nickel perovskite nanoclusters-based composite catalyst for effective oxidation of urea in alkaline medium. *Synth. Met.* **2020**, *266*, 116372. [CrossRef]

62. Al-Kadhi, N.S.; Hefnawy, M.A.; SNafee, S.; Alamro, F.S.; Pashameah, R.A.; Ahmed, H.A.; Medany, S.S. Zinc Nanocomposite Supported Chitosan for Nitrite Sensing and Hydrogen Evolution Applications. *Polymers* **2023**, *15*, 2357. [CrossRef]
63. Al-Kadhi, N.S.; Hefnawy, M.A.; Alamro, F.S.; Pashameah, R.A.; Ahmed, H.A.; Medany, S.S. Polyaniline-Supported Nickel Oxide Flower for Efficient Nitrite Electrochemical Detection in Water. *Polymers* **2023**, *15*, 1804. [CrossRef]
64. Hefnawy, M.A.; Fadlallah, S.A.; El-Sherif, R.M.; Medany, S.S. Systematic DFT studies of CO-Tolerance and CO oxidation on Cu-doped Ni surfaces. *J. Mol. Graph. Model.* **2023**, *118*, 108343. [CrossRef]
65. Liu, J.; Chen, X.; Shao, Z.; Zhou, P. Preparation and characterization of chitosan/Cu (II) affinity membrane for urea adsorption. *J. Appl. Polym. Sci.* **2003**, *90*, 1108–1112. [CrossRef]
66. Zhang, J.-J.; Bao, W.-W.; Li, M.-Y.; Yang, C.-M.; Zhang, N.-N. Ultrafast formation of an FeOOH electrocatalyst on Ni for efficient alkaline water and urea oxidation. *Chem. Commun.* **2020**, *56*, 14713–14716. [CrossRef]

polymers

MDPI

Article

Two-Way Reversible Shape Memory Behavior of Chitosan/Glycerol Film Triggered by Water

Shuozi Li [1], Hu Lyu [2], Yujia Wang [1], Xianzhi Kong [2,*], Xiangxian Wu [1], Lina Zhang [1], Xiaojuan Guo [1] and Dawei Zhang [1,*]

1 Engineering Research Center of Advanced Wooden Materials, Ministry of Education, Northeast Forestry University, Harbin 150040, China; 15387045795@163.com (S.L.); wangyujia990508@163.com (Y.W.); w18773337677@163.com (X.W.); 13853065099@163.com (L.Z.); gxj15776632498@163.com (X.G.)

2 Institute of Petrochemistry, Heilongjiang Academy of Sciences, Harbin 150036, China; 15546002614@163.com

* Correspondence: kongxianzhi2023@hipc.org.cn (X.K.); zhangdawei@nefu.edu.cn (D.Z.)

Abstract: Reversible shape memory polymers (SRMPs) have been identified as having great potential for biomedical applications due to their ability to switch between different shapes responding to stimuli. In this paper, a chitosan/glycerol (CS/GL) film with a reversible shape memory behavior was prepared, and the reversible shape memory effect (SME) and its mechanism were systematically investigated. The film with 40% glycerin/chitosan mass ratio demonstrated the best performance, with 95.7% shape recovery ratio to temporary shape one and 89.4% shape recovery ratio to temporary shape two. Moreover, it shows the capability to undergo four consecutive shape memory cycles. In addition, a new curvature measurement method was used to accurately calculate the shape recovery ratio. The suction and discharge of free water change the binding form of the hydrogen bonds inside the material, which makes a great reversible shape memory impact on the composite film. The incorporation of glycerol can enhance the precision and repeatability of the reversible shape memory effect and shortens the time used during this process. This paper gives a hypothetical premise to the preparation of two-way reversible shape memory polymers.

Keywords: reversible shape memory; chitosan; composite film; hydrogen bonds

1. Introduction

Shape memory polymers, which have been widely used as a kind of smart material, can recover from the presupposed temporary shape to the permanent shape upon external stimulus [1–4]. However, the unidirectional nature of one-way shape memory deformation had restricted the application of materials in various fields. Derived from it, the researchers' interest has gradually focused on the new reversible shape memory polymers (RSMPs), which could achieve the shape interconversion between two shapes while receipting the external stimuli [5–8].

Currently, the thermo-responsive bidirectional RSMPs have occupied the research's highlights [9–13]. For semi-crystalline polymers, the directional crystallization of anisotropic polymer networks would be responsible for the thermal-triggered two-way SME [14–16]. Due to the relevance between the thermal effect and mechanical behaviors of semi-crystalline SMPs, the effect of thermal stimuli on the SMPs within a certain temperature range could be anticipated and controlled.

Solution actuation [17–22], as a triggering pattern through changing the polymer structure since Huang [23], develops relatively slower because the effect of the solution on the RSMPs is all-directional and complex. The solvent-driven shape memory effect can be classified into two modes: water-driven and organic solvent-driven [24–26]. Various solvents can trigger different shape recovery responses, similar to the temperature memory effect, and the use of diverse solvents can enable multi-staged and multiple shape memory recovery mechanisms [27,28]. Therefore, how to improve the recovery speed, accuracy and stability of RSME of polymers, especially of natural polymer materials upon the solution, has become an investigated hotspot. The SME of shape-memory materials is always associated with their crystallization properties [29]. The entry of solution molecules swells the molecular chains in amorphous regions, and the movement of the segments eventually leads to macroscopic deformation, which can be achieved through the disruption and reforming of intermolecular hydrogen bonds [30,31].

Chitosan (CS), as the only biodegradable cationic polysaccharide originated from the partial deacetylation of chitin from the shells of crustaceans, has attracted great research interest due to its wide availability, biocompatibility and biodegradability [32–36]. The use of chitosan-based biomaterials is integral in skin tissue engineering primarily [37]. Its antimicrobial and hemostatic properties allow chitosan to promote wound healing and prevent infections [38,39]. The hydrogen bonds formed between the chain segments benefit from the amino and hydroxyl groups; the protonation of amino groups allows electrostatic interactions, which could be conducive to the film formation in an acidic medium [40,41].

In this paper, chitosan/glycerol (CS/GL) films were prepared, and a novel curvature method was employed to characterize the shape programming rate and recovery rate of RSMPs. Glycerol (GL) is dispersed within chitosan molecules primarily through hydrogen bonding and the van der Waals force. This dispersion enhances the structure of the chitosan network, allowing for a more stable CS/GL complex in solution. Compared to polymers, such as polyvinyl alcohol [42], glycerol's smaller molecular size results in improved solubility and biocompatibility. The RSME was analyzed by various characterization techniques including dynamic mechanical analysis (DMA), thermogravimetric analysis (TGA), Fourier transform infrared (FTIR), differential scanning calorimetry (DSC) and X-ray diffraction (XRD). Through the implementation of a water–air bidirectional triggering mode, we demonstrated the ability of RSMPs to exhibit completely reversible shape memory behavior under conditions of water absorption and loss, with increased glycerol content resulting in a greater shape recovery rate and a shorter response time of RSMPs. Through our investigation, it was found that the stronger the mobility of chitosan chains, the stronger the RSME. The binding mode of water molecules was also closely related to different stages of RSME. Furthermore, we emphasized the key roles of intermolecular hydrogen bonding in shaping the mechanical properties of RSME.

2. Materials and Methods

2.1. Materials

CS (degree of deacetylation 90.3%, viscosity 100 mPa·s) was purchased from Zhejiang Golden Shell Pharmaceutical Co. (Zhejiang, China). GL (purity 99%) was commercially available from Tianjin Hengxing Chemical Reagent Co. (Tianjin, China). Acetic acid (purity 99.5%) and NaOH (purity 99.0%) were both obtained from Tianjin Kermel Chemical Reagent Co. (Tianjin, China). Distilled water was bought from Harbin Wenjing (Harbin, China) distilled water factory. Ethanol (purity 99.7%) was obtained from Tianjin Zhiyuan Chemical Reagent Co. (Tianjin, China). Unless otherwise stated, all the materials were of analytical grade and all the materials were used without further treatment.

2.2. Preparation of CS/GL Film

Figure 1a. shows the preparation method of CS/GL films. The powdered CS was dissolved in 2% weight percent acetic acid and stirred evenly to obtain 3% wt CS solution. Glycerol was added to the CS solution and stirred for 1 h to obtain the 10%, 20%, 30%, and 40% GL/CS (mass ratio) solutions. The resulting solutions were poured into a homemade film-shaping mold and allowed to dry at 50 °C and −0.09 MPa for 72 h to fabricate uniform CS/GL films.

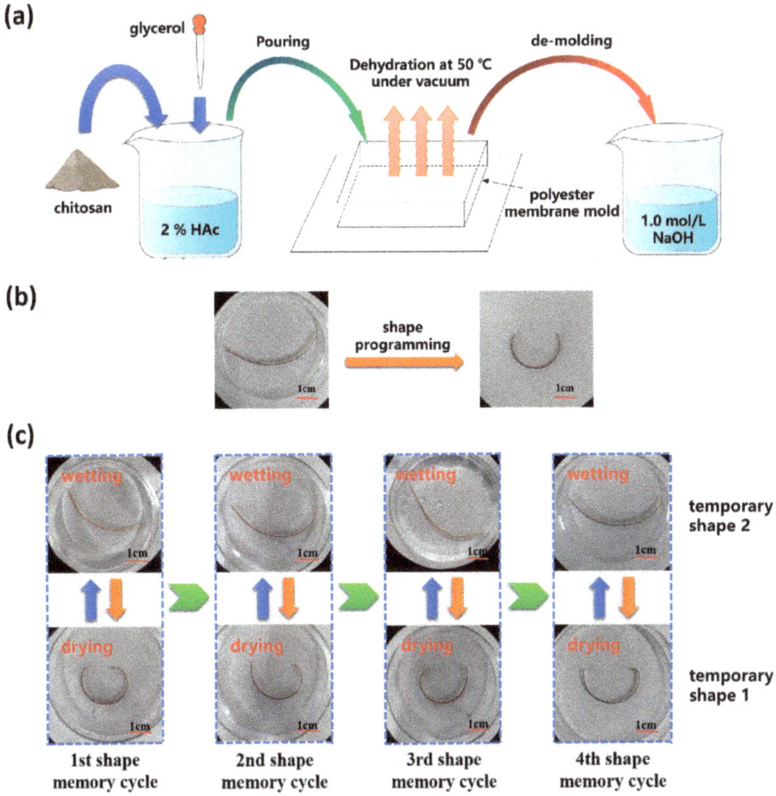

Figure 1. Preparation method of CS/GL films with RSME and the shape memory process of RSMPs. (**a**) Preparation method of CS/GL films; (**b**) thermal programming of RSMPs (under 40% GL mass ratio) to obtain a curve shape (temporary shape 1); and (**c**) RSME between temporary shape 1 shape and temporary shape 2 can be repeated for many times (RSMPs under 40% GL mass ratio).

2.3. Shape Memory Test

To prepare RSMPs for testing, rectangular samples with a width of 3 mm and a length of 50 mm were cut from the prepared films. These samples were immersed in a 1.0 mol/L NaOH solution for 30 s and then washed with distilled water. The prepared samples were thermally programmed on the inner surface of the mold to obtain its temporary shape 1 in an oven at 80 °C for 1 h, as shown in Figure 1b.

A curvature determination method was utilized to characterize and evaluate the RSME in the process of shape programming and shape recovery, as depicted in Figure 2. The shape programming ratio (R_p) was employed to evaluate the degree of thermal programming of the material, which was determined by dividing $\overline{K_p}$ by $\overline{K_{p0}}$:

$$R_p = \frac{\overline{K_p}}{\overline{K_{p0}}} \times 100\% \quad (1)$$

where $\overline{K_p}$ represents the mean curvature of the sample after thermal molding, while $\overline{K_{p0}}$ denotes the inner surface curvature of the mold used for fixing the sample.

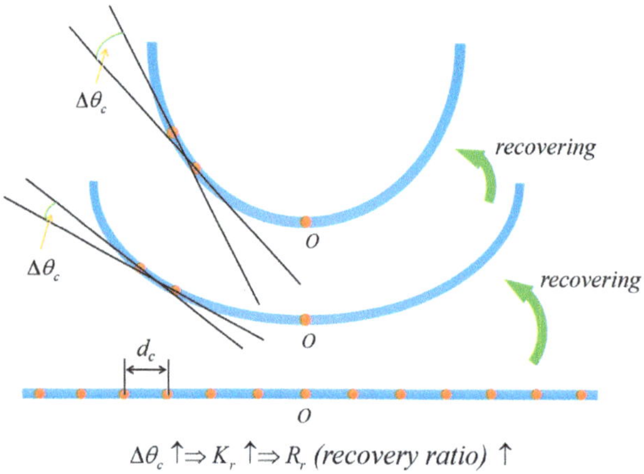

Figure 2. The recovery ratio measured by curvature was used to characterize differences in the degree of deformation across different shape memory processes. Curvature-determining method used for testing RSME.

After thermal shape programming, the sample was cooled adequately at room temperature and placed in water to elicit a response. After a certain period, the sample obtained temporary shape 2 (flattened shape). The shape recovery ratio of temporary shape 2 (R_{r2}) was determined by $\overline{K_{r0}}$ and $\overline{K_{r2}}$:

$$R_{r2} = (1 - \frac{\left|\overline{K_{r2}} - \overline{K_{r0}}\right|}{\overline{K_{r0}}}) \times 100\% \tag{2}$$

where $\overline{K_{r0}}$ represents the mean curvature of the sample during the first extension in water and $\overline{K_{r2}}$ represents the average curvature when temporary shape 2 was obtained again after undergoing the first shape memory cycle.

The sample with temporary shape 2 was removed from the water environment and exposed to air. After the sample underwent a dehydration process in the air (the abbreviation "in air" is used later in the text), it regained its temporary shape 1, determined during thermal programming. During this process, the rate of water loss depends to some extent on the temperature and air humidity. In this study, the samples were exposed to the air for dehydration in a closed laboratory with a temperature of 25 °C and a relative humidity of 41%. At this point, the shape recovery ratio R_{r1} was determined by dividing $\overline{K_{r1}}$ by $\overline{K_p}$:

$$R_{r1} = \frac{\overline{K_{r1}}}{\overline{K_p}} \times 100\% \tag{3}$$

where $\overline{K_{r1}}$ denotes the mean curvature of the sample after shape recovery to temporary shape 1.

During an entire cycle of shape memory transformation, the deformation state of the RSMP was recorded using a camera every 3 s. One shape recovery cycle is defined as follows: the temporary shape 1 transforms into the temporary shape 2 in response to water stimulation, and then the temporary shape 2 transforms back to the temporary shape 1 after dehydration in the air. The material's training through thermal programming is independent of the shape recovery cycle. In this paper, the shape recovery time is defined as the time required for the material to undergo directional shape memory deformation and deform to another stable shape after being stimulated by the external environment from a stable shape. The judgment method for stable shape is: if the deviation between the shape recovery ratios measured within 60 s after this state is less than 0.2%, the material in this state is determined to have a stable shape.

Methods to calculate the mean curvature of RSMPs are shown below. Set a positive integer n greater than 3; the RSMPs sample was divided into (n + 1) segments with equal lengths by n points. The distance between adjacent points was denoted as d and the angles between tangents ($\Delta\theta_c$, c = 1, 2, 3 ... n) were calculated. The mean curvature was defined as follows:

$$\overline{K} = \frac{1}{n}\sum_{c=1}^{n}\frac{\Delta\theta_c}{d}\ (c = 1,\ 2,\ 3\ \ldots\ n) \tag{4}$$

It should be noted that the larger the value of n, the more significant the measured mean curvature is. In this paper, the value of n was uniformly set to 9.

2.4. Characterization and Measurements

Free water content and thermal stability of samples were performed on a thermal analyzer (TG 209 F3, Netzsch) under an N_2 flow at a heating rate of 10 °C/min. Dynamic mechanical behaviors of samples were carried out using a DMA-242 E at a heating rate of 5 °C/min with 10 μm maximum tensile strain from −30 °C to 100 °C. Fourier transform infrared (FTIR) spectra (TENSORIIBruker, Ettlingen, Germany) were selected to characterize the chemical structure of RSMPs and changes in hydrogen bonding interactions at different shape memory stages, covering the range of 4000–400 cm^{-1} with a resolution of 4 cm^{-1}. The crystallinity of RSMPs at different stages of RSME was investigated by X-ray diffraction (XRD) analysis performed with a Rigaku D/max 2200 VPC (Tokyo, Japan). XRD patterns were examined within the diffraction angle from 5° to 65° with a scanning speed of 5°/min. Differential scanning calorimetry (DSC) analysis was conducted on a Netzsch 204 DSC instrument under an N_2 flow. The samples were heated up from 30 °C to 200 °C with the warming rate 5 °C/min throughout the whole process.

3. Results and Discussion

3.1. Two-Way Reversible Water-Triggered Shape Memory Behaviors

As illustrated in Figure 3a, the shape memory effect of RSMPs with 10% and 20% GL ratio was not entirely satisfactory, evidenced by R_{r1} values of only 83.8% and 87%, respectively. However, when increasing the GL ratio to 30% and 40%, the R_{r1} value could reach 95.0% and 95.7%, indicating that the materials have excellent RSME at the macroscopic level. In contrast, there was no significant difference in the R_{r2} value of RSMPs undergoing a transformation from temporary shape one to temporary shape two for various GL concentrations. Remarkably, as shown in Table 1, irrespective of varying glycerol content, the R_p value of the materials exceeded 98%, cementing CS's position as an exceptional matrix material for shape memory materials. As shown in Figure 3b, the shape recovery time of RSMPs during both stages exhibited an increasing reduction with a greater GL ratio, thereby affirming an escalated responsiveness to environmental stimuli. It is noteworthy that the transition time of RSMPs to temporary shape one was longer than that to temporary shape two, which mainly depends on the rate of water absorption and desorption.

The CS/GL composite film exhibits remarkable reversibility in the process of bidirectional shape memory, which is manifested by consistent recovery time and consistent recovery ratio across multiple shape memory cycles (Figure 1c). Meanwhile, this reversible bidirectional shape memory process can be forcibly interrupted at any stage and redriven in both directions (to temporary shape one or temporary shape two). The thermally programmed shape (temporary shape one) of the composite film was obtained during water loss in air, regardless of whether the material was fully soaked in water during the previous process. Similarly, regardless of whether the composite film was completely dehydrated in air, it eventually recovered to temporary shape two in the aqueous solution. Mechanistically, the polymer deformation process could be attributed to a relaxation behavior [43,44], which was always synchronized with real-time changes in the polymer network for RSME.

The high degree of reversibility of RSME was largely attributed to the reversibility of the structural changes in polymer networks.

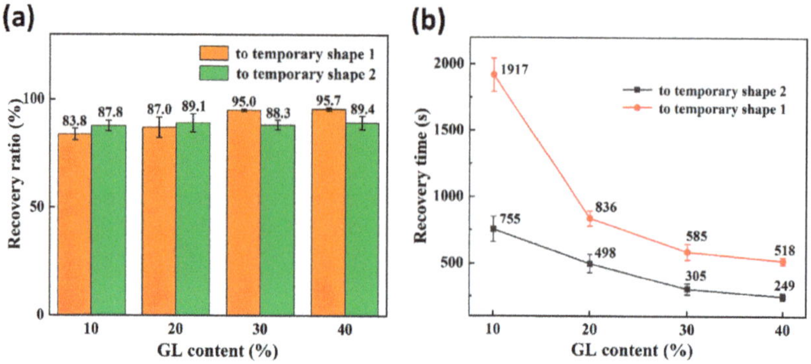

Figure 3. Characterization of the two shape recovery processes in a shape memory cycle. (**a**) The recovery ratio of the reversible shape memory polymers (RSMPs) under various GL ratios; (**b**) the recovery time of RSMPs under various GL ratios.

Table 1. Programming ratio of RSMPs under various GL ratios.

GL Content	Programming Ratio	Standard Deviation
10%	98.7%	0.21%
20%	98.8%	0.21%
30%	99.1%	0.25%
40%	98.9%	0.26%

3.2. Thermogravimetric (TG) Analysis

Thermogravimetric (TG) curves of RSMPs with a 30% GL ratio (named 30-RSMPs) at different stages of RSME (after shape programming and in the first shape memory cycle) are shown in Figure 4a. Figure 4a′ represents two curves from Figure 4a (in water and in air), with the TG curve in air having the intercept corresponding to 5% of water content subtracted. Figure 4b displays the TG curves of samples with a 40% GL ratio (named 40-RSMPs) at different RSME stages; Figure 4b′ is the curve obtained by adjusting the intercept of 3% water content between the two curves (in water and in air) shown in Figure 4b, respectively. It can be observed that the curves in Figure 4 follow the typical pattern of thermal degradation of natural polymers [45,46]. Most of the free water would be removed in the heating process (<100 °C), while the dehydration of combined water was a gradual process during the heating process. When the temperature reached 250 °C, the curve dropped sharply, indicating the significant thermal degradation of RSMPs. Comparing Figure 4a′,b′, it could be observed that after compensating for the difference in water content, the TG curves of the water-absorbed sample and the dehydrated sample (>100 °C) are almost identical, indicating that the water triggering the RSME of the composite films exists in the form of free water. Moreover, the difference in water content has no influence on the internal structure of the material. The change in crystallinity was usually accompanied by the change of thermal stability for natural polymer materials [47,48], and the parallelism of TG curves between two shapes indicated that the entry and exit of water did not affect the crystallization region of CS. It is evident that the water molecules lost by RSMPs between temporary shape one and temporary shape two did not bind with chitosan, but rather formed hydrogen-bonded chains to occupy the gaps in the chitosan chain segments, which mainly consisted of free water. In contrast, the TG curve of the thermally-programmed material did not exhibit such "parallelism" with the temporary shape one and temporary shape two curves. Therefore, it can be confirmed that the water molecules absorbed by the material during the transition from thermal programming to the shape memory period were linked with chitosan through hydrogen bonding, which is referred to as bound water.

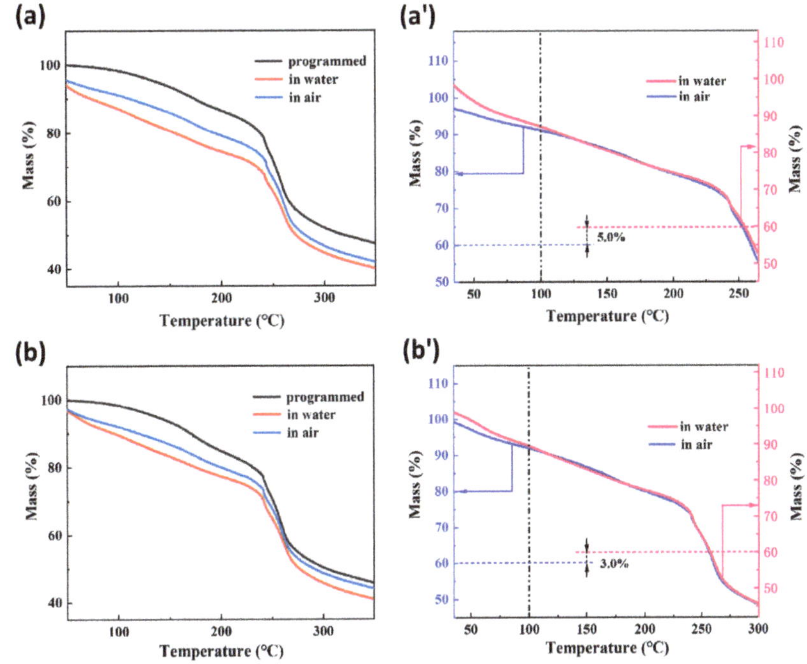

Figure 4. Results of TG analysis after shape programming and in the first shape memory cycle. (**a**) TG curves of 30-RSMPs; (**a′**) parallel determination of the two curves from (**a**); (**b**) TG curves of 40-RSMPs; (**b′**) parallel determination of the two curves from (**b**).

3.3. Differential Scanning Calorimetry (DSC) Analysis

Figure 5a,b display the results of DSC tests conducted on 30-RSMPs and 40-RSMPs during the first shape memory cycle. The 30-RSMPs DSC curve exhibits a significant endothermic peak at 91 °C (in water) and 64 °C (in air), while the 40-RSMPs DSC curve exhibits similar peaks at 85 °C (in water) and 75 °C (in air). The peak difference for 40-RSMPs is only 10 °C, which is significantly less than 30-RSMPs' 27 °C, thus confirming that the recovery time (to temporary shape one and to temporary shape two) for 40-RSMPs is faster than for 30-RSMPs. The shape recovery speed is regulated by the glycerol content in the film. With an increase in the glycerol content, the water absorption and water loss is fast, thus recovery speed of the film is increased. The absorption or loss water was in the form of free water; the results observed were consistent with the findings from the thermogravimetric (TG) analysis.

3.4. Dynamic Mechanical Behaviors of CS/GL Composite Films

Figure 6a,b show the results of the DMA (dynamic mechanical analysis) test of 30-RSMPs at different stages of RSME (after shape programming and in the first shape memory cycle), while Figure 6c,d shows the results of 40-RSMPs at different stages of RSME. In Figure 6a,c, the storage moduli gradually decreased with the increase in temperature. Compared with these two groups, the storage modulus was negatively correlated with water content, which proved that the water triggering of RSMP could be interpreted as a process of reducing storage moduli. As shown in Figure 5b, the storage moduli of CS networks exhibit a sharp decrease at temperatures corresponding to this transition caused by water molecule diffusion. In an environment with air exposure, water molecules gradually diffuse from the material into the surrounding air, causing a reduction in the available space for polymer chain mobility. This results in increased entanglements between the polymer chains, which in turn leads to a corresponding increase in the strength of the samples. Accordingly, the secondary transition temperature of RSMPs after programming, immersing in water, and exposing to air were T_p = 76.0 °C, T_w = 6.1 °C and T_a = 56.7 °C [49,50], as well as the transition temperature in Figure 5d being T_p = 74.2 °C, T_w = 16.9 °C and T_a = 47.3 °C, respectively. The secondary transition temperature decreased when the RSMP was immersed in water while increased in air condition. This suggests that the triggering effect of water molecules on shape memory behaviors occurs within the amorphous region of CS, which is likely due to the presence of reversible hydrogen bonds [51–53]. When water infiltrates the CS molecules, the hydrogen bonds between the CS chains are disrupted, causing the chain segments to become mobile and resulting in an earlier appearance of the tanδ peak. After water actuation, the tanδ peak of 40-RSMPs was higher than that of 30-RSMPs. In addition, the tanδ peak of both RSMPs with 30% and 40% GL ratio increased after the water immersion. The tanδ peak in 30-RSMPs increased from 0.285 to 0.307 and from 0.272 to 0.364 in 40-RSMPs. Under the premise that 40-RSMPs had a shorter water absorbing time, the swelling effect of water molecules is more effective on the 40-RSMPs, thus significantly increasing the tanδ peak.

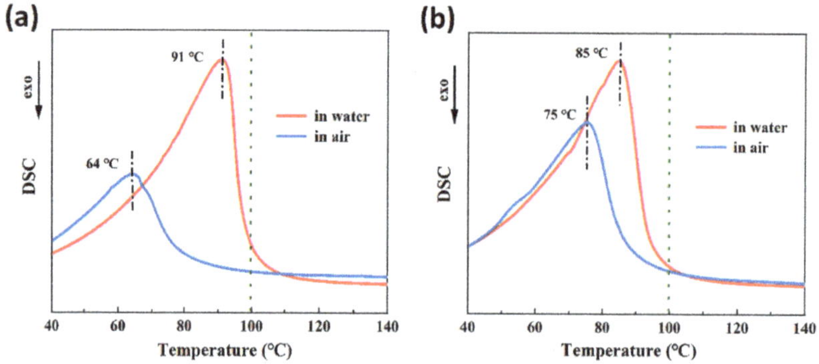

Figure 5. Differential scanning calorimetry (DSC) analysis in the first shape memory cycle. (**a**) DSC curves of 30-RSMPs in the first shape memory cycle; (**b**) DSC curves of 40-RSMPs in the first shape memory cycle.

3.5. *Fourier Transform Infrared (FTIR) Spectra*

Figure 7a,b showed the FTIR test spectra of RSMPs at different stages of RSME (after shape programming and in the first shape memory cycle) with 30% and 40% GL ratio. According to the curve analysis in the two plots, for the programmed RSMPs, the two absorption peaks at 3274 cm^{-1} and 3297 cm^{-1} were assigned to the overlapping results of O−H and N−H stretching vibrations. In response to water stimulation, the overlapping peak moved in low wavenumbers to 3247 cm^{-1} and 3283 cm^{-1}, respectively. Moreover, when the SMP responded in the air environment, the overlapping peaks moved again to the high wavenumber of 3259 cm^{-1} and 3288 cm^{-1} but were still lower than the programmed wavenumber. This was strong evidence for the formation of hydrogen bond-based supramolularly reversible cross linking inside RSMP networks. The absorption peak at 2850 cm^{-1} was assigned to the C−H of methylene at the carbon atom at position 6 (C6) and its intensity decreased delicately with the formation of hydrogen bonds between water and hydroxyl groups at C6. The elongation vibration peak of C=O corresponded to the RSMPs of 1640 cm^{-1} and 1641 cm^{-1} in two plots, with a significantly enhanced peak strength after increasing the water content, which was also associated with the base energy to form reversible hydrogen bonds with water molecules. Therefore, the process of hydrogen bond re-establishment and the swelling effects of solution molecules could be important sources of reproducibility and reversibility of RSME.

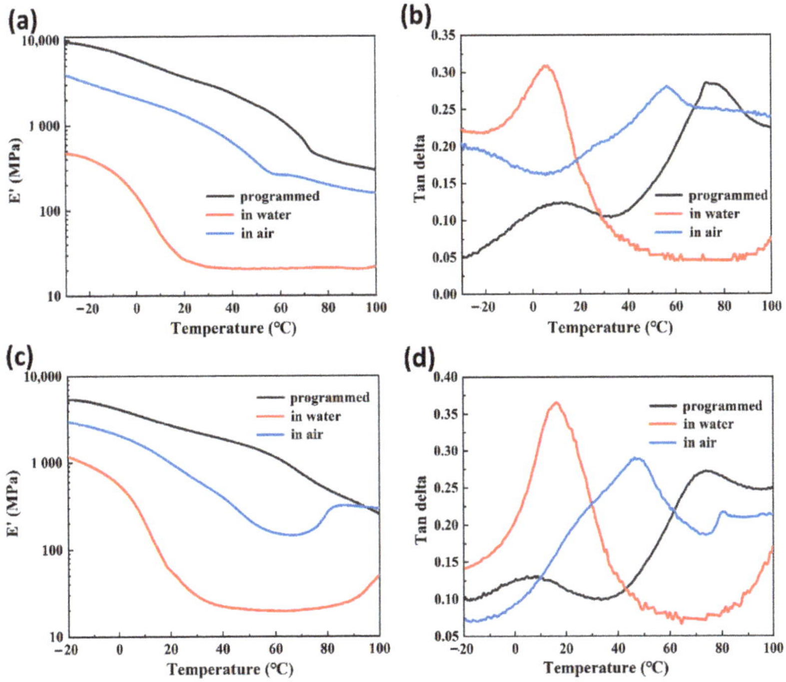

Figure 6. Dynamic mechanical analysis after shape programming and in the first shape memory cycle. (**a**) Storage modulus curves of 30-RSMPs; (**b**) tanδ curves of 30-RSMPs; (**c**) storage modulus curves of 40-RSMPs; and (**d**) tanδ curves of 40-RSMPs.

3.6. X-ray Diffraction (XRD) Analysis

The crystalline properties (after shape programming and in the first shape memory cycle) at different stages of RSME of 30-RSMPs and 40-RSMPs were investigated through XRD analysis. As shown in Figure 8a, diffraction peaks were observed at around $2\theta = 10°$, $2\theta = 15°$ and $2\theta = 24°$, while at around $2\theta = 7°$, $2\theta = 15°$ and $2\theta = 21°$ in Figure 8b, respectively, indicating the semi-crystalline nature of the CS.

The crystallinity of the 30-RSMPs and 40-RSMPs was found to be 39.1% and 35.7% (shown in Table 2), respectively, which represents the highest achievable degree of crystallinity for RSMPs without free water. However, as the material absorbs water, the intensity of these diffraction peaks significantly reduces, resulting in a smoother diffraction pattern and a broader XRD spectrum presenting a smooth, bun-like peak. Therefore, determining the crystallinity of 30-RSMPs and 40-RSMPs at this stage is no longer meaningful. Upon exposure to air, the diffraction peaks gradually became narrower and more pronounced, and the crystallinity of RSMPs increased to some extent. These findings suggest that water content is an important factor that affects the crystalline properties of the RSMPs. The entry of water molecules disrupts the hydrogen bonding interactions between CS chains, promoting the migration of CS chain segments. As water molecules penetrate the gaps between chitosan chains, hydrogen bonds are formed between water molecules and CS chains, opening the pores in the gaps between chitosan chains, effectively enhancing the flexibility of polymer chains and reducing the crystallinity of RSMPs. It is noteworthy that the crystallinity of 30-RSMPs was consistently higher than that of 40-RSMPs, indicating the significant plasticizing effect of GL on CS. Increased GL content decreased the crystallinity of CS, promoting the motion of polymer chain segments, which is consistent with the experimental results obtained earlier.

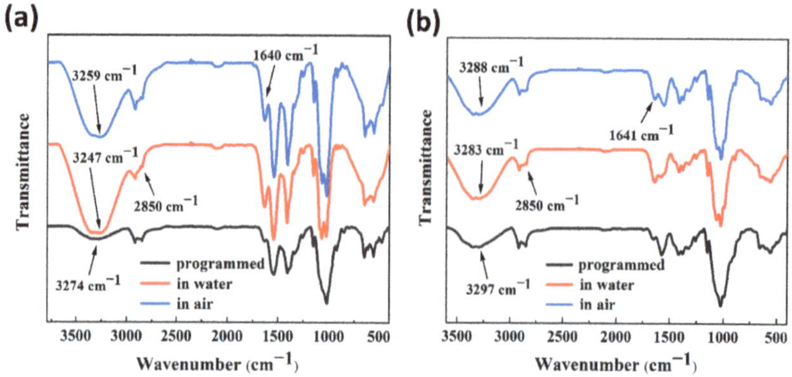

Figure 7. FTIR test spectra after shape programming and in the first shape memory cycle. (**a**) FTIR spectra of 30% GL ratio RSMPs; (**b**) FTIR spectra of 40% GL ratio RSMPs.

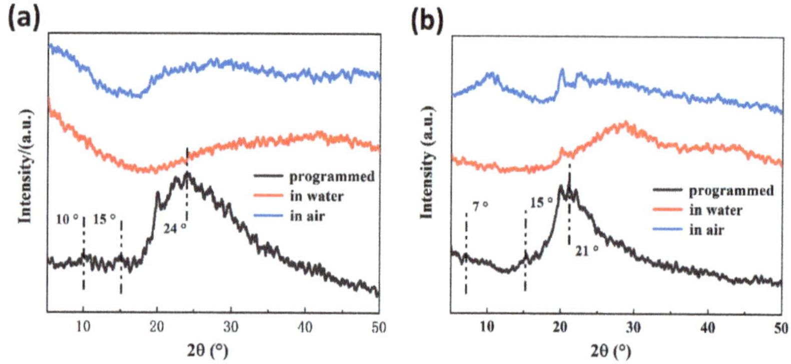

Figure 8. XRD spectra of RSMPs after shape programming and in the first shape memory cycle. (**a**) XRD spectra of 30-RSMPs at different stages of RSME; (**b**) XRD spectra of 40-RSMPs at different stages of RSME.

A schematic model is proposed through the above discussion to better understand the water triggered reversible shape memory mechanism of CS/GL films, as shown in Figure 9. RSMPs of a specific shape were thermally programmed by bending loading to obtain and store stress and fix it into a temporary shape one. At this point, chitosan segments are connected by hydrogen bonds and contain a small amount of bound water. When RSMPs are stimulated by water, they obtain temporary shape two and chitosan molecules absorb a large amount of water. The water disrupts the hydrogen bonds between chitosan–chitosan and forms new hydrogen bonds with the amino and hydroxyl groups of chitosan, which becomes bound water. At the same time, the water molecules form a hydrogen bond chain, swelling the chitosan matrix. This water exists in the form of free water. Afterward, if the RSMPs are placed in an air environment, the water will naturally diffuse and evaporate, and the RSMPs will regain their temporary shape one. Free water, which is swelling between chitosan segments, is evaporated and removed, and the stress stored between segments is released, causing them to arrange towards temporary shape one. After a shape memory cycle, the RSMPs can be dissolved in water and exposed to air to enter another shape memory cycle. During a shape memory cycle, temporary shape one can be further dehydrated to lose all free water and some bound water to obtain sufficient mechanical strength (such as grabbing and hanging a ball in Figure 9), but it needs to reabsorb enough bound water to re-enter the shape memory cycle.

Table 2. The crystallinity of RSMPs attached to the XRD spectra, "-" means the crystallinity that cannot be accurately calculated.

RSMPs	Crystallinity (Programmed)	Crystallinity (in Air)	Crystallinity (in Water)
30-RSMPs	39.1%	-	-
40-RSMPs	35.7%	-	-

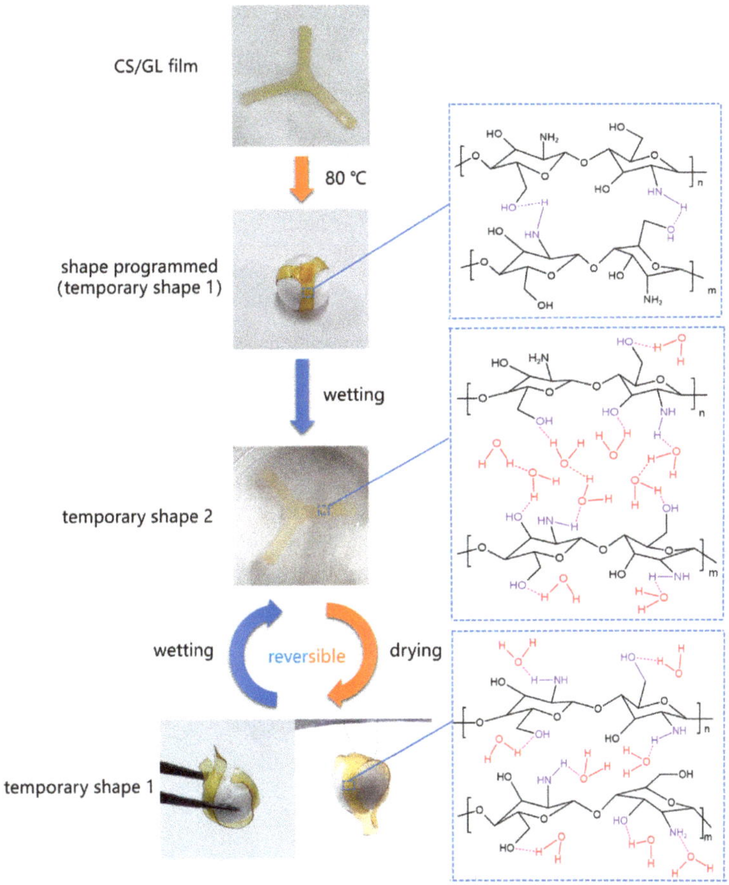

Figure 9. Schematic illustration of the reversible shape memory behavior of CS/GL films under 40% GL ratio triggered by water.

4. Conclusions

This paper provides a comprehensive understanding of the reversible shape memory behavior of chitosan/glycerol (CS/GL) films triggered by water. The systematic investigation of the binding mode and mechanism between chitosan and water molecules in different shape memory stages sheds light on the intricate interplay between water and hydrogen bonding in the film. Specifically, when CS/GL films are thermally programmed at a specific shape and store stress, their CS segments are connected by hydrogen bonds and contain a small amount of bound water, leading to temporary shape one. Upon contact with water, the CS molecules absorb a large amount of water, disrupting the CS–CS hydrogen bonds and forming new CS–water hydrogen bonds with amino and hydroxyl groups. Furthermore, the free water molecules form a hydrogen bond chain, swelling the CS matrix. When the RSMPs are placed back in the air environment, the water naturally diffuses and evaporates, causing them to revert to temporary shape one, due to the release of the stress stored between the CS segments and further removal of free water. Moreover, the GL content was found to significantly impact the material's shape memory performance, with higher concentrations improving the flexibility and response rate of the films. These findings not only contribute to our fundamental understanding of the shape memory behavior of CS/GL composite films but also hold promising implications for their potential application in various fields where reversible shape changes are desired.

Author Contributions: Conceptualization, S.L., X.W. and D.Z.; data curation, S.L. and H.L.; funding acquisition, S.L. and Y.W.; investigation, S.L., Y.W. and X.G.; methodology, S.L. and X.K.; supervision, S.L.; writing—original draft, S.L. and D.Z.; writing—review and editing, S.L., D.Z. and L.Z. All authors have read and agreed to the published version of the manuscript.

Funding: This research was supported by the National Undergraduate Training Program for Innovations, grant number 202110225125.

Institutional Review Board Statement: Not applicable.

Data Availability Statement: Not applicable.

Acknowledgments: We appreciate the Laboratory of Polymer Materials and Science of the College of Materials Science and Engineering at Northeast Forestry University for providing advanced equipment and an experimental environment for our research.

Conflicts of Interest: The authors declare no conflict of interest.

References

1. Bian, Q.; Fu, L.; Li, H. Engineering shape memory and morphing protein hydrogels based on protein unfolding and folding. *Nat. Commun.* **2022**, *13*, 137. [CrossRef] [PubMed]
2. Wang, K.; Zhu, X.X. Two-way reversible shape memory polymers containing polydopamine nanospheres: Light actuation, robotic locomotion, and artificial muscles. *ACS Biomater. Sci. Eng.* **2018**, *4*, 3099–3106. [CrossRef] [PubMed]
3. Wan, X.; He, Y.; Liu, Y.; Leng, J. 4D printing of multiple shape memory polymer and nanocomposites with biocompatible, programmable and selectively actuated properties. *Addit. Manuf.* **2022**, *53*, 102689. [CrossRef]
4. Aberoumand, M.; Soltanmohammadi, K.; Rahmatabadi, D.; Soleyman, E.; Ghasemi, I.; Baniassadi, M.; Abrinia, K.; Bodaghi, M.; Baghani, M. 4D printing of polyvinyl chloride (PVC): A detailed analysis of microstructure, programming, and shape memory performance. *Macromol. Mater. Eng.* **2023**, *308*, 2200677. [CrossRef]
5. Hui, J.; Xia, H.; Chen, H.; Qiu, Y.; Fu, Y.; Ni, Q.Q. Two-way reversible shape memory polymer: Synthesis and characterization of benzoyl peroxide-crosslinked poly (ethylene-co-vinyl acetate). *Mater. Lett.* **2020**, *258*, 126762. [CrossRef]
6. Wang, Z.; Song, W.; Ke, L.; Wang, Y. Shape memory polymer composite structures with two-way shape memory effects. *Mater. Lett.* **2012**, *89*, 216–218. [CrossRef]
7. Wang, K.; Jia, Y.G.; Zhao, C.; Zhu, X.X. Multiple and two-way reversible shape memory polymers: Design strategies and applications. *Prog. Mater. Sci.* **2019**, *105*, 100572. [CrossRef]
8. Jin, B.; Song, H.; Jiang, R.; Song, J.; Zhao, Q.; Xie, T. Programming a crystalline shape memory polymer network with thermo-and photo-reversible bonds toward a single-component soft robot. *Sci. Adv.* **2018**, *4*, eaao3865. [CrossRef] [PubMed]
9. Zare, M.; Prabhakaran, M.P.; Parvin, N.; Ramakrishna, S. Thermally-induced two-way shape memory polymers: Mechanisms, structures, and applications. *Chem. Eng. J.* **2019**, *374*, 706–720. [CrossRef]
10. Xu, Z.; Fan, Z.Y.; Wei, D.W.; Bao, R.Y.; Wang, Y.; Ke, K.; Liu, Z.Y.; Yang, M.B.; Yang, W. Tunable reversible deformation of semicrystalline polymer networks based on temperature memory effect. *Polymer* **2021**, *232*, 124157. [CrossRef]
11. Liang, R.; Yu, H.; Wang, L.; Lin, L.; Nan, W.; Naveed, K.R. Highly tough hydrogels with the body temperature-responsive shape memory effect. *ACS Appl. Mater. Interfaces* **2019**, *11*, 43563–43572. [CrossRef] [PubMed]
12. Lendlein, A.; Gould, O.E.C. Reprogrammable recovery and actuation behaviour of shape-memory polymers. *Nat. Rev. Mater.* **2019**, *4*, 116–133. [CrossRef]
13. Rahmatabadi, D.; Aberoumand, M.; Soltanmohammadi, K.; Soleyman, E.; Ghasemi, I.; Baniassadi, M.; Abrinia, K.; Zolfagharian, A.; Bodaghi, M.; Baghani, M. A New Strategy for Achieving Shape Memory Effects in 4D Printed Two-Layer Composite Structures. *Polymers* **2022**, *14*, 5446. [CrossRef] [PubMed]

14. Song, H.; Fang, Z.; Jin, B.; Pan, P.; Zhao, Q.; Xie, T. Synergetic chemical and physical programming for reversible shape memory effect in a dynamic covalent network with two crystalline phases. *ACS Macro Lett.* **2019**, *8*, 682–686. [CrossRef] [PubMed]
15. Yang, L.; Zhao, H.; Xie, Y.; Ouyang, Y.; Ruan, Y.; Chen, J.; Weng, W.; He, X.; Xia, H. Optically reconfigurable shape memory metallo-polymer mediated by a carbolong complex and radically exchangeable covalent bond. *Polym. Chem.* **2022**, *13*, 1844–1851. [CrossRef]
16. Yan, W.; Rudolph, T.; Noechel, U.; Gould, O.; Behl, M.; Kratz, K.; Lendlein, A. Reversible actuation of thermoplastic multiblock copolymers with overlapping thermal transitions of crystalline and glassy domains. *Macromolecules* **2018**, *51*, 4624–4632. [CrossRef]
17. Dumlu, H.; Marquardt, A.; Zirdehi, E.M.; Varnik, F.; Shen, Y.; Neuking, K.; Eggeler, G. A mechanical analysis of chemically stimulated linear shape memory polymer actuation. *Materials* **2021**, *14*, 481. [CrossRef] [PubMed]
18. Liang, R.; Yu, H.; Wang, L.; Amin, B.U.; Wang, N.; Fu, J.; Xing, Y.; Shen, D.; Ni, Z. Triple and two-way reversible shape memory polymer networks with body temperature and water responsiveness. *Chem. Mater.* **2021**, *33*, 1190–1200. [CrossRef]
19. Ren, Y.; Zhang, Z.Z.; Xia, W.; Zhou, Q.; Son, X. Water-responsive shape memory PLLA via incorporating PCL-(PMVS-s-PAA)-PCL-PTMG-PCL-(PMVS-s-PAA)-PCL. *Eur. Polym. J.* **2021**, *147*, 110252. [CrossRef]
20. Zou, F.; Chen, S. A Novel 2, 6-Diaminopyridine-base Polymer with Thermo-/Water-responsive Shape Memory Effect. *Mater. Today Proc.* **2019**, *16*, 1548–1553. [CrossRef]
21. Wang, Y.; Yu, X.; Cheng, Z.; Zhi, C.; Liu, Y.; Liu, Y. Switchable shape memory wetting surface based on synergistic regulation of surface chemistry and microstructure. *Compos. Part A-Appl. S.* **2021**, *149*, 106579. [CrossRef]
22. Peng, S.; Sun, Y.; Ma, C.; Duan, G.; Liu, Z.; Ma, C. Recent advances in dynamic covalent bond-based shape memory polymers. *e-Polymers* **2022**, *22*, 285–300. [CrossRef]
23. Huang, W.M.; Yang, B.; An, L.; Li, C.; Chan, Y.S. Water-driven programmable polyurethane shape memory polymer: Demonstration and mechanism. *Appl. Phys. Lett.* **2005**, *86*, 114105. [CrossRef]
24. Du, H.; Zhang, J. Solvent induced shape recovery of shape memory polymer based on chemically cross-linked poly (vinyl alcohol). *Soft Matter* **2010**, *6*, 3370–3376. [CrossRef]
25. Lu, H.; Huang, W.; Wu, X.; Ge, Y.; Zhang, F.; Zhao, Y.; Geng, J. Heating/ethanol-response of poly methyl methacrylate (PMMA) with gradient pre-deformation and potential temperature sensor and anti-counterfeit applications. *Smart Mater. Struct.* **2014**, *23*, 067002. [CrossRef]
26. Qiu, Y.; Xi, J.; Wanyan, Q.; Wu, D. Selectively sensing capacities of biocompatible shape memory materials based on cross-linked poly (L-malic acid): Visual discrimination of the solvents with similar structures. *ACS Appl. Polym. Mater.* **2020**, *2*, 1672–1681. [CrossRef]
27. Xiao, R.; Huang, W. Heating/solvent responsive shape-memory polymers for implant biomedical devices in minimally invasive surgery: Current status and challenge. *Macromol. Biosci.* **2020**, *20*, 2000108. [CrossRef]
28. Xiao, R.; Guo, J.; Safranski, D.L.; Nguyen, T.D. Solvent-driven temperature memory and multiple shape memory effects. *Soft Matter* **2015**, *11*, 3977–3985. [CrossRef]
29. Zhao, Q.; Qi, H.J.; Xie, T. Recent progress in shape memory polymer: New behavior, enabling materials, and mechanistic understanding. *Prog. Polym. Sci.* **2015**, *49*, 79–120. [CrossRef]
30. Zhang, D.; Yang, S.; Zhang, K.; Zhou, G.; Jiang, Z.; Gu, J. Shape memory effect of chitosan/glycerol composite film in mixed water/ethanol solution. *J. Appl. Polym. Sci.* **2019**, *136*, 47037. [CrossRef]
31. Han, Y.; Hu, J.; Jiang, L. Collagen skin, a water-sensitive shape memory material. *J. Mater. Chem. B* **2018**, *6*, 5144–5152. [CrossRef] [PubMed]
32. Wu, M.; Zhang, X.; Zhao, Y.; Yang, C.; Jing, S.; Wu, Q.; Brozena, A.; Miller, J.T.; Libretto, N.J.; Wu, T.; et al. A high-performance hydroxide exchange membrane enabled by Cu^{2+}-crosslinked chitosan. *Nat. Nanotechnol.* **2022**, *17*, 629–636. [CrossRef] [PubMed]
33. Rajabloo, Z.; Farahpour, M.R.; Saffarian, P.; Jafarirad, S. Biofabrication of ZnO/Malachite nanocomposite and its coating with chitosan to heal infectious wounds. *Sci. Rep.* **2022**, *12*, 11592. [CrossRef] [PubMed]
34. Karimi-Maleh, H.; Ayati, A.; Davoodi, R.; Tanhaei, B.; Karimi, F.; Malekmohammadi, S.; Orooji, Y.; Fu, L.; Sillanpaa, M. Recent advances in using of chitosan-based adsorbents for removal of pharmaceutical contaminants: A review. *J. Clean. Prod.* **2021**, *291*, 125880. [CrossRef]
35. Liu, D.; Shu, G.; Jin, F.; Qi, J.; Xu, X.; Du, Y.; Yu, H.; Wang, J.; Sun, M.; You, Y.; et al. ROS-responsive chitosan-SS31 prodrug for AKI therapy via rapid distribution in the kidney and long-term retention in the renal tubule. *Sci. Adv.* **2020**, *6*, eabb7422. [CrossRef]
36. Huq, T.; Khan, A.; Brown, D.; Dhayagude, N.; He, Z.; Ni, Y. Sources, production and commercial applications of fungal chitosan: A review. *J. Bioresour. Bioprod.* **2022**, *7*, 85–98. [CrossRef]
37. Madni, A.; Kousar, R.; Naeem, N.; Wahid, F. Recent advancements in applications of chitosan-based biomaterials for skin tissue engineering. *J. Bioresour. Bioprod.* **2021**, *6*, 11–25. [CrossRef]
38. Sun, S.; Deng, P.; Peng, C.; Ji, H.; Mao, L.; Peng, L. Selenium-Modified Chitosan Induces HepG2 Cell Apoptosis and Differential Protein Analysis. *Cancer Manag. Res.* **2022**, *14*, 3335–3345. [CrossRef]
39. Lu, J.; Chen, Y.; Ding, M.; Fan, X.; Hu, J.; Chen, Y.; Li, J.; Li, Z.; Liu, W. A 4arm-PEG macromolecule crosslinked chitosan hydrogels as antibacterial wound dressing. *Carbohydr. Polym.* **2022**, *277*, 118871. [CrossRef]

40. Zhang, H.; Kong, M.; Jiang, Q.; Hu, K.; Ouyang, M.; Zhong, F.; Qin, M.; Zhuang, L.; Wang, G. Chitosan membranes from acetic acid and imidazolium ionic liquids: Effect of imidazolium structure on membrane properties. *J. Mol. Liq.* **2021**, *340*, 117209. [CrossRef]
41. Tu, G.; Li, S.; Han, Y.; Li, Z.; Liu, J.; Liu, X.; Li, W. Fabrication of chitosan membranes via aqueous phase separation: Comparing the use of acidic and alkaline dope solutions. *J. Membr. Sci.* **2022**, *646*, 120256. [CrossRef]
42. Chen, Y.; Li, J.; Lu, J.; Ding, M.; Chen, Y. Synthesis and properties of poly (vinyl alcohol) hydrogels with high strength and toughness. *Polym. Test.* **2022**, *108*, 107516. [CrossRef]
43. Di Prima, M.A.; Lesniewski, M.; Gall, K.; McDowell, D.L.; Sanderson, T.; Campbell, D. Thermo-mechanical behavior of epoxy shape memory polymer foams. *Smart Mater. Struct.* **2007**, *16*, 2330. [CrossRef]
44. Lei, M.; Yu, K.; Lu, H.; Qi, H.J. Influence of structural relaxation on thermomechanical and shape memory performances of amorphous polymers. *Polymer* **2017**, *109*, 216–228. [CrossRef]
45. Mohamed, A.E.; Elgammal, W.E.; Eid, A.M.; Dawaba, A.M.; Ibrahim, A.G.; Fouda, A.; Hassan, S.M. Synthesis and characterization of new functionalized chitosan and its antimicrobial and in-vitro release behavior from topical gel. *Int. J. Biol. Macromol.* **2022**, *207*, 242–253. [CrossRef]
46. Riaz, A.; Lei, S.; Akhtar, H.M.S.; Wan, P.; Chen, D.; Jabbar, S.; Abid, M.; Hashim, M.M.; Zeng, X. Preparation and characterization of chitosan-based antimicrobial active food packaging film incorporated with apple peel polyphenols. *Int. J. Biol. Macromol.* **2018**, *114*, 547–555. [CrossRef]
47. Vijay, R.; Singaravelu, D.L.; Vinod, A.; Sanjay, M.R.; Siengchin, S.; Jawaid, M.; Khan, A.; Parameswaranpillai, J. Characterization of raw and alkali treated new natural cellulosic fibers from Tridax procumbens. *Int. J. Biol. Macromol.* **2019**, *125*, 99–108. [CrossRef]
48. Sisti, L.; Kalia, S.; Totaro, G.; Vannini, M.; Negroni, A.; Zanaroli, G.; Celli, A. Enzymatically treated curaua fibers in poly (butylene succinate)-based biocomposites. *J. Environ. Chem. Eng.* **2018**, *6*, 4452–4458. [CrossRef]
49. Wu, S. Secondary relaxation, brittle–ductile transition temperature, and chain structure. *J. Appl. Polym. Sci.* **1992**, *46*, 619–624. [CrossRef]
50. Johari, G.P.; Goldstein, M. Viscous liquids and the glass transition. III. Secondary relaxations in aliphatic alcohols and other nonrigid molecules. *J. Chem. Phys.* **1971**, *55*, 4245–4252. [CrossRef]
51. Yu, Y.; Tyrikos-Ergas, T.; Zhu, Y.; Fittolani, G.; Bordoni, V.; Singhal, A.; Fair, R.J.; Grafmuller, A.; Seeberger, P.H.; Delbianco, M. Systematic hydrogen-bond manipulations to establish polysaccharide structure–property correlations. *Angew. Chem.* **2019**, *131*, 13261–13266. [CrossRef]
52. Li, T.; Zhang, C.; Xie, Z.; Xu, J.; Guo, B. A multi-scale investigation on effects of hydrogen bonding on micro-structure and macro-properties in a polyurea. *Polymer* **2018**, *145*, 261–271. [CrossRef]
53. Xing, K.; Chatterjee, S.; Saito, T.; Gainaru, C.; Sokolov, A.P. Impact of hydrogen bonding on dynamics of hydroxyl-terminated polydimethylsiloxane. *Macromolecules* **2016**, *49*, 3138–3147. [CrossRef]

polymers

Article

Crude Enzyme Concentrate of Filamentous Fungus Hydrolyzed Chitosan to Obtain Oligomers of Different Sizes

Cleidiane Gonçalves e Gonçalves [1], Lúcia de Fátima Henriques Lourenço [2], Hellen Kempfer Philippsen [3], Alberdan Silva Santos [4], Lucely Nogueira dos Santos [1] and Nelson Rosa Ferreira [1,2,*]

1 Graduate Program in Food Science and Technology, Federal University of Pará, Belem 66075-110, PA, Brazil
2 Faculty of Food Engineering, Technology Institute, Federal University of Pará, Belem 66075-110, PA, Brazil
3 Faculty of Biology, Socioenvironmental and Water Resources Institute, Federal Rural University of the Amazon, Campus Belém, Belem 66077-830, PA, Brazil
4 Faculty of Chemistry, Institute of Exact and Natural Sciences, Federal University of Pará, Belem 66075-110, PA, Brazil
* Correspondence: nelson.ufpa@gmail.com

Abstract: Chitosan is a non-cytotoxic polysaccharide that, upon hydrolysis, releases oligomers of different sizes that may have antioxidant, antimicrobial activity and the inhibition of cancer cell growth, among other applications. It is, therefore, a hydrolysis process with great biotechnological relevance. Thus, this study aims to use a crude enzyme concentrate (C_{EC}) produced by a filamentous fungus to obtain oligomers with different molecular weights. The microorganism was cultivated in a liquid medium (modified Czapeck—with carboxymethylcellulose as enzyme inducer). The enzymes present in the C_{EC} were identified by LC-MS/MS, with an emphasis on cellobiohydrolase (E.C 3.2.1.91). The fungus of the *Aspergillus* genus was identified by amplifying the ITS1-5.8S-ITS2 rDNA region and metaproteomic analysis, where the excreted enzymes were identified with sequence coverage greater than 84% to *A. nidulans*. Chitosan hydrolysis assays compared the C_{EC} with the commercial enzyme (Celluclast 1.5 L®). The ability to reduce the initial molecular mass of chitosan by 47.80, 75.24, and 93.26% after 2.0, 5.0, and 24 h of reaction, respectively, was observed. FTIR analyses revealed lower absorbance of chitosan oligomers' spectral signals, and their crystallinity was reduced after 3 h of hydrolysis. Based on these results, we can conclude that the crude enzyme concentrate showed a significant technological potential for obtaining chitosan oligomers of different sizes.

Keywords: chitosan; hydrolysis; filamentous fungus; enzyme

1. Introduction

Chitosan is a linear polysaccharide obtained by the partial deacetylation of chitin under alkaline conditions, which contains copolymers of D-glucosamine (deacetylated units) and N-acetyl-D-glucosamine (acetylated units) interconnected by β-glycosidic bonds (1→4) [1], and has the unique properties of biocompatibility, biodegradability, bioactivity, and non-toxicity.

In addition to the positive aspects of chitosan, it is necessary to meet technological conditions and specific needs that are enhanced depending on the size of the oligomer obtained through hydrolysis. The size of the oligomer is directly proportional to its respective molecular weight [2]. Based on molecular weight, chitosan can be grouped into low molecular weight (<100 kDa), medium molecular weight (100–1000 kDa), and high molecular weight (>1000 kDa) [3].

Compared to chitosan, its oligomers have a lower molecular weight, better water solubility, and greater physiological activities which involve: antimicrobial and antioxidant activity [4–6]; hypocholesterolemic properties [7]; antimutagenicity [8]; decrease in acrylamide formation in glucose/fructose-asparagine solutions [9]; and inhibition of tumor cell growth [10].

The standard method of depolymerization of chitosan is through acid hydrolysis. However, this method has some disadvantages, including the difficulty in obtaining oligosaccharides with a low degree of polymerization and in controlling the extent of hydrolysis, which often results in hydrolysates with a high amount of monosaccharides [11]. Furthermore, the reaction conditions require high temperatures and high concentrations of reagents, which often result in chemically modified oligosaccharides forming [12].

The use of enzymes in chitosan hydrolysis has received attention because it presents less variation in obtaining oligosaccharides than chemical hydrolysis [13]. In addition, the enzymatic method has advantages over chemical reactions because the enzymes act under milder conditions, have high specificity, and do not modify the structure of the glucose ring [14]. However, this method has the disadvantages of higher cost, limited availability of some enzymes, and slow action in viscous solutions, requiring a low substrate concentration and a more significant amount of enzymes [13].

Chitosanases and chitinases are specific enzymes responsible for the hydrolysis of chitosan and chitin, respectively. Chitosanases (E.C 3.2.1.132) are glycosyl hydrolases that catalyze the endohydrolysis of β-1,4-glycosidic bonds of chitosan. Chitinases can be classified according to their mode of action into endochitinases (E.C 3.2.1.14) and exochitinases (E.C 3.2.1.52), which catalyze the internal and external hydrolysis of chitosan [15].

Non-specific or promiscuous enzymes are also capable of hydrolyzing chitosan. These enzymes are cellulases, hemicellulases, pectinases, papain, pepsin, proteases, bromelain, and lysozyme, which are less expensive and can cleave the β-1,4-glycosidic bond of chitosan [5,16–20].

Cellulases are industrially essential enzymes and have traditionally been obtained from submerged fermentation by controlling extrinsic and intrinsic factors such as temperature and pH [21]. In this context, filamentous fungi are preferred for producing these enzymes rather than yeast or bacteria [22].

The search for new enzyme-producing sources and new enzymatic depolymerization mechanisms are objects of considerable scientific and technological interest. Thus, this study shows the possibility of using an efficient enzymatic extract obtained with few downstream steps capable of hydrolyzing chitosan to obtain chitooligosaccharides with varied sizes that present physiological activities.

2. Materials and Methods

2.1. Chitosan

Commercial chitosan (Polymar, Fortaleza, Ceará, Brazil) (cód. PB2112) (obtained from crustacean exoskeleton) with a certified degree of deacetylation of 85% was used.

2.2. Commercial Enzyme

Celluclast 1.5 L® (Novozymes, Araucária, Paraná, Brazil), produced by Trichoderma reesei, was used as a standard in the hydrolysis of chitosan.

2.3. Microorganism

The filamentous fungus (code MIBA0664) belongs to the mycotheque of the Laboratory of Systematic Research in Biotechnology and Molecular Biodiversity of the Federal University of Pará (Brazil). This microorganism was isolated from a xylophagous mollusk using the procedure described by Ferreira et al. [23]. The strain was previously selected among others in the laboratory due to its potential endoglucanase activity.

2.4. Identification of the Microorganism

The fungal biomass used in the identification was obtained from cultivation in liquid medium Czapek Dox (Sigma-Aldrich, St. Louis, MO, USA, EUA) with the addition of 10% yeast extract and subsequent incubation in shaken flasks at 120 rpm/30 °C for five days. DNA was extracted according to the manufacturer's instructions for the genomic DNA extraction and purification kit (Axygen®, ref. AP-MN-MS-GDNA-50) (Union City,

CA, USA, EUA). The primer pair ITS-1f (5′-TCC GTA GGT GAA CCT GCG G-3′ and primer ITS-4r (5′-TCC TCC GCT TAT TGA TAT GC-3′) [24,25]. PCR was performed in a thermocycler (model TX96, Amplitherm®, State of São Paulo, Brazil) programmed at 95 °C/5′, 35 cycles of 94 °C/1′, 55.5 °C/2′, 72 °C/2′ (denaturation, annealing, and extension stages, respectively), and final extension at 72 °C/10′ [25].

Sequence annotation was performed using Geneious® (version 9.1.5) and compared to GenBank. The sequence > 98% was considered to identify the filamentous fungus species.

2.5. Crude Enzyme Concentrate (C_{EC})

Initially, two culture media were evaluated for better enzymatic activity, established as endoglucanase activity [26]. Two media were used, modified Czapeck (C_{zm}): carboxymethylcellulose (CMC) 10.0 g/L, sodium nitrate 3.0 g/L, potassium phosphate 1.0 g/L, potassium chloride 0.5 g/L, magnesium sulfate 0.5 g/L, and ferrous sulfate 0.01 g/L; and CPY: carboxymethylcellulose (CMC) 10.0 g/L, peptone 1.0 g/L, and yeast extract 20.0 g/L.

Cultivation was carried out in 500 mL conical flasks containing 250 mL of C_{zm} or CPY media. Ten discs of fungal mycelium fragments 5 mm in diameter were added and incubated in each flask at 30 °C, under constant agitation at 120 rpm. Endoglucanase activity was evaluated after seven days, and the assays were performed in triplicate.

After selecting the culture medium, a scale-up was performed, where five 1000-mL conical flasks containing 500 mL of C_{zm} and 20 mycelia discs (5 mm) were incubated under the same conditions used previously. Aliquots were withdrawn at 24-h intervals to assess endoglucanase activity. The final culture was filtered through a quantitative paper disc and a Buchner funnel under a vacuum to remove the biomass. The liquid phase was concentrated by lyophilization until 90% of the initial volume was reduced.

2.6. Protein Content

The amount of total protein present in the C_{EC} and the commercial enzyme was determined according to Bradford methodology [27] with modifications. Thus, 200 μL of C_{EC} or commercial enzyme was added to 1800 μL of Bradford reagent solution (0.1 mg/mL) with a reaction time of 5 min. The reading was performed in a spectrophotometer (Thermo, GENESYS, Fisher Scientific, Arendalsvägen, Göteborg, Sweden) at 595 nm. The blank was prepared under the same conditions, replacing the CEC with deionized water.

2.7. Endoglucanase Activity Assay (CMCase Activity)

The enzymatic activity was determined by quantifying reducing sugars [28], with modifications. In a 1.5 mL microtube, 500 μL of the enzymatic extract and 1000 μL of carboxymethylcellulose (CMC) at 0.5% diluted in sodium citrate buffer were added (0.1 M; pH 4.8). The reaction was carried out at 50 °C for 10 min in a Thermomixer (Thermomixer ® compact, Eppendorf) at 400 rpm. After the established time, 100 μL of the reaction mixture was added to 200 μL of water and 300 μL of DNS reagent and incubated at 100 °C for 5 min. After the mixture was cooled, 1000 μL of distilled water was added. The blank control sample was evaluated under the same conditions, replacing the CMC with distilled water. The difference with the blank control sample determined the released reducing sugars (*Ra*). The reading was performed in a spectrophotometer at 540 nm. Enzyme activity was calculated using Equations (1) and (2):

$$Enzymatic\ activity\left(\frac{\mathrm{U}}{\mathrm{mL}}\right) = \frac{Ra\left(\frac{\mathrm{mg}}{\mathrm{mL}}\right) \times \mathrm{V_r(mL)}}{\mathrm{T_h(min)} \times 0.18 \times \mathrm{V_e(mL)}} \quad (1)$$

$$Enzymatic\ activity\left(\frac{\mathrm{U}}{\mathrm{mg}}\right) = \frac{Ra\left(\frac{\mathrm{mg}}{\mathrm{mL}}\right) \times \mathrm{V_r(mL)}}{\mathrm{T_h(min)} \times 0.18 \times \mathrm{P_c(mg)}} \quad (2)$$

where: *Ra*: released reducing sugars; V_r: reaction volume; T_h: hydrolysis time; V_e: enzymatic extract volume; P_c: Protein content.

2.8. Identification of Proteins in Crude Enzyme Concentrate

2.8.1. Protein Precipitation

The enzymatic extract concentrated by lyophilization was subjected to precipitation [29]. Proteins were precipitated by adding methanol, chloroform, and water in a ratio of 1:4:3 ($v/v/v$), respectively. The solution was vortexed, followed by centrifugation at 12.000$\times$ g for 5 min and 25 °C. The upper phase was discarded, and then methanol was added in a ratio of 3 mL to 1 mL of the initial volume of the sample. The solution was centrifuged at 12.000$\times$ g for 5 min and 25 °C. The supernatant was discarded, and the pellet was dried at room temperature 25 °C. The obtained pellet was washed with 50 mM bicarbonate and ammonium in Amicon® Ultra 3 K MWCO (MilliporeSigma in the U.S. and Canada) 5 times, with subsequent centrifugation at 12.000$\times$ g for 1 hour.

2.8.2. Protein Digestion

After washing with 50 mM bicarbonate and ammonium in Amicon®Ultra 3 K MWCO (MilliporeSigma in the U.S. and Canada), protein digestion was performed [30]. Each protein sample was homogenized with 50 mL of ammonium bicarbonate (50 mM), 10 mM DTT (dithiothreitol), and 0.25% RapiGest SF (Waters Corp., Milford, MA, USA). Digestion occurred from incubating trypsin samples with protein (1:50) at 37 °C for 16 h. Possible interferences in the samples were removed with the addition of formic acid (0.1%) and incubated at 60 °C for 60 min. Insoluble detergents were removed from samples by centrifugation at 10.000$\times$ g for 30 min. The supernatant was collected and frozen at −80 °C for subsequent identification of proteins by LC/MS.

2.8.3. Identification by LC-MS/MS

Proteins were identified using the nanoElute nanoflow chromatographic system (Bruker Daltonics, Bremen, Germany), coupled online to a mass spectrometer (Hybrid Trapped Ion Mobility Spectrometry (TIMS), model Quadrupole Time-of-Flight (timsTOF Pro 2, Bruker Daltonics, Bremen, Germany). An aliquot (1 µL) sample, equivalent to 200 ng of digested peptides, was injected into a Bruker FIFTEEN C18 column (1.9 µm, 150 mm × 75 µm), from Bruker. The column was coupled online with a CaptiveSpray ion source (Bruker Daltonik GmbH). A typical RP gradient (Solvent A: 0.1% AF, 99.9% H_2O; Solvent B: 0.1% AF, 99.9% CH_3CN) was established in a liquid chromatography nanoflow system and separated at a flow rate of 500 nL/min. The column temperature was maintained at 50 °C. The chromatographic run lasted 60 min (30% of Solvent B for 55 min and 95% at 56 min; maintained at this percentage of Solvent B for another 4 min). The temperature of the capillary ion transfer line was set to 180 °C. Ion accumulation and mobility separation were obtained with an input potential ramp from −160 V to −20 V within 123 s. During the acquisition, to enable the PASEF method, the accumulation parallel to the fragmentation of the ions, the precursor m/z and mobility information was first derived from a Tims-MS full scan experiment, with an m/z range of 100–1700. Monocharged precursors were excluded for their position in the m/z-ion plane of mobility, and precursors that reached the target value of 20,000 a.u. were dynamically excluded for 0.4 min. The operational mode of the TIMS-TOF, MS, and PASEF were controlled and synchronized with the aid of the instrumental control software OtofControl 5.1 by Bruker Daltonik.

2.8.4. Data Processing and Search Parameters

Data processing, protein identification, and relative quantification were performed using PEAKS Studio Software, Version 10.6, Bioinformatics Solutions Inc., Waterloo, ON, Canada. The processing parameters included carbamidomethylation of cysteine with fixed amino acid modification, methionine oxidation, and N-terminal acetylation, which were considered variable variations. Trypsin was used as a proteolytic enzyme, with a maximum of 2 possible cleavage errors. The ion mass shift tolerance for peptides and fragments was adjusted to 20 ppm and 0.05 Da, respectively. A maximum false positive rate (FDR) of 1% was used for peptide and protein identification, considering at least one unique peptide for identification as a criterion. All proteins were identified with a confidence level ≥ 95%, using the PEAKS Software algorithm and searching within the *Aspergillus* sp. database, using the UniProt database (http://www.uniprot.proteomes/ (accessed on 10 December 2022)).

The identified proteins were grouped into functional categories according to their functions and molecular weights using the database (http://www.uniprot.proteomes/ (accessed on 11 December 2022)).

2.8.5. Identification of Microorganisms Based on Peptides

The identified peptides were analyzed using a bioinformatics tool for metaproteomics (https://unipept.ugent.be/ (accessed on 11 December 2022)) to verify the taxonomy of the studied microorganism. After identifying the microorganism, the proteins identified were analyzed against the microorganism database (https://blast.ncbi.nlm.nih.gov/ (accessed on 11 December 2022)).

2.9. *Enzymatic Hydrolysis of Chitosan*

2.9.1. Crude Enzyme Concentrate

Hydrolysis was performed according to the method by Roncal et al. [18], with modifications. Chitosan hydrolysis was performed using a 1% solution (*w*/*v*) in sodium acetate buffer (pH 4.5; 0.1 M). From the C_{EC}, 0.14 mg of protein/g of chitosan was added. Thus, 1000 µL of C_{EC} was added in 100 mL of sodium acetate buffer, then 1000 mg of chitosan was added and incubated at 45 °C under agitation (120 rpm) at different times: 1; 2; 3; 4; 5; and 24 h. The hydrolysis reaction was stopped by heating the mixture at 100 °C for 5 min to inactivate the enzyme. The content was adjusted to pH 7, resulting in a chitosan precipitate. The precipitate was filtered and dried at 60 °C.

2.9.2. Commercial Enzyme

Hydrolysis was performed according to the method by Roncal et al. [18], with modifications. Chitosan hydrolysis was performed using a 1% solution (*w*/*v*) in sodium acetate buffer (pH 4.5; 0.1 M). From the commercial enzyme Celuclast 1.5 L, 0. 51 mg of protein/g of chitosan was added. Thus, 20 µL of Celuclast was added in 100 mL of sodium acetate buffer, then 1000 mg of chitosan was added and incubated at 45 °C under agitation (120 rpm) at different times: 0.5; 1; 2; and 3 h. The hydrolysis reaction was stopped by heating the mixture at 100 °C for 5 min to inactivate the enzyme. The content was adjusted to pH 7, resulting in a chitosan precipitate. The precipitate was filtered and dried at 60 °C.

2.10. *Characterization of Chitosan and Oligomers*

2.10.1. Molecular Mass Determination

The molecular mass of chitosan (Qt) and hydrolyzed chitosan (Qh) was determined by viscometry according to the method described by Garcia et al. [31], with modifications. Qt and Qh samples were prepared with a concentration of 0.005 g/mL in acetic acid buffer solution (0.3 M) and sodium acetate (0.2 M), pH 4.5, and kept at 30 °C, 120 rpm, for 24 h.

To determine the intrinsic viscosity, [η], the Qt and Qh solutions were diluted at concentrations of 0.004, 0.003, 0.002, and 0.001 g/mL, and the solution flow times were determined in a Canon Fensk capillary viscometer (Schott AVS 350) at 25 °C. The specific viscosity (η_{sp}) was determined using Equation (3):

$$\eta_{sp} = (t - t_0)/t_0 \quad (3)$$

where: t is the flow time of the chitosan solution and t_0 is the flow time of the solvent.

The reduced viscosity (η_{red}) was obtained through the relationship between specific viscosity and chitosan concentration (C), as in Equation (4):

$$\eta_{red} = \eta_{sp}/C \quad (4)$$

Intrinsic viscosity [η] is defined as reduced viscosity, extrapolated to a chitosan concentration (C) of zero, as in Equation (5):

$$[\eta] = (\eta_{sp}/C)_{c\to 0} = (\eta_{red})_{c\to 0} \quad (5)$$

From the intrinsic viscosity, the molecular mass of chitosan was calculated using the Mark–Houwink equation, as in Equation (6):

$$[\eta] = KM_w{}^{\alpha} \quad (6)$$

where: M_w is the viscosity average molecular weight, and K and α are constants that depend on the chitosan polydispersion and the solvent system used. The values of these constants were previously determined to be K = 0.074 and α = 0.76.

2.10.2. FTIR Deacetylation Degree

The Qt and Qh samples were analyzed in an infrared spectrometer (Agilent, model-Cary 360, Santa Clara, CA, USA, EUA) with Total Attenuated Reflectance (FTIR-ATR) and zinc selenide crystal (ZnSe), in the range of 4000 cm^{-1} to 650 cm^{-1}, resolution of 4 cm^{-1} and 32 scans.

The degree of Qt and Qh deacetylation was determined by FTIR-AT by calculating the areas of the infrared spectral bands corresponding to the functional groups of amine (1350 cm^{-1}) and CH_2 (1465 cm^{-1}), according to the methodology described by Barragán et al. [32]. The areas were calculated by the Spectragryph software (v. 1.2.14/2020) using the integration with baseline function [33]. The degree of acetylation (DA) was determined using Equation (7):

$$A_{1350}/A_{1465} = 0.3822 + 0.0313\ GA \quad (7)$$

where: DA: degree of acetylation; A_{1350}: area under the curve of the infrared spectrum band with a wavenumber of 1350 cm^{-1}; and A_{1465}: area under the curve of the infrared spectral band with wavenumber 1465 cm^{-1}. Values 0.3822 and 0.0313 were obtained by linear regression [34].

The degree of deacetylation (DD) was determined using Equation (8):

$$DD = 100 - DA \quad (8)$$

3. Results and Discussion

3.1. Identification of the Microorganism

The strain was identified as *Aspergillus* sp. (GenBank ID accession number MT135987) with greater than 97% similarity. In several studies, it was observed that the genus *Aspergillus* stands out in the production of enzymes that act in the hydrolysis of chitosan, such as cellulases [35], hemicellulase [18], and pectinase [36] by *Aspergillus niger*; chitin deacetylase by *Aspergillus nidulans* [37] chitosanase by *Aspergillus fumigatus* [38].

After identifying the enzymes in the secreted extract of *Aspergillus* sp, it was possible to perform a metaproteomic analysis based on peptides. The analysis allowed the identification of the microorganism belonging to the species *A. nidulans* (Figure 1).

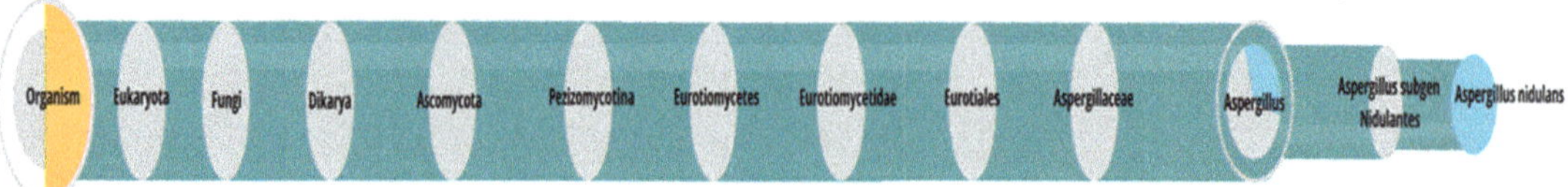

Figure 1. Dendogram based on peptides built by Unipept (https://unipept.ugent.be/, accessed on 11 December 2022). Phylogenetic tree based on the list of peptides identified using the lowest common ancestor (LCA) method.

3.2. Selection of the Culture Medium

The structural similarity between cellulose and chitosan—D-glucose polymers linked by β-1,4-glycosidic bonds—enabled carboxymethylcellulose to be an inducing substrate in both culture media (C_{zm} and CPY) in order to produce enzymes, which will later be used in the hydrolysis of chitosan. Figure 1 shows that the microorganism produced enzymes capable of hydrolyzing carboxymethylcellulose (CMCase activity), indicated by endoglucanase activity.

Among the culture media studied, the C_{zm} showed the highest enzymatic activity (Figure 2). In addition, an increase in enzymatic activity is observed up to the eighth day of cultivation, followed by a decrease in subsequent days, defining the eighth day in C_{zm} medium for the production of C_{EC}.

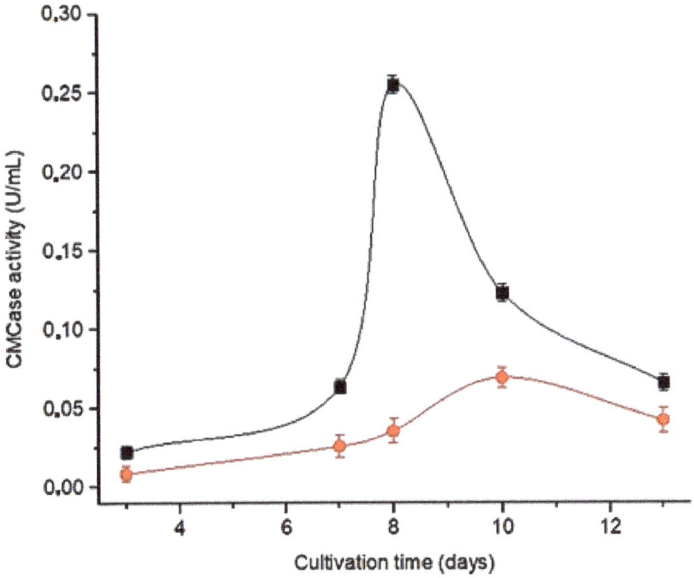

Figure 2. Enzyme activity profile (CMCase) in different culture media for *Aspergillus nidulans*: ■—modified czapek culture medium (C_{zm}); •—medium cellulose peptone yeast extract (CPY).

3.3. Crude Enzyme Concentrate (CEC)

Table 1 shows the total activity (U/mL) and specific activity (U/mg) for the crude enzyme and the crude enzyme concentrate, respectively. The crude enzyme produced by the microorganism in C_{zm} medium during eight days of cultivation showed a specific enzymatic activity of 13.3 U/mg. After concentration by lyophilization, the crude enzyme showed an activity of 15.9 U/mg. The total protein content increased eight times after lyophilization, and there was a 19% increase in the specific activity. There was no significant increase in the specific activity. This may be related to the fact that no purification step was performed, only concentration, which enables more significant competitive inhibition by the saccharides present in the medium. Otherwise, this culture was characterized by low protein production, which contributed to good results of a specific activity.

Table 1. Enzyme activity after extract concentration by lyophilization.

	Volume (mL)	Total Activity (U/mL)	Total Protein (mg/mL)	Specific Activity (U/mg)
Crude enzyme	500.0	0.22 ± 0.031	0.0165 ± 0.001	13.3 ± 0.97
Crude enzyme concentrate	25.0	2.23 ± 0.19	0.14 ± 0.013	15.9 ± 1.38

3.4. Identification of Proteins in Crude Enzyme Concentrate

The fungus was grown in the C_{zm} culture medium added with chitosan. The proteins secreted by the microorganism in the culture medium were precipitated and submitted to LC-MS/MS analysis to verify the possible secretion of enzymes involved in the hydrolysis of chitosan by the fungus. Specific enzymes involved in the hydrolysis of chitosan are called chitosanases. They carry out hydrolysis from the cleavage of glycosidic bonds s β-1,4 [39]. These enzymes are usually extracellular and microorganisms such as *Bacillus subitilis* [40,41], and *Janthinobacterium* [42] showed capacity in the production of chitosanases. Among fungi, the genus *Trichoderma* [43] and *Aspergillus* [44,45] have stood out in this production.

Enzymes for polymers structurally similar to chitosan (such as cellulose and pectin) can also be used for the hydrolysis of chitosans, as they have chitosanolytic activity [5]. Regardless of whether it is a specific enzyme or not, chitosan is basically hydrolyzed by the action of glycosidic hydrolases (GH) [39]. In chitin and chitosan, acetyl groups remaining from GlcNAc play the nucleophilic role of the hydrolysis reaction by GH; thus, non-specific GHs can cleave the glycosidic bonds of chitosan [39].

Analysis by LC-MS/MS in the present study allowed for the identification of enzymes that hydrolyze chitosan. The peptides were blasted against the *Aspergilus* database. The taxonomic analysis based on identified peptides indicated that *A. nidulans* is the study microorganism (Figure 2); therefore, the identified proteins that did not belong to *A. nidulans/Emericella nidulans* were blasted against the *A. nidulans* database, which allowed the identification of several proteins, listed in Table 2. All identified proteins showed sequence coverage greater than 90% of *A. nidulans* proteins (Table 2). The accession numbers Q8NK02, Q5AUX2, Q5BA61, and Q9HGI3, which refer to the identification of *Emericella nidulans*, were not compared against *A. nidulans* because they are the same microorganism, only differentiated by the sexual or teleomorphic form.

Among the identified proteins were identified chitinases (E.C 3.2.1.14) and (E.C 3.2.1.52), responsible for catalyzing the cleavage of chitin and chitosan [46,47]. Depending on the type of cleavage, chitinases can be classified as endo- or exo-chitinases. Endochitinases (E.C 3.2.1.14) catalyze the internal hydrolysis of chains at random points along the polysaccharide, and hydrolyze the ends of the reducing or non-reducing polymeric chain, producing low molecular weight N-acetylglucosamine multimers [48], while exochitinases (E.C 3.2.1.52) act by randomly cleaving internal chitin sites. Exochitinases, in turn, are divided into two subgroups: chitobiosidases, which cleave chitin from the non-reducing end, releasing diacetylchitobiose (N-acetylglucosamine dimer); and 1-4-β-glucosaminidase, which cleave chitin oligomers released by endochitinases and chitobiosidases producing GlcNAc monomers. The β-hexosaminidase enzyme identified in the present work secreted by *A. nidulans* performs the hydrolysis of terminal non-reducing N-acetyl-D-hexosamine residues in N-acetyl-β-D-hexosaminides and can be used for the enzymatic synthesis of complex type sugar chains containing GlcNAc and GalNAc as components [48].

Chitinases are widely studied for different applications, such as biocontrol [49,50], as an antitumor agent [51], and bioethanol production [52], among others. The secretion of chitinases by *Aspergillus* spp. Is already described. It was observed that *A. flavus* secretes thermostable chitinases in a medium with high salt concentration [53], *A. niger* secretes thermostable chitinases that act in the biocontrol of *Candida* and *Galleria mellonella* [54], and *A. nidulans* secretes chitinases, cellulases, hemicellulases, esterases, and lipases when cultivated in sorghum straw [55].

Table 2. Proteins identified by LC-MS/MS against the *Aspergillus* sp. and *A. nidulans* database.

Data on Proteins Identified by LC-MS/MS against the *Aspergillus* sp. Database					Data on Proteins Identified by LC-MS/MS against the *Aspergillus nidulans* Database		
Access Number—Proteins Identified against the *Aspergillus* sp. Database	**Peptide**	**Mass**	**Ppm**	***m/z***	**Accession Number**	**Coverage (%)**	**Catalytic Class (E.C)**
A1DMA5—Probable 1,4-beta-D-glucan cellobiohydrolase A OS = *Neosartorya fischeri*	VIANSVSNVADVSGNSISSDFC(+57.02)TAQK	26.692.603	16.7	8.907.712	Q5B2Q4—Probable 1,4-beta-D-glucan cellobiohydrolase A; flags: Precursor [*Aspergillus nidulans* FGSC A4]	99%	Glycosidase, Hydrolase (3.2.1.91)
Q8NK02—1 4-beta-D-glucan cellobiohydrolase B OS = *Emericella nidulans*	YGTGYC(+57.02)DSQC(+57.02)PR LNFVTQSQQK	14.625.605 11.916.248	0.9 −1.9	7.322.845 5.968.185	Q8NK02—1 4-beta-D-glucan cellobiohydrolase B OS = *Emericella nidulans*	100%	Glycosidase, Hydrolase (3.2.1.91)
Q5AUX2—Alpha-L-arabinofuranosidase axhA-2 OS = *Emericella nidulans*	ANSGATWTDDISHGDLVR	19.138.867	2.8	6.389.697	Q5AUX2—Alpha-L-arabinofuranosidase axhA-2 OS = *Emericella nidulans*	100%	Glicosidase, Hidrolase (3.2.1.55)
Q5BA61 Pectin lyase B OS = *Emericella nidulans*	SLVGEGSSGVIK	11.316.135	1.5	5.668.149	Q5BA61 Pectin lyase B OS = *Emericella nidulans*	100%	Liase (4.2.2.10)
E9QRF2 Endochitinase B1 OS = *Neosartorya fumigata*	IVLGMPLYGR	11.176.317	−0.6	5.598.228	G5EAZ3—Endochitinase B; AltName: Full = Chitinase B [*Aspergillus nidulans* FGSC A4]	90%	Glycosidase, Hydrolase (3.2.1.14)
Q9HGI3—Beta-hexosaminidase OS = *Emericella nidulans* OX = 162425 GN = nagA PE = 1 SV = 1	HISWGHSGPKPLSDVSLRTERDTDDSILTNAWNR	38.598.989	−5.7	12.876.329	Q9HGI3—Beta-hexosaminidase OS = *Emericella nidulans* OX = 162425 GN = nagA PE = 1 SV = 1	100%	Glycosidase, Hydrolase (3.2.1.52)

In the pool of enzymes secreted by the microorganism, also identified were the enzyme cellobiohydrolase (E.C3.2.1.91), also known as exoglucanase. Two cellobiohydrolases were identified in the present work: Probable 1,4-β-D-glucan cellobiohydrolase A (Q5B2Q4) and Probable 1,4-β-D-glucan cellobiohydrolase B (Q8NK02). These enzymes release cellobiose units from cellulose's reducing and non-reducing ends [56]. Cellobiohirolases can also hydrolyze chitosan, producing low molecular weight oligomers. In addition, it was verified that the hydrolysis product of chitosan through cellobiohydrolase is similar to that of chitosanase [56], with the formation of chitooligosaccharides that may have relevant bioactive effects, such as maintaining the immunological indices of patients undergoing chemotherapy [57], as well as antimicrobial and antioxidant activity [5].

Another identified enzyme was α-L-arabinofuranosidase (E.C 3.2.1.55). This enzyme releases L-arabinose and is involved in the hydrolysis of oligosaccharides and hemicelluloses [58]. Pectin lyase (E.C. 4.2.2.10), also identified, catalyzes the cleavage of the α1-4 glycosidic bond in pectic acid and pectin [59]. In addition to pectin hydrolysis, several studies have found that pectinases can be used in chitosan hydrolysis [20,60,61].

3.5. Chitosan Hydrolysis

The commercial enzyme (C_E), Cellulast 1.5 L®, was used as a parameter to verify the efficiency in the hydrolysis of chitosan using the C_{EC} obtained in this study. C_E can cleave the β-1,4-glycosidic bond of chitosan [62], that is, the same region where chitosanase acts.

Figure 3 shows a decrease in molecular weight with increasing hydrolysis time for the two enzyme concentrates. The molecular weight of chitosan hydrolyzed by C_{EC} and C_E after 1 h was 85.20 and 71.14 kDa; at 2 h, it was 56.8 and 57.86 kDa, respectively.

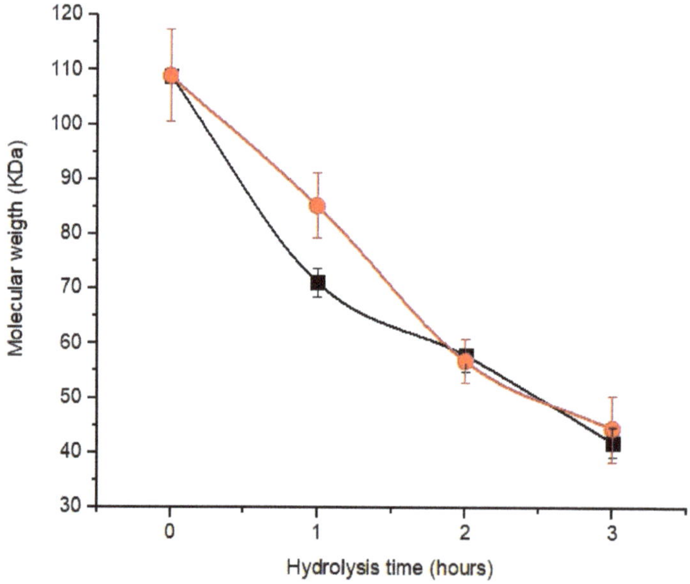

Figure 3. Chitosan hydrolysis profile to obtain oligomers as a function of reaction time and different enzyme concentrates: ■—C_E (commercial enzyme); •—C_{EC} (crude enzyme concentrate).

Although the enzyme concentrates showed similar responses during hydrolysis, the amount of protein in C_E (0.51 mg of protein/g of chitosan) was higher than that of C_{EC} (0.14 mg of protein/g of chitosan). However, after 2 h of hydrolysis, there was no significant difference ($p < 0.05$) between the samples.

Due to the lower concentration of enzymes present in the C_{EC} compared to the C_E, we can preliminarily infer that the C_{EC} showed greater efficiency in the hydrolysis of chitosan, highlighting the presence of endo-chitinases (E.C 3.2.1.14), exo-chitinases (E.C 3.2.1.52), and cellobiohydrolase (E.C 3.2.1.91) identified in this study.

3.6. Effect of Hydrolysis Time Using C_{EC}

Figure 4 presents the molecular weight of chitosan after different hydrolysis times using the C_{EC}. Chitosan with an initial molecular weight of 108.94 kDa showed a reduction of 47.80% (56.87 kDa) after 2 h of hydrolysis, 75.24% (26.86 kDa) after 5 h, and in 24 h, it was possible to reach a reduction of 93.26% (7.23 kDa). Thus, a significant decrease in molecular weight was observed, and thus, preliminarily a good correlation with the hydrolysis time to obtain the oligomers of interest.

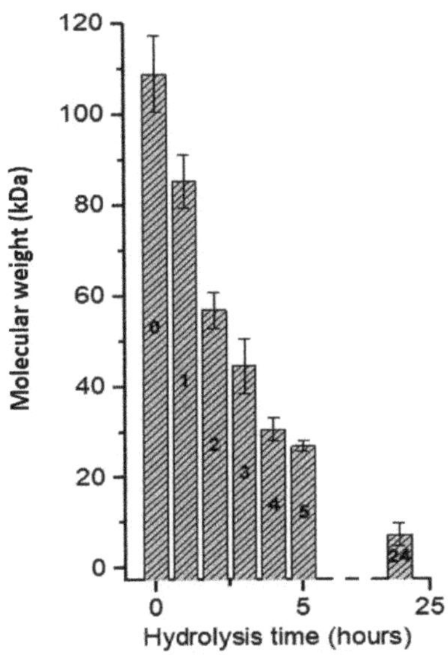

Figure 4. Chitosan hydrolysis as a function of reaction time using *A. nidulans* crude enzyme concentrate.

We believe that the efficiency of C_{EC} is due to the presence of enzymes that have hydrolytic action, especially for obtaining chitooligosaccharides [46,56,62,63].

3.7. FTIR Analysis

FTIR analyzed the chemical structure of chitosan and its oligomers through the absorption bands (Figure 5). A pattern between the spectra with different hydrolysis times was observed at the 3435 cm^{-1} peak (a) due to the O–H stretching vibration merged with the N–H stretching band. The 2870 cm^{-1} bands (b) were attributed to the elongation of the C–H groups [4].

The spectra also showed characteristic bands for amide groups, including the characteristic band of C=O (amide I) elongation at 1648 cm^{-1} in the chitosan spectrum, with a displacement of this peak in hydrolyzed chitosan to 1660 cm^{-1} (c). At 1589 cm^{-1} (d), it showed bending vibrations of N–H coupled to stretching vibrations of C–N (amide II), and at 1330 cm^{-1} (e), characteristic of amide III is observed [7]. Furthermore, absorption bands were observed at 1150 cm^{-1}, corresponding to the asymmetric stretching of the C–O–C bond (f). At 1067 and 1024 cm^{-1}, the bands correspond to the vibration involving the C–O stretching (g) [64] and at 893 cm^{-1} (h), there is the absorption β-1,4 glycosidic linkages [16].

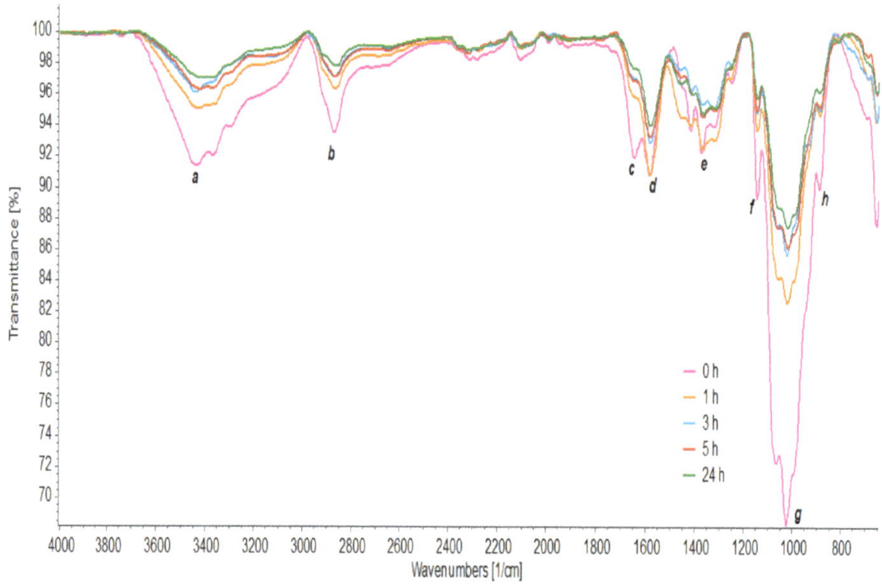

Figure 5. FT-IR spectrum of chitosan hydrolyzed with crude enzyme concentrate at times of 0 h, 1 h, 3 h, 5 h, and 24 h.

From the results, it is observed that the characteristics of chitosan were identified in the spectra, confirming its chemical identity before and after hydrolysis. In all spectra, the wavenumber values did not suffer significant displacements. However, other differences were observed between the spectra. Lower spectral intensities are observed in hydrolyzed chitosan because the functional groups vibrate with greater freedom of movement. The signal strength at 3465 cm^{-1} is due to more O–H groups, while the signal strength at 3364 cm^{-1} is associated with a more significant number of N–H units [4].

The absorption band at 3435 cm^{-1} shifted to a lower number of waves at 5 h and 24 h of hydrolysis, indicating that the crystalline order of chitosan was altered [16]. Furthermore, the absorption intensity relative to the C–H stretching band at 1380 cm^{-1} decreased at times of 3 h, 5 h, and 24 h, which indicated that intermolecular and intramolecular hydrogen bonds were weakened and its crystallinity was reduced [64].

3.8. Deacetylation Degree of Chitosan (DG)

The DG was determined using the FTIR spectra through the relationship between the integration values of the peaks at 1350 cm^{-1} and 1465 cm^{-1}, according to Equation (8). This interaction showed the highest linear correlation compared to others in the infrared spectrum for different crustacean biopolymers [34]; Table 3 shows the DG for different hydrolysis times.

Table 3. Degree of deacetylation about hydrolysis time using crude enzyme concentrate.

Hydrolysis Time	Integrated Area (A_{1345})	Integrated Area (A_{1460})	DG (%)	CV (%)
Without hydrolysis	1.051	1.062	80	0.49
1 h	1.030	1.154	83	3.23
3 h	1.175	1.150	79	1.74
5 h	0.991	1.010	80	0.49
24 h	1.220	1.991	80	0.49

A_{1345}: integrated area value for the 1345 cm^{-1} peak; A_{1460}: integrated area value for the 1460 cm^{-1} peak; DG: degree of deacetylation; and CV: coefficient of sampling variation.

Table 3 shows a shift of 5 cm^{-1} in the values at the peaks of the integration areas. Thus, the values of 1345 cm^{-1} and 1460 cm^{-1} were considered and not 1350 cm^{-1} and 1465 cm^{-1} as recommended in Equation (8). Variations of this magnitude are likely to occur due to the degree of purity of the commercial sample compared to a standard sample. However, this fact did not influence the vibrational characteristics of the functional groups.

The DG values show no variation in polymer deacetylation when comparing chitosan before and after hydrolysis. This result confirms the idea of the specificity of enzymatic hydrolysis, in this case, in glycosidic bonds against conventional chemical hydrolysis with the possibility of obtaining unwanted values of GD.

4. Conclusions

The fungus used to produce C_{EC} was identified as *Aspergillus* sp. (GenBank ID accession number MT135987), and through metaproteomic analysis, the microorganism was identified as *Aspergillus nidulans*.

The enzyme concentrate produced by the fungus showed enzymes identified as cellobiohydrolases, chitinases, alpha-L-arabinofuranosidase, and pectin lyase. These enzymes, belonging to the glycosyl hydrolase class, show activity in the hydrolysis of chitosan, reported in other studies.

The C_{EC} showed greater efficiency in chitosan hydrolysis than the commercial enzyme. This is because, although the commercial cellulase has certified efficiency in the hydrolysis of cellulose, the concentrate may have presented a synergistic effect that catalyzed the hydrolysis of the inner and outer chain of chitosan.

The oligomers obtained by hydrolysis using C_{EC} showed lower absorbance of spectral signals due to the functional groups vibrating with greater freedom of movement. In addition, their crystallinity was reduced, and their degree of deacetylation was maintained.

The C_{EC} showed efficiency in the hydrolysis of chitosan to obtain oligomers of interest with few steps and lower costs.

Author Contributions: Conceptualization, C.G.e.G. and N.R.F.; funding acquisition, N.R.F. and L.d.F.H.L.; investigation, C.G.e.G.; methodology, C.G.e.G. and H.K.P.; project administration, L.d.F.H.L.; resources, A.S.S.; supervision, N.R.F.; validation, A.S.S. and H.K.P.; visualization, C.G.e.G.; writing—original draft, C.G.e.G.; writing—review and editing, L.N.d.S., H.K.P. and N.R.F. All authors have read and agreed to the published version of the manuscript.

Funding: The publication in this journal was funded by Pró-Reitoria de Pesquisa e Pós-Graduação/UFPA (PROPESP/UFPA, Edital PAPQ 02/2023).

Institutional Review Board Statement: Not applicable.

Informed Consent Statement: Not applicable.

Data Availability Statement: The data presented in this study are available on request from the corresponding author.

Acknowledgments: The authors acknowledge the Programa de Pós-graduação em Ciência e Tecnologia de Alimentos (PPGCTA/UFPA) and Pró-Reitoria de Pesquisa e Pós-Graduação (PROPESP/UFPA) for providing the infrastructure and publication support.

Conflicts of Interest: The authors declare no conflict of interest.

References

1. Hamed, I.; Özogul, F.; Regenstein, J.M. Industrial applications of crustacean by-products (chitin, chitosan, and chitooligosaccharides): A review. *Trends Food Sci. Technol.* **2016**, *48*, 40–50. [CrossRef]
2. Naveed, M.; Phil, L.; Sohail, M.; Hasnat, M.; Baig, M.M.F.A.; Ihsan, A.U.; Shumzaid, M.; Kakar, M.U.; Khan, T.M.; Akabar, M.D.; et al. Chitosan oligosaccharide (COS): An overview. *Int. J. Biol. Macromol.* **2019**, *129*, 827–843. [CrossRef] [PubMed]
3. Santoso, J.; Adiputra, K.C.; Soerdirga, L.C.; Tarman, K. Effect of acetic acid hydrolysis on the characteristics of water soluble chitosan. *IOP Conf. Ser. Earth Environ. Sci.* **2020**, *414*, 012021. [CrossRef]
4. Águila-Almanza, E.; Salgado-Delgado, R.; Vargas-Galarza, Z.; García-Hernández, E.; Hernández-Cocoletzi, H. Enzymatic depolimerization of chitosan for the preparation of functional membranes. *J. Chem.* **2019**, *2019*, 5416297. [CrossRef]

5. Laokuldilok, T.; Potivas, T.; Kanha, N.; Surawang, S.; Seesuriyachan, P.; Wangtueai, S.; Phimolsiripol, Y.; Regenstein, J.M. Physicochemical, antioxidant, and antimicrobial properties of chitooligosaccharides produced using three different enzyme treatments. *Food Biosci.* **2017**, *18*, 28–33. [CrossRef]
6. Yin, N.; Du, R.; Zhao, F.; Han, Y.; Zhou, Z. Characterization of antibacterial bacterial cellulose composite membranes modified with chitosan or chitooligosaccharide. *Carbohydr. Polym.* **2020**, *229*, 4. [CrossRef]
7. Xu, W.; Mohan, A.; Pitts, N.L.; Udenigwe, C.; Mason, B. Bile acid-binding capacity of lobster shell-derived chitin, chitosan and chitooligosaccharides. *Food Biosci.* **2020**, *33*, 100476. [CrossRef]
8. Chang, S.-H.; Wu, C.-H.; Tsai, G.-J. Effects of chitosan molecular weight on its antioxidant and antimutagenic properties. *Carbohydr. Polym.* **2018**, *18*, 1026–1032. [CrossRef]
9. Chang, Y.-W.; Zeng, X.Y.; Sung, W.-C. Effect of chitooligosaccharide and different low molecular weight chitosans on the formation of acrylamide and 5-hydroxymethylfurfural and Maillard reaction products in glucose/fructose-asparagine model systems. *LWT* **2020**, *119*, 108879. [CrossRef]
10. Qin, C.; Zhou, B.; Zeng, L.; Zhang, Z.; Liu, Y.; Du, Y.; Xiao, L. The physicochemical properties and antitumor activity of cellulase-treated chitosan. *Food Chem.* **2004**, *84*, 107–115. [CrossRef]
11. Xing, R.; Liu, Y.; Li, K.; Yu, H.; Liu, S.; Yang, Y.; Chen, X.; Li, P. Monomer composition of chitooligosaccharides obtained by different degradation methods and their effects on immunomodulatory activities. *Carbohydr. Polym.* **2017**, *157*, 1288–1297. [CrossRef]
12. Kim, S.-K.; Rajapakse, N. Enzymatic production and biological activities of chitosan oligosaccharides (COS): A review. *Carbohydr. Polym.* **2005**, *62*, 357–368. [CrossRef]
13. Sánchez, Á.; Mengíbar, M.; Rivera-Rodríguez, G.; Moerchbacher, B.; Acosta, N.; Heras, A. The effect of preparation processes on the physicochemical characteristics and antibacterial activity of chitooligosaccharides. *Carbohydr. Polym.* **2017**, *157*, 251–257. [CrossRef]
14. Li, J.; Du, Y.; Yang, J.; Feng, T.; Li, A.; Chen, P. Preparation and characterisation of low molecular weight chitosan and chito-oligomers by a commercial enzyme. *Polym. Degrad. Stab.* **2005**, *87*, 441–448. [CrossRef]
15. Aranaz, I.; Alcántara, A.R.; Civera, M.C.; Arias, C.; Elorza, B.; Caballero, A.H.; Acosta, N. Chitosan: An overview of its properties and applications. *Polymers* **2021**, *13*, 3256. [CrossRef]
16. Li, M.; Han, J.; Xue, Y.; Dai, Y.; Liu, J.; Gan, L.; Xie, R.; Long, M. Hydrogen peroxide pretreatment efficiently assisting enzymatic hydrolysis of chitosan at high concentration for chitooligosaccharides. *Polym. Degrad. Stab.* **2019**, *164*, 177–186. [CrossRef]
17. Pan, A.-D.; Zeng, H.-Y.; Foua, G.B.; Alain, C.; Li, Y.-Q. Enzymolysis of chitosan by papain and its kinetics. *Carbohydr. Polym.* **2016**, *135*, 199–206. [CrossRef]
18. Roncal, T.; Oviedo, A.; Armentia, I.L.; Fernández, L.; Villarán, M.C. High yield production of monomer-free chitosan oligosaccharides by pepsin catalyzed hydrolysis of a high deacetylation degree chitosan. *Carbohydr. Res.* **2007**, *342*, 2750–2756. [CrossRef] [PubMed]
19. Santos-Moriano, P.; Kidibule, P.E.; Alleyne, E.; Ballesteros, A.O.; Heras, A.; Fernandez-Lobato, M.; Plou, F.J. Efficient conversion of chitosan into chitooligosaccharides by a chitosanolytic activity from *Bacillus thuringiensis*. *Process Biochem.* **2018**, *73*, 102–108. [CrossRef]
20. Wang, H.; Cui, H.; Wang, X.; Lin, C.; Xia, S.; Hayat, K.; Hussain, S.; Tahir, M.U.; Zhang, X. Metal complexed-enzymatic hydrolyzed chitosan improves moisture retention of fiber papers by migrating immobilized water to bound state. *Carbohydr. Polym.* **2020**, *235*, 115967. [CrossRef]
21. Sirohi, R.; Singh, A.; Tarafdar, A.; Shahi, N.C.; Verma, A.K.; Kushwaha, A. Cellulase production from pre-treated pea hulls using *Trichoderma reesei* under submerged fermentation. *Waste Biomass Valoriz.* **2019**, *10*, 2651–2659. [CrossRef]
22. Mrudula, S.; Murugammal, R. Production of cellulase by *Aspergillus Niger* under submerged and solid state fermentation using coir waste as a substrate. *Braz. J. Microbiol.* **2011**, *42*, 1119–1127. [CrossRef]
23. Ferreira, N.R.; Sarquis, M.I.M.; Gobira, R.M.; Souza, M.G.; Santos, A.S. The multivariate statistical selection of fungal strains isolated from *Neoteredo reynei*, with the high hydrolytic potential to deconstruct celulose. *Food Res. Int.* **2019**, *122*, 402–410. [CrossRef]
24. Chan, L.G.; Cohen, J.L.; Bell, J.M. Conversion of agricultural streams and food-processing by-products to value-added compounds using filamentous fungi. *Annu. Rev. Food Sci. Technol.* **2018**, *9*, 503–523. [CrossRef] [PubMed]
25. Chen, Y.C.; Eisner, J.D.; Kattar, M.M.; Rassoulian-Barrett, S.L.; Lafe, K.; Bui, U.; Limaye, A.P.; Cookson, B.T. Polymorphic internal transcribed spacer region 1 DNA sequences identify medically important yeasts. *J. Clin. Microbiol.* **2001**, *39*, 4042–4051. [CrossRef]
26. Peláez, R.D.R.; Wischral, D.; Cunha, J.R.B.; Mendes, T.D.; Pacheco, T.F.; Siqueira, F.G.d.; Almeida, J.R.M.d. Production of Enzymatic Extract with High Cellulolytic and Oxidative Activities by Co-Culture of Trichoderma reesei and Panus lecomtei. *Fermentation* **2022**, *8*, 522. [CrossRef]
27. Bradford, M.M. A rapid and sensitive method for the quantitation of microgram quantities of protein utilizing theprincipleof protein-dyebinding. *Anal. Biochem.* **1976**, *72*, 248–254. [CrossRef]
28. Miller, G.L. Use of dinitrosalicylic acid reagent for determination of reducing sugar. *Anal. Chem.* **1959**, *31*, 426–428. [CrossRef]
29. Wessel, D.; Flugge, U.I. A method for the quantitative recovery of protein in dilute solution in the presence of detergents and lipids. *Anal. Biochem.* **1984**, *138*, 141–143. [CrossRef]
30. Chapman, J.R. (Ed.) *Mass Spectrometry of Proteins and Peptides*; Humana Press: Totowa, NJ, USA, 2000.

31. Garcia, L.G.; Guedes, G.M.; Silva, M.L.; Castelo-Branco, D.S.; Sidrim, J.J.; Cordeiro, R.; Rocha, M.F.; Vieira, R.S.; Brilhante, R.S. Effect of the molecular weight of chitosan on its antifungal activity against Candida spp. in planktonic cells and biofilm. *Carbohydr. Polym.* **2018**, *195*, 662–669. [CrossRef]
32. Barragán, C.A.R.; Fornué, E.D.; Ortega, J.A.A. Determinación del grado de desacetilación de quitosana mediante titulación potenciométrica, FTIR y Raman. *J. CIM* **2016**, *4*, 769–777.
33. Santos, L.N.; Santos, A.S.; Dantas, K.d.G.F.; Ferreira, N.R. Adsorption of Cu (II) Ions Present in the Distilled Beverage (Sugar Cane Spirit) Using Chitosan Derived from the Shrimp Shell. *Polymers* **2022**, *14*, 573. [CrossRef]
34. Brugnerotto, J.; Lizardi, J.; Goycoolea, F.M.; Argüelles-Monal, W.; Desbrières, J.; Rinaudo, M. An infrared investigation in relation with chitin and chitosan characterization. *Polymer* **2001**, *42*, 3569–3580. [CrossRef]
35. Xie, Y.; Wei, Y.; Hu, J.G. Depolymerization of Chitosan with a Crude Cellulase Preparation from *Aspergillus niger*. *Appl. Biochem. Biotechnol.* **2010**, *160*, 1074–1083. [CrossRef]
36. Kittur, F.S.; Kumar, A.B.V.; Varadaraj, M.C.; Tharanathan, R.N. Chitooligosaccharides-preparation with the aid of pectinase isozyme from *Aspergillus niger* and their antibacterial activity. *Carbohydr. Res.* **2005**, *340*, 1239–1245. [CrossRef] [PubMed]
37. Liu, Z.; Gay, L.M.; Tuveng, T.R.; Agger, J.W.; Westereng, B.; Mathiesen, G.; Horn, S.J.; Vaaje-Kolstad, G.; van Aalten, D.M.F.; Eijsink, V.G.H. Structure and function of a broad-specificity chitin deacetylase from *Aspergillus nidulans* FGSC A4. *Sci. Rep.* **2017**, *7*, 1746. [CrossRef]
38. Cheng, C.-Y.; Chang, C.-H.; Wu, Y.-J.; Li, Y.-K. Exploration of glycosyl hydrolase family 75, a chitosanase from *Aspergillus fumigatus*. *J. Biol. Chem.* **2006**, *28*, 3137–3144. [CrossRef] [PubMed]
39. Poshina, D.N.; Raik, S.V.; Poshin, A.N.; Skorik, Y.A. Accessibility of chitin and chitosan in enzymatic hydrolysis: A review. *Polym. Degrad. Stab.* **2018**, *156*, 269–278. [CrossRef]
40. Su, P.C.; Hsueh, W.C.; Chang, W.S.; Chen, P.T. Enhancement of chitosanase secretion by Bacillus subtilis for production of chitosan oligosaccharides. *J. Taiwan Inst. Chem. Eng.* **2017**, *79*, 49–54. [CrossRef]
41. Thadathil, N.; Velappan, S.P. Recent developments in chitosanase research and its biotechnological applications: A review. *Food Chem.* **2014**, *150*, 392–399. [CrossRef]
42. Johnsen, M.G.; Hansen, O.C.; Stougaard, P. Isolation, characterization and heterologous expression of a novel chitosanase from Janthinobacterium sp. strain 4239. *Microb. Cell Factories* **2010**, *9*, 1–9. [CrossRef] [PubMed]
43. da Silva, L.C.; Honorato, T.L.; Franco, T.T.; Rodrigues, S. Optimization of Chitosanase Production by *Trichoderma koningii* sp. Under Solid-State Fermentation. *Food Bioprocess Technol.* **2012**, *5*, 1564–1572. [CrossRef]
44. Xia, W.; Yu, X. Purification and characterization of two types of chitosanase from *Aspergillus* sp. CJ22-326. *Food Res. Int.* **2005**, *38*, 315–322. [CrossRef]
45. Embaby, A.M.; Melika, R.R.; Hussein, A.; El-Kamel, A.H.; Marey, H.S. Biosynthesis of chitosan-Oligosaccharides (COS) by nonaflatoxigenic *Aspergillus* sp. strain EGY1 DSM 101520: A robust biotechnological approach. *Process Biochem.* **2018**, *64*, 16–30. [CrossRef]
46. Qu, T.; Zhang, C.; Qin, Z.; Fan, L.; Jiang, L.; Zhao, L. A novel GH family 20 β-N-acetylhexosaminidase with both chitosanase and chitinase activity from *Aspergillus oryzae*. *Front. Mol. Biosci.* **2021**, *8*, 684086. [CrossRef]
47. Kidibule, P.E.; Santos-Moriano, P.; Plou, F.J.; Fernández-Lobato, M. Endo-chitinase Chit33 specificity on different chitinolytic materials allows the production of unexplored chitooligosaccharides with antioxidant activity. *Biotechnol. Rep.* **2020**, *27*, e00500. [CrossRef] [PubMed]
48. Kim, S.; Matsuo, I.; Ajisaka, K.; Nakajima, H.; Kitamoto, K. Cloning and characterization of the nagA gene that encodes beta-n-acetylglucosaminidase from *Aspergillus nidulans* and its expression in Aspergillus oryzae. *Biosci. Biotechnol. Biochem.* **2002**, *66*, 2168–2175. [CrossRef]
49. Patil, N.S.; Jadhav, J.P. Significance of *Penicillium ochrochloron* chitinase as a biocontrol agent against pest *Helicoverpa armigera*. *Chemosphere* **2015**, *128*, 231–235. [CrossRef] [PubMed]
50. Han, P.; Yang, C.; Liang, X.; Li, L. Identification and characterization of a novel chitinase with antifungal activity from 'Baozhu' pear (*Pyrus ussuriensis* Maxim.). *Food Chem.* **2016**, *196*, 808–814. [CrossRef]
51. Le, B.; Yang, S.H. Microbial chitinases: Properties, current state and biotechnological applications. *World J. Microbiol. Biotechnol.* **2019**, *35*, 144. [CrossRef]
52. Purushotham, P.; Podile, A.R. Synthesis of Long-Chain Chitooligosaccharides by a Hypertransglycosylating Processive Endochitinase of *Serratia proteamaculans* 568. *J. Bacteriol.* **2016**, *194*, 4260–4271. [CrossRef]
53. Al Abboud, M.A.; Al-Rajhi, A.M.H.; Shater, A.-R.M.; Alawlaqi, M.M.; Mashraqi, A.; Selim, S.; Al Jaouni, S.K.; Ghany, T.M.A. Halostability and Thermostability of Chitinase Produced by Fungi Isolated from Salt Marsh Soil in Subtropical Region of Saudi Arabia. *BioResources* **2022**, *17*, 4763–4780. [CrossRef]
54. Abdel-Moneim, A.M.E.; El-Saadony, M.T.; Shehata, A.M.; Saad, A.M.; Aldhumri, S.A.; Ouda, S.M.; Mesalam, N.M. Antioxidant and antimicrobial activities of *Spirulina platensis* extracts and biogenic selenium nanoparticles against selected pathogenic bacteria and fungi. *Saudi J. Biol. Sci.* **2022**, *29*, 1197–1209. [CrossRef] [PubMed]
55. Saykhedkar, S.; Ray, A.; Ayoubi-Canaan, P.; Hartson, S.D.; Prade, R.; Mort, A.J. A time course analysis of the extracellular proteome of *Aspergillus nidulans* growing on sorghum stover. *Biotechnol. Biofuels* **2012**, *5*, 52. [CrossRef] [PubMed]
56. Tegl, G.; Öhlknecht, C.; Vielnascher, R.; Rollett, A.; Hofinger-Horvath, A.; Kosma, P.; Guebitz, G.M. Cellobiohydrolases produce different oligosaccharides from chitosan. *Biomacromolecules* **2016**, *17*, 2284–2292. [CrossRef] [PubMed]

57. Ma, J.-X.; Qian, L.; Zhou, Y. Stimulation effect of chitosan on the immunity of radiotherapy patients suffered from lung câncer. *Int. J. Biol. Macromol.* **2015**, *72*, 195–198. [CrossRef]
58. Squillaci, G.; Esposito, A.; La Cara, F.; Morana, A. Free and immobilized thermophilic alpha-L-arabinofuranosidase for arabinose production. *Arch. Ind. Biotechnol.* **2017**, *1*, 27–34.
59. Bibi, F.; Irshad, M.; Anwar, Z.; Bhatti, K.H.; Raza, A. Improved catalytic functionalities of purified pristine and chitosan-immobilized polygalacturonase, and pectin lyase. *Chem. Eng. Res. Des.* **2019**, *128*, 146–154. [CrossRef]
60. Kittur, F.S.; Kumar, A.B.V.; Tharanat Purushotham and Podile, R.N. Low molecular weight chitosans/preparation by depolymerization with *Aspergillus niger* pectinase, and characterization. *Carbohydr. Res.* **2003**, *33*, 1283–1290. [CrossRef]
61. Tishchenko, G.; Šimůnek, J.; Brus, J.; Netopilík, M.; Pekárek, M.; Walterová, Z.; Koppová, I.; Lenfeld, J. Low-molecular-weight chitosans: Preparation and characterization. *Carbohydr. Polym.* **2011**, *86*, 1077–1081. [CrossRef]
62. Lin, S.B.; Lin, Y.C.; Chen, H.H. Low molecular weight chitosan prepared with the aid of cellulase, lysozyme and chitinase: Characterisation and antibacterial activity. *Food Chem.* **2009**, *116*, 47–53. [CrossRef]
63. Beer, B.; Bartolome, M.J.; Berndorfer, L.; Bochmann, G.; Guebitz, G.M.; Nyanhongo, G.S. Controlled enzymatic hydrolysis and synthesis of lignin cross-linked chitosan functional hydrogels. *Int. J. Biol. Macromol.* **2020**, *161*, 1440–1446. [CrossRef] [PubMed]
64. Luo, W.; Han, Z.; Zeng, X.; Yu, S.; Kennedy, J.F. Study on the degradation of chitosan by pulsed electric fields treatment. *Innov. Food Sci. Emerg. Technol.* **2010**, *11*, 587–591. [CrossRef]

polymers

MDPI

Article

Polyols and Polyurethane Foams Based on Water-Soluble Chitosan

Anna Maria Strzałka [1] and Jacek Lubczak [2,*]

1 Doctoral School of Engineering and Technical Sciences, Rzeszow University of Technology, Al. Powstańców Warszawy 6, 35-959 Rzeszów, Poland
2 Faculty of Chemistry, Rzeszów University of Technology, Al. Powstańców Warszawy 6, 35-959 Rzeszów, Poland
* Correspondence: jml@prz.edu.pl

Abstract: At present, majority of polyols used in the synthesis of polyurethane foams are of petrochemical origin. The decreasing availability of crude oil imposes the necessity to convert other naturally existing resources, such as plant oils, carbohydrates, starch, or cellulose, as substrates for polyols. Within these natural resources, chitosan is a promising candidate. In this paper, we have attempted to use biopolymeric chitosan to obtain polyols and rigid polyurethane foams. Four methods of polyol synthesis from water-soluble chitosan functionalized by reactions of hydroxyalkylation with glycidol and ethylene carbonate with variable environment were elaborated. The chitosan-derived polyols can be obtained in water in the presence of glycerol or in no-solvent conditions. The products were characterized by IR, ^{1}H-NMR, and MALDI-TOF methods. Their properties, such as density, viscosity, surface tension, and hydroxyl numbers, were determined. Polyurethane foams were obtained from hydroxyalkylated chitosan. The foaming of hydroxyalkylated chitosan with 4,4′-diphenylmethane diisocyanate, water, and triethylamine as catalysts was optimized. The four types of foams obtained were characterized by physical parameters such as apparent density, water uptake, dimension stability, thermal conductivity coefficient, compressive strength, and heat resistance at 150 and 175 °C. It has been found that the obtained materials had most of the properties similar to those of classic rigid polyurethane foams, except for an increased thermal resistance up to 175 °C. The chitosan-based polyols and polyurethane foams obtained from them are biodegradable: the polyol is completely biodegraded, while the PUF obtained thereof is 52% biodegradable within 28 days in the soil biodegradation oxygen demand test.

Keywords: water-soluble chitosan; hydroxyalkylation; polyols; polyurethane foams

1. Introduction

At present, most polyurethane foams (PUF) are obtained from petrochemical substrates [1–3]. The fact that petroleum resourcesare decreasing makes it necessary to look for resources of a biological origin. Polyols based on renewable natural resources are easily available, comparatively not expensive, and biodegradable. The constant increase of foam products on one hand and environmental requirements on the other hand imposes on the searchfor biodegradable polyols from plant oils [4,5] or carbohydrates such asstarch and cellulose [6–11]. Chitosan also belongs to the latter group of compounds and was not used untilnow to obtain polyols. Chitosan can be obtained by D-deacetylation of chitin (Scheme 1). Chitin is apolymer that is present in marine invertebrates and in the skeleton material in crustaceans and insects [12].

Chitosan can be considered as a natural resource because this polymer is the building material of the fungal cell wall. It can be easily isolated by extraction. Chitosans are a group of polymers of variable degrees of molecular weight and deacetylation. These physicochemical characteristics of fungal chitosan such as molecular weight and degree of degradation can be better controlled compared to chitosan obtained from crustacean sources [13]. Fungal chitosan has many advantages in biomedical applications due to its molecular properties. It is easier to obtain very low or high molecular mass polymer from fungal chitosan compared to shellfish chitosan. Fungal chitosan can be used as a potential carrier of drug and non-viral gene delivery systems [13].

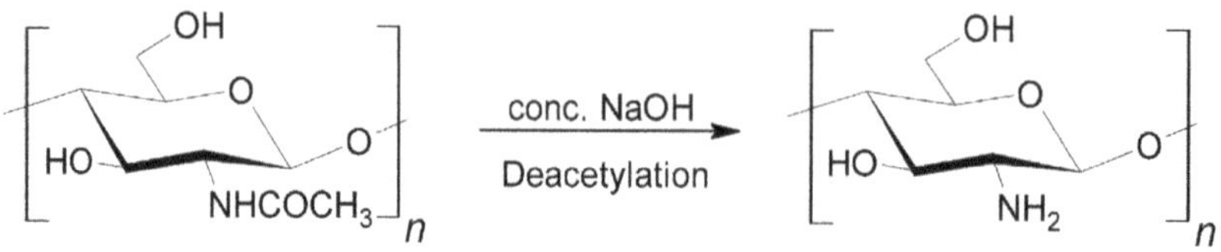

Scheme 1. Deacetylation of chitin.

The chitosan chain comprises the units of β-(1,4)-D-glucoseamine and N-acetyl-D-glucoseamine (Scheme 2) [14]. The distributions of deacetylated and acetylated subunits within the polymeric chain influence the chemical properties of chitosan. There are various kinds of chitosan commercially available, which differ in molecular weight and deacetylation degree. Generally, chitosan is a biodegradable, biocompatible, and non-toxic polymer with some antigen properties [15]. It is used in tissue engineering and as an ingredient of drug delivery systems, diet supplements, cosmetics, in plant cultivation, and in environmental technologies [16–20], and also to separate dyes and heavy-metal ions [21,22].

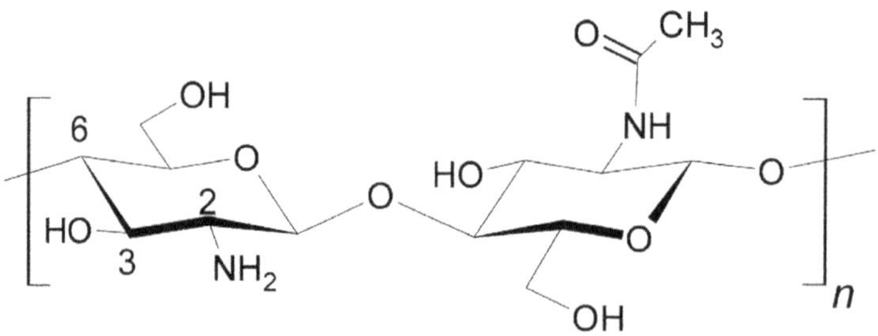

Scheme 2. Structural formula of chitosan.

Chitosan used in industry is not soluble in water, although it dissolves in aqueous solutions of organic acids such as formic, acetic, and citric at pH levels below 6.5 [12]. In order to convert chitosan, the amine groups at C-2 and/or primary and secondary hydroxyl groups at C-4 and C-6, respectively, can be used. Thus O- and N-functionalized derivatives with carboxymethyl, acyl, sulfone, and alkyl derivatives were obtained as well as coordination compounds with metal ions [23]. Due to the lack of systemic toxicity, biocompatibility, and facile biodegradation, many derivatives of these polymers have found their application in medicine, tissue engineering, and pharmacy [24–27]. In recent years, chitosan-based material shave found a further application in skin tissue engineering [28]. Furthermore, the chitosans with free amine groups are applicable in wastewater treatment [24]; for example, binary coagulation system graphene oxide/chitosan was used for polluted surface water treatment [29,30].

Amine and hydroxyl groups of chitosans provide a path to obtain polyols suitable for further use to obtain polyurethane foams (PUF) in a similar way as starch or cellulose derivatives [31,32]. Chitosan-derived polyols were not reported, except for in [33], although hydroxyalkylation of chitosan methods were elaborated [33–35]. Additionally, chitosan-filled PUFs were described. Thus, the hydroxyalkylation of chitosan in 15% aqueous NaOH with ethylene, propylene, and butylene oxides, and then further reaction with oxirane, led to a hydroxyalkylated derivative of chitosan. This product was grafted on collagen or nisin to obtain the sorbent materials suitable for pharmaceutic and medical applications [35,36].

Similar hydroxyalkylation with glycidol was also reported [37]. Generally, the chitosan hydroxyalkylated with epoxides can be further substituted with carboxyl groups [38,39]. The application of chitosan to obtain polyols was proposed by Fernandes et al. [33]. They have activated chitin and chitosan with potassium hydroxide and treated them with propylene oxide to obtain polyol, which was suitable for further conversion to polyurethanes and polyesters. However, the obtained polyols had high viscosity, which rendered them immiscible with diisocyanate to obtain PUF. They were also contaminated with the catalyst. As mentioned earlier, the chitosan-derived polyols were not used as a component to obtain PUFs, although it has been shown that composites of PUF and chitosan were an elastomeric product. The PUFs with added chitosan were also biodegradable, which promised their application as sorbents and biomaterials [40–47].

In this paper, for the first time, chitosan was used to obtain polyols and rigid polyurethane foams. Four methods of polyol synthesis were elaborated, namely by hydroxyalkylation of water-soluble chitosan with glycidol and ethylene carbonate in water and in glycerol, as well as without a solvent. The new series of rigid polyurethane foams were obtained, which are susceptible for biodegradation.

2. Materials and Methods

2.1. Materials

The following materials were used in this work: water-soluble chitosan, degree of deacetylation, DD = 85.8%, viscosity molecular weight, Mv~14, 1250 Da (CS, Biosynth-Carbosynth, Staad, Switzerland), glycidol (GL, pure 98%, Sigma-Aldrich, Taufkirchen, Germany), ethylene carbonate (EC, pure $\geq$ 99%, Fluka, Buchs, Switzerland), potassium carbonate (anal. grade 100%, POCH, Gliwice, Poland), polymeric diphenylmethane 4,4′-diisocyanate (pMDI, Merck, Darmstadt, Germany), triethylamine (TEA, anal. grade $\geq$ 99%, Fluka, Buchs, Switzerland), surfactant Silicon L-6900 (pure, Momentive, Wilton, CT, USA), and glycerol (GLYC, anal. grade 99.5–100%, POCH, Gliwice, Poland).

2.2. Synthesis of Polyols

2.2.1. Synthesis 1: Polyol (CS + H_2O + GL) + EC

Here, 6 g of CS, 60 g of GL, and 45 g of water were placed in a three-necked round-bottom flask equipped with a reflux condenser, mechanical stirrer, and a thermometer. The mixture was heated at 105–110 °C until complete reaction of GL (determined by the epoxide number). Then, water was distilled under reduced pressure (p = 30 mm Hg, up to temperature 150 °C). The product was a clear gelatinous liquid (the semi-product was CS + H_2O + GL). To this semi-product, EC (75 g) and 0.5 g of potassium carbonate (catalyst) were added. The mixture was reheated up to 160 °C and maintained until all the EC was reacted.

2.2.2. Synthesis 2: Polyol (CS + GLYC + GL) + EC

Here, 6 g of CS, 60 g of GL, and 45 g of GLYC were placed in a three-necked round-bottom flask equipped with a reflux condenser, mechanical stirrer, and a thermometer. The mixture was heated at 160 °C until complete reaction of GL. Afterward, to this semi-product (CS + GL + GLYC), EC (105 g) and potassium carbonate (0.5 g) were added, and the mixture was heated at 180 °C until EC was completely consumed.

2.2.3. Synthesis 3: Polyol (CS + GL) + EC

CS (6 g) and GL (90 g) were placed in a three-necked round-bottom flask, equipped with a reflux condenser, mechanical stirrer, and a thermometer. The mixture was heated at 190 °C until complete reaction of GL. Then, to the semi-product (CS + GL), EC (135 g) and potassium carbonate (0.5 g) were added, and the mixture was heated at 170 °C.

2.2.4. Synthesis 4: Polyol (CS + GL + EC)

CS (6 g), GL (75 g), and EC (90 g) were placed in the flask and heated at 140–145 °C, until chitosan was dissolved. Then, the temperature was increased to 160 °C and the mixture was refluxed. During stepwise consumption of GL, the temperature increased to 190 °C and the mixture was maintained at this temperature for ca 0.5 h until GL was fully consumed. Then, the mixture was cooled down to 140 °C, potassium carbonate (1 g) was added, and the mixture was kept at 140 °C until EC was completely reacted.

2.3. *Analytical Methods*

The deacetylation degree of chitosans was determined according to the results of the elemental analysis, as it was described in [48]. Molecular mass was determined by the viscosimetric method at 30 °C using the Mark–Houvink equation, as described in [49]:

$$[\eta] = k\, M_{\upsilon}{}^{\alpha} \tag{1}$$

where: $[\eta]$ is intrinsic viscosity, M_{υ} is the viscosity-average mass weight, and k, α are constants that are characteristic for a particular polymer–solvent system at a specific temperature:

$$k = (1.64 \cdot 10^{-30}) \cdot (DD^{14})\ [cm^3/g] \tag{2}$$

$$\alpha = (-1.02 \cdot 10^{-2}) \cdot (DD) + 1.82 \tag{3}$$

where DD is the % degree of deacetylation.

The reaction of the mixture of CS with GL was monitored by epoxide number determination using hydrochloric acid in dioxane [50]. Specifically, 25 cm^3 of hydrochloric acid solution in dioxane (1.6 cm^3 in 100 cm^3 dioxane)was added into a 0.5 g mass sample. Excess of HCl was then titrated with 0.2 M NaOH in methanol in the presence of *o*-cresol red as an indicator. The progress of the reaction of hydroxyalkylation with EC was monitored using the barium hydroxide method described in [51]. In particular, the samples of 0.1–0.5 g of mass were treated with 2.5 cm^3 of 0.15 M $Ba(OH)_2$ and then titrated with 0.1 M HCl in the presence of 0.2% thymoloftalein in alcohol. Finally, the hydroxyl number (HN) of polyol was determined by acylation with acetate anhydride in dimethylformamide [52]. Thus, 1 g of sample was heated with a 20 cm^3 acetylating mixture (acetic anhydride and dimethylformamide at a 23:77 v:v ratio) for 1 h at 100 °C. Excess of anhydride was titrated with 1.5 M $NaOH_{aq}$ in the presence of phenolphthalein. The ^{1}H-NMR spectra of reagents were recorded using a500 MHz Bruker UltraShield instrument in DMSO-d_6 and D_2O with hexamethyldisiloxane as an internal standard. IR spectra were registered on an ALPHA FT-IR BRUKER spectrometer in KBr pellets or by the ATR technique. The samples were scanned 25 times, in the range from 4000 to 450 cm^{-1} at a 2 cm^{-1} resolution. MALDI-TOF (Matrix-Associated Laser Desorption Ionization Time of Flight) spectra of polyols were obtained on a Voyager-Elite Perceptive Biosystems (US) mass spectrometer working in linear mode with delayed ion extraction, equipped with a nitrogen laser working at 352 nm. The method of laser desorption from gold nanoparticles (AuNPET LDI MS) was applied [53]. The observed peaks corresponded to the molecular K^+ (from catalyst) ions. The samples were diluted with methanol to 0.5 mg/cm^3.

2.4. *Physical Properties of Polyol*

Density, viscosity, and surface tension of polyol were determined with a pycnometer, Höppler viscometer (type BHZ, Prüfgeratewerk, Germany), and by the detaching ring method, respectively.

2.5. *Polyurethane Foams*

Foaming of polyol was performed in500 cm^3 cups at room temperature. The foams were prepared from 10 g of polyol, to which 0.30–0.39 g of surfactant (Silicon L-6900) and 0.08–0.27 g of TEA as a catalyst and water (2–3%) as a blowing agent were added.

After homogenization, the polymeric diphenylmethane 4,4′-diisocyanate was added in the amount of 11.0–18.5 g. The commercial isocyanate containing 30% of tri-functional isocyanates was used. The mixture was vigorously stirred until creaming began. The materials were then conditioned at room temperature for 3 days. The samples for further studies were cut from the obtained foam.

2.6. Properties of Foams

The apparent density [54], water absorption [55], dimensional stability at 150 °C [56], thermal conductivity coefficient (IZOMET 2104, Bratislava, Slovakia), and compressive strength [57] of PUF were measured. The apparent density of PUFs was calculated as the ratio of PUF mass to the measured volume of the PUF sample in a cube of a 50 mm edge length. Water volume absorption was measured on cubic samples of 30 mm edge lengths by full immersion of PUF in water and mass measurement after 5 min, 3 h, and 24 h. Dimensional stability was tested on samples of 100 mm × 100 mm × 25 mm in size. The thermal conductivity coefficient was measured at 20 °C after 72 h of PUF conditioning. The needle was inserted 8 cm deep into a cylindrical PUF sample 8 cm in diameter and 9 cm high. Compressive strength was determined using burden causing 10% compression of PUF height related to the initial height (in accordance with the PUF growing direction). The thermal resistance of modified foams was determined by both static and dynamic methods. In the static method, the foams were heated at 150 and 175 °C with continuous measurement of mass loss and determination of mechanical properties before and after heat exposure. The 100 × 100 × 100 mm cubic samples were used to determine the static thermal resistance and compressive strength. In the dynamic method, thermal analyses of foams were performed in a ceramic crucible in a20–600 °C temperature range, with about 100 mg of sample, under air atmosphere with a Thermobalance TGA/DSC 1 derivatograph, Mettler, with a 10 °C/min heating rate. Topological pictures of PUFs were recorded for cross-sections of PUF samples. The pictures were analyzed with a Panthera microscope (prod. Motic, Wetzlar, Germany) with 4 × /0.13, 10 × /0.30 lenses and worked up with Motic Multi-Focus Professional 1.0 software, enabling merging and manipulation of images with adjustable lensing planes.

2.7. Biodegradation of Polyol and Foam

The biodegradation of polyol and the PUF obtained from it was tested by the use of the OxiTop Control S6 instrument (WTW-Xylem, Rye Brook, NY, USA). The respirometric method was used to measure the oxygen demand necessary for aerobic biodegradation of polymeric materials in soil. The measurement of consumed oxygen was presented using the value of biochemical oxygen demand (*BOD*), which is the number of milligrams of captured oxygen per mass unit of the tested polyurethane material. The instrument was composed of six 510 cm^3 glass bottles, equipped with rubber quiversand measuring heads, which were used to measure the *BOD*. They allowed to measure the pressure in the range of 500 to 1350 hPa with an accuracy of 1% at a temperature of 5 to 50 °C. The communication between the measuring heads and the user was performed with Achat OC computer software (WTW-Xylem, Rye Brook, NY, USA), which was applied to interpret the obtained measurement results.

The biodegradation tests were performed according to the norm [58]. For a biodegradation test, the sieved and dried gardening soil was used with the following parameters: 5% humidity (according to ISO 11274-2019 [59]), pH = 6 (according to ISO 10390-2005 [60]), and particle diameters < 2 nm. The measurement was carried out in a system consisting of 200 mg of the tested sample (oligomer or foam), 200 g of soil, and 100 g of distilled water. The samples were homogenized in bottled, rubber quivers containing two pastilles of solid NaOH, and were mounted and sealed with measuring heads for six samples. The set was incubated at 20 ± 0.2 °C for 28 days. The current oxygen consumption was determined within 2–3-day intervals for the samples and 2 references: positive and negative, plus a blank, which was the soil and water only. The starch was used as the positive sample, while polyethylene was the negative sample. BOD was determined for every sample, taking into account the *BOD* of the tested system reduced by the BOD of the soil and the concentration of the tested compound in the soil using the following formula:

$$BODs = \frac{BOD_x - BOD_g}{c} \quad (4)$$

where: *S*—number of measurements (in days), BOD_S—biochemical oxygen demand of the analyzed sample within S days (mg/dm^3), *BODx*—biochemical oxygen demand of the measuring system (bottle with sample and soil) (mg/dm^3), *BODg*—biochemical oxygen demand of the soil without a sample (mg/dm^3), and c—sample concentration in the tested system (mg/dm^3). The degree of biodegradation of the polyol or the foam based on it was determined using the formula:

$$D_t = \frac{BOD_S}{TOD} \cdot 100\% \quad (5)$$

where *Dt* is the biodegradation degree of the sample (%) and *TOD* is the theoretical oxygen demand (mg/dm^3).

The theoretical oxygen demand was calculated using the formula provided in norm ISO17556-2019 [59]. It has been assumed that in oxygen conditions, the carbon is converted into CO_2, hydrogen into H_2O, and nitrogen into NH_3.

For the compounds of known *C, H, N,* and *O* percentages and total mass of the sample, the *TOD* value can be calculated from the following equation:

$$TOD = \frac{16 \cdot [2C + 0.5 \cdot (H - 3N) - O]}{m} \quad (6)$$

where *C, H, N,* and *O* are the mass fractions of elements in the biodegraded material, and *m* is the sample mass of the material (g).

3. Results and Discussion

3.1. Obtaining of Polyols

The known methods of CS hydroxyalkylations with oxiranes are difficult to perform. Oxiranes are low boiling, toxic, and flammable liquids, cancerogenic, and form an explosive mixture with air. Therefore, using them requires high-pressure reactors. Common ways to convert CS require preliminary treatment of CS with NaOH in alcohol, followed by a reaction with an oxirane [33]. Side products are formed in the reaction of alcohols with alkylene oxides. Thus, in order to use CS as a substrate for PUF, it needs to obtain a semi-product: a liquid polyol suitable to react with diisocyanate. Chitosans of high molecular weight were not good candidates for such a conversion because of their low solubility. Therefore, the water-soluble CS was chosen for a reaction with GL. Our earlier experience [10,11] on hydroxyalkylation of starch and cellulose showed that those sparingly soluble polymers could be successfully hydroxyalkylated by preliminary heating the substrates in water with GL, which enabled to obtain a better soluble substrate which was further converted by a reaction with EC to obtain polyols. We applied the elaborated method to convert the CS in water with GL at a slightly elevated temperature (40 °C). CS itself is a product of chitin deacetylation in which the degree of deacetylation (DD) is 85%. That rendered the CS water-soluble and enabled to convert it in the reaction with GL. The reaction mixture was then gradually heated up to reflux (ca. 100 °C). The product analysis by determination of the epoxide number indicated that GL was consumed in the reaction with water to yield GLYC (Scheme 3).

The amount of water distilled from the reaction mixture was 26.4 mass % instead of theoretically 27.4 mass % if all the GL were to react with water. Thus, this semi-product contained GLYC, which was isolated after the initial removal of water, and identified by IR, refraction index, and MALDI-TOF of polyol (vide infra). We concluded that unreacted and dissolved CS was present in the post-reaction mixture. The obtained semi-product was semisolid resin, not miscible with isocyanates, and was then further liquefied by hydroxyalkylation with EC according to Scheme 4.

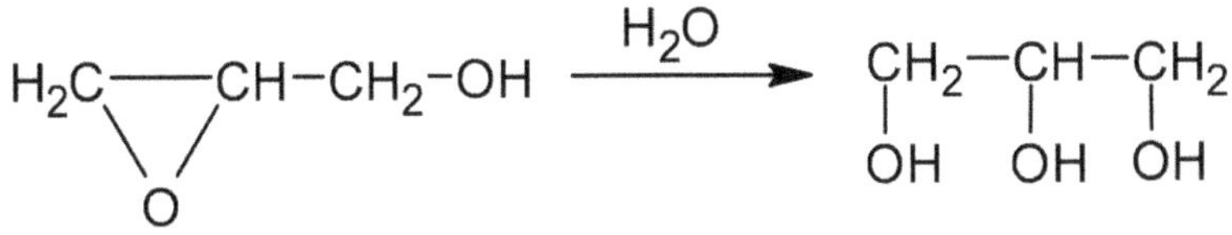

Scheme 3. The reaction of GL with water.

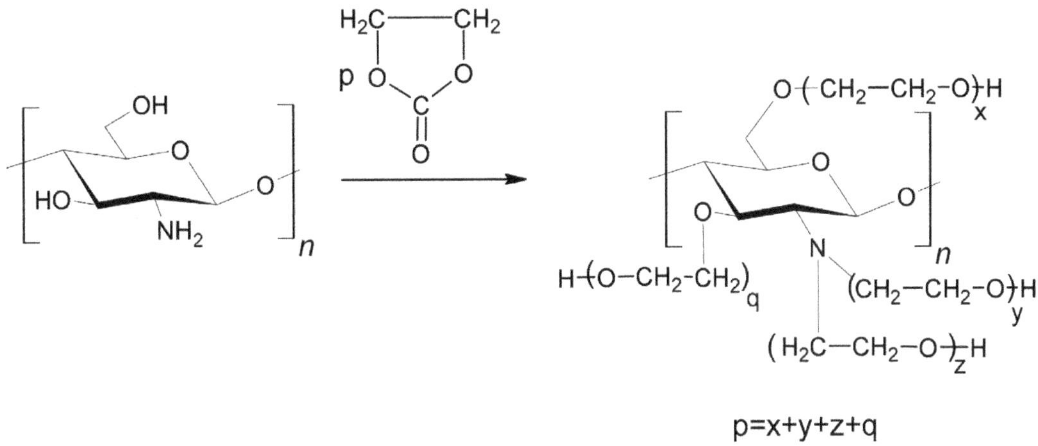

Scheme 4. Reaction of CS with EC.

The amount of EC was minimized in order to obtain polyol of low viscosity, miscible with diisocyanate, pMDI. Preliminary experiments of the direct reaction between CS and EC in the presence of the K_2CO_3 catalyst at 180 °C led to carbonization of the polymer. We also found that water was not a necessary solvent to dissolve CS, and GLYC could be used as a solvent for CS and hydroxyalkylation could be performed with GL, and further with EC. Thus, the way to obtain the polyol could be simplified due to avoiding water removal. A general scheme of chitosan hydroxyalkylation is presented in Scheme 5. This remains valid for other methods of polyol synthesis further described below.

Obtained liquid polyols contain side products, namely the products of the reaction between glycerol and glycidol (see Table 1), for example those in Scheme 6.

Further attempts indicated that the synthetic pathway could be simplified by straightforward hydroxyalkylation of CS with excess GL, without using GLYC. The obtained semi-product required consecutive hydroxyalkylation with EC in order to obtain a final product of low viscosity.

Finally, we performed the one-pot synthesis of polyol by introducing CS, GL, and EC into a reaction flask. We have previously found that CS does not react with EC without a catalyst. Thus, the EC acted as a solvent for CS, while GL reacted with CS. After consuming all GL, the catalytic amount of K_2CO_3 was added to trigger the reaction with EC.

All tested methods resulted in the formation of liquid polyols, miscible with pMDI.

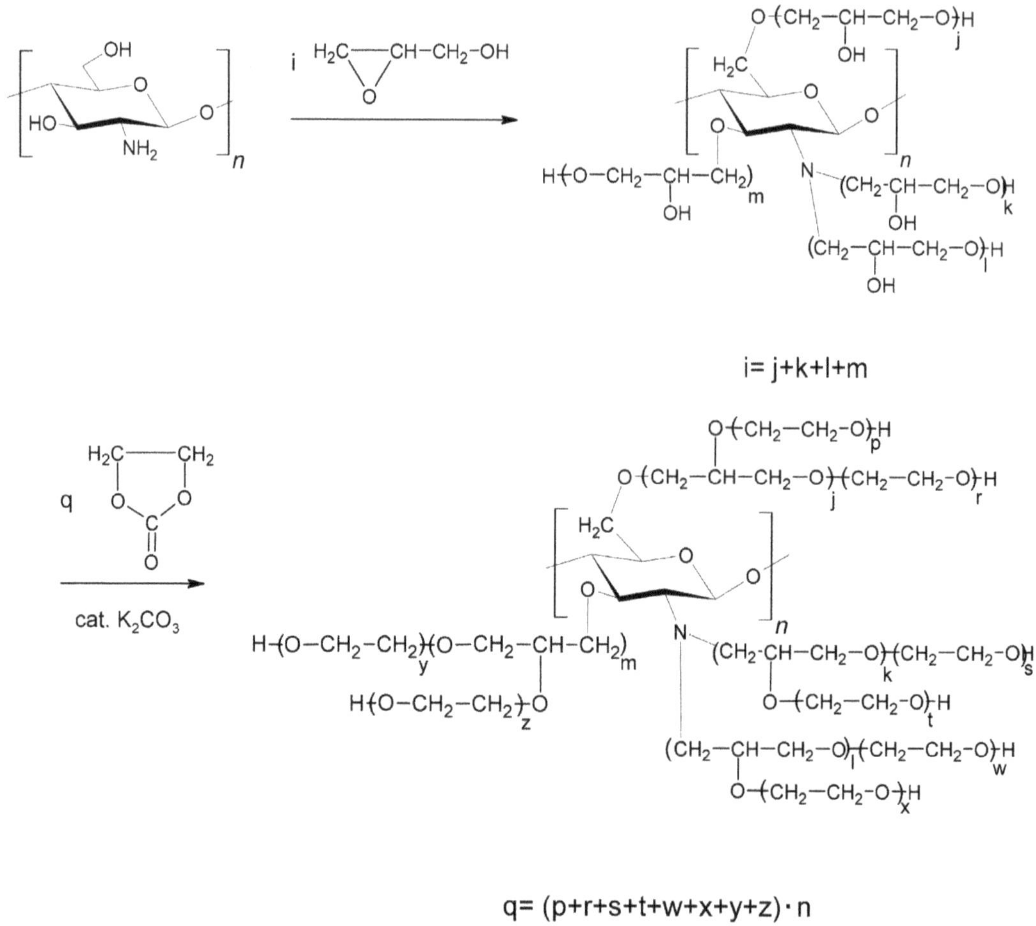

Scheme 5. Hydroxyalkylation of CS with GL and EC.

3.2. Composition and Structure of Polyols

The progress of the reaction was monitored by spectroscopes IR and ^{1}H-NMR and the MALDI-TOF technique. The spectra of obtained polyols were compared with those of the starting CS (Figure 1). The IR spectrum of CS showed a broad band centered at 3400 cm^{-1}from hydroxyl and amine groups' stretching vibrations, as well as deformation bands at 1420 cm^{-1} and 1630 cm^{-1}, respectively. The band centered at ca. 1030 cm^{-1} was attributed to valence of ether C-O-C vibrations. The presence of acetylamine groups was demonstrated by I and II amide bands at 1630 cm^{-1} and 1520 cm^{-1} (overlapped with the amine deformation band of chitosan). The IR spectra of the obtained polyols are presented in Figure 2. The IR spectra of all polyols were similar due to the similarity of the chemical structural fragments. The increase in intensity of the C-O-C band at 1030 cm^{-1} was observed as well as methine and methylene bands (2900 cm^{-1}, 1400–1300 cm^{-1}), which derive from glycidol ring opening and the incorporation into polyol. The presence of a carbonyl band at 1750 cm^{-1} indicated that ester bonds are present in polyol, especially those obtain data temperature lower than 180 °C. In such conditions, carbonate groups are able to incorporate into the polyol structure, as it was observed before in [51].

Table 1. Interpretation of MALDI-TOF spectrum of polyol (CS + GL) + EC.

Entry	Signal Position, m/z	Relative Intensity of Signal (%)	Molecular Ion Structure	Calc. Molecular Weight (g/mol)
1	96.929	43.84	$GL + Na^+$	97.027
2	112.895	100.00	$GL + K^+$	113.000
3	114.898	44.54	$GL + H_2O + Na^+$	115.037
4	119.934	11.95	$GL + OE + H^+$	119.071
5	196.334	30.58	$GL + 2OE + CH_3OH$	194.115
6	210.700	4.12	$2GL + H_2O + OE$	210.110
7	230.217	6.74	$2GL + OE + K^+$	231.063
8	257.024	9.33	$2GL + 2OE + Na^+$	259.116
9	274.287	7.92	$2GL + 2OE + K^+$	275.090
10	287.078	8.44	$3GL + OE\text{-}H_2O + K^+$	287.090
11	305.108	15.14	$3GL + OE + K^+$	305.100
12	333.149	11.86	$3GL + 2OE + Na^+$	333.153
13	349.134	23.33	$H_2O + 3GL + 2OE + Na^+$	351.163
14	363.142	13.82	$4GL + OE + Na^+$	363.163
15	393.992	16.36	$2GL + 6OE - H_2O$	394.220
16	473.184	9.06	$6GL + OE - H_2O + H^+$	471.244
17	590.878	16.40	$7GL + 2OE - H_2O + H^+$	589.307
18	627.395	4.04	$7GL + 2OE + Na^+$	629.300
19	671.165	3.40	$7GL + 3OE + Na^+$	673.326
20	707.320	3.34	$8GL + 3OE - H_2O + H^+$	707.370

Description of abbreviations used in the table: GL—glycidol, OE—oxyethylene group from EC, H_2O—water, K^+—potassium ion from the catalyst K_2CO_3, Na^+—sodium ion, CH_3OH—methanol.

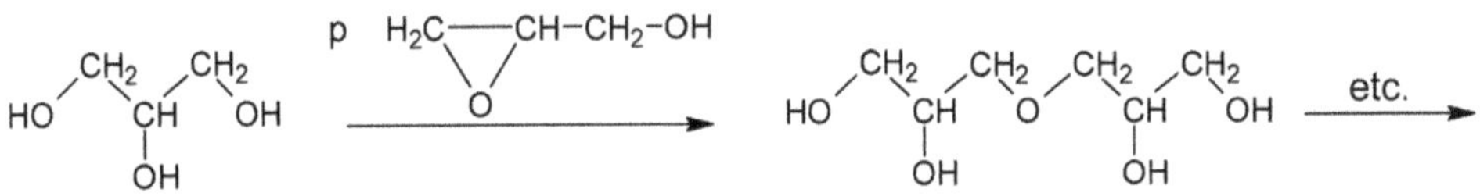

Scheme 6. Hydroxyalkylation of glycerol with GL.

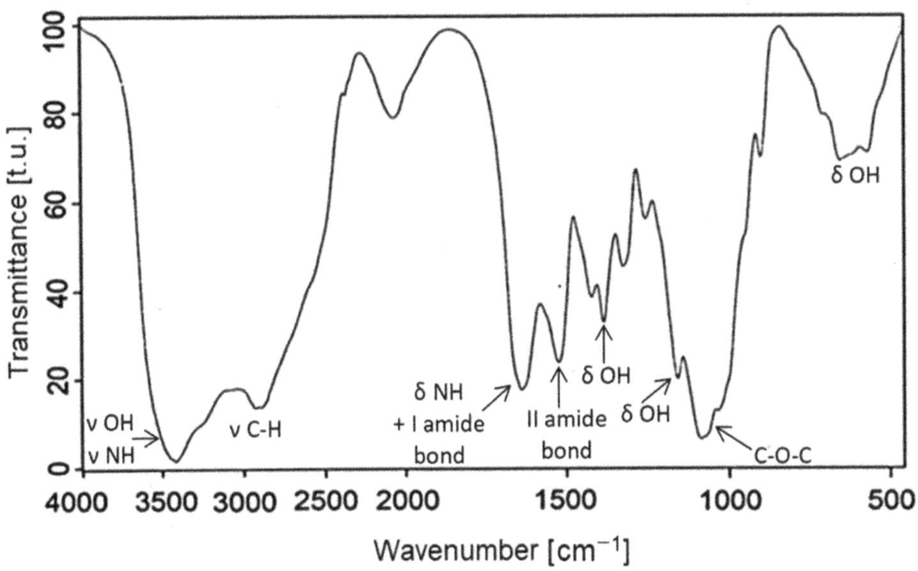

Figure 1. IR spectrum of water-soluble chitosan.

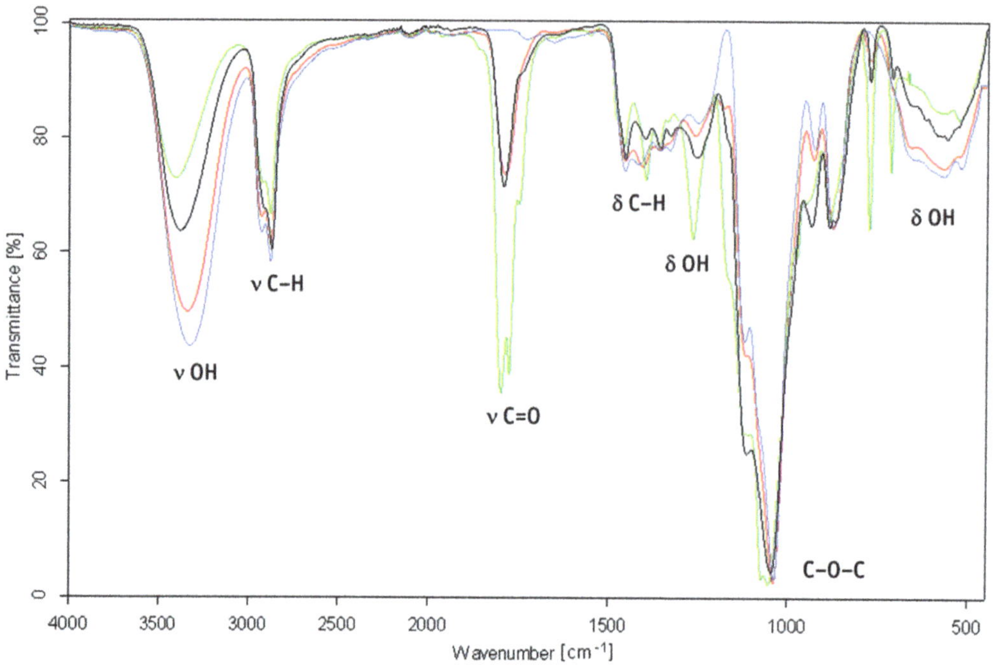

Figure 2. IR spectra of polyols (CS + H_2O + GL) + EC (red), (CS + GLYC + GL) + EC (blue), (CS + GL) + EC (green), and (CS + GL + EC) (black).

In the ^{1}H-NMR spectra of CS (Figure 3), the amine group protons yielded resonances at ca. 8.2 ppm, while primary and secondary hydroxyl group protons were present at 4.8 ppm and 5.5 ppm. Between the latter, the characteristic resonance of C_1H from a chitosan ring was present. The C_3-C_6 methine proton resonance multiplets were present at 3.5–3.9 ppm regions, while the C_2H signal overlapped with a water residual broad signal. Methyl resonance from acetylamine was observed at 1.9 ppm [14]. The ^{1}H-NMR-obtained polyols are shown in Figure 4. In these spectra, the amine proton resonances disappeared due to hydroxyalkylation of amine groups.

The region 3.2–3.5 ppm was considerably modified, and the additional resonances from methylene and methine protons appeared, which evidenced the GL and EC ring opening and incorporation of their structural fragments into the polyol. The chitosan hydroxyl resonances (previously observed at 4.8 and 5.5 ppm) disappeared, while new hydroxyl proton resonances grew within the 4.3–4.6 ppm region. No considerable differences were observed between polyol obtained from various chitosans and variable conditions of the reaction. Using mass spectrometry in the MALDI-TOF technique, the side products formed in polyol syntheses were identified. The illustrative example of the results is shown in Table 1 for (CS + GL) + EC polyol obtained by method 3. Low molecular weight peaks corresponded to not-reacted GL (Table 1, entries 1, 2). There were also the peaks corresponding to the product of hydroxyalkylation of GL and its oligomers with EC with elimination of CO_2 (Table 1, e.g., entries 4, 5, 7–9, 11, 14, 18, and Schemes 7 and 8).

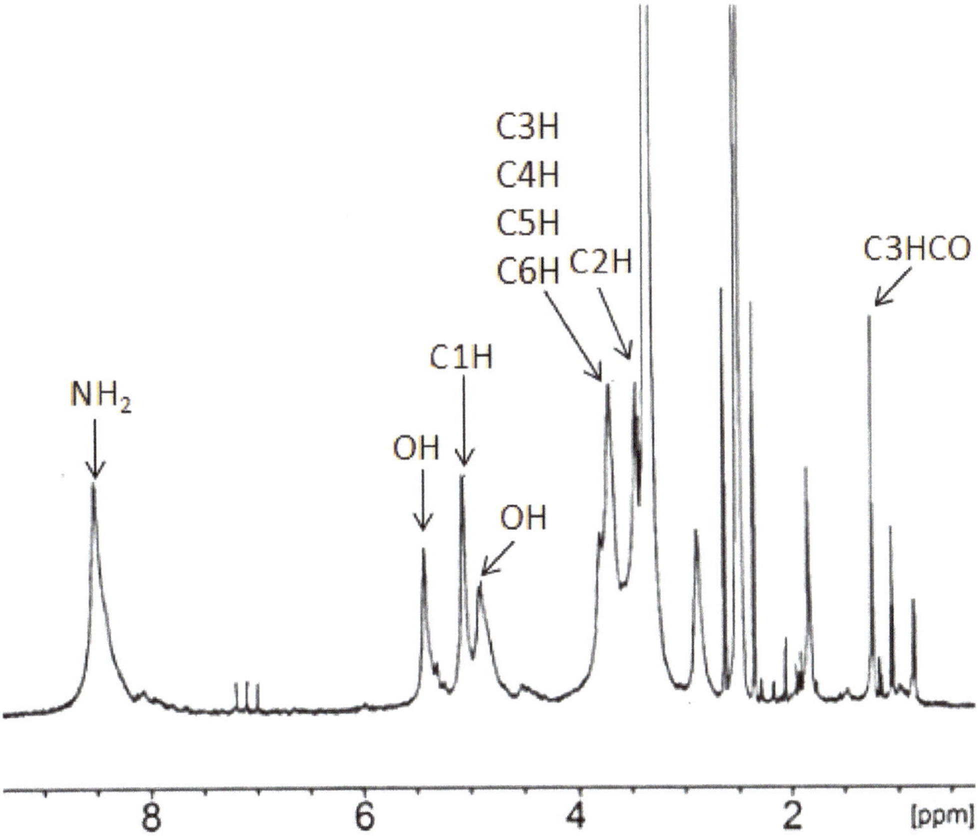

Figure 3. ^{1}H-NMR spectrum of water-soluble chitosan.

Side products can also be dehydrated in the reaction conditions (Table 1, for example entries 10, 15–17, and Scheme 9).

The MALDI-TOF spectrum of polyol (CS + GL + H_2O) + EC obtained in water confirmed the presence of GLYC, which was formed in the reaction between water and GL (Table 2, entry 5). It can be further hydroxyalkylated with GL (Table 2, entries 9, 13, 14, 21, 26). The obtained oligomers can then react with EC (Table 2, entries 11, 15, 17–20, 22–25). The MALDI-TOF spectrum of polyol (CS + GL + GLYC) + EC contained similar peaks as the previous case because GLYC was added into the reacting system (Table 3). In both spectra, the products of oligomerization of GL and the GL + EC reaction were present, similarly to the spectrum of the aforementioned polyol (CS + GL) + EC.

The obtained polyols are liquids miscible with pMDI. Their hydroxyl numbers and physical properties, such as density, viscosity, and surface tension, were determined. The results are collected in Table 4. The temperature dependencies of the physical parameters of the studied polyols are typical of those used to produce PUFs (Figure 5) [61]. Polyols obtained in water or in GLYC showed a lower density and viscosity, which were caused by the presence of a reactive solvent in the system. The hydroxyl numbers within 409–654 mg KOH/g suggested that the obtained polyols can be used to obtain rigid PUFs. The high values of hydroxyl numbers of polyols (CS + GL + H_2O) + EC and (CS + GL + GLYC) + EC were due to the presence of products of water and glycerol hydroxyalkylation (high functionality, high number of hydroxyl groups).

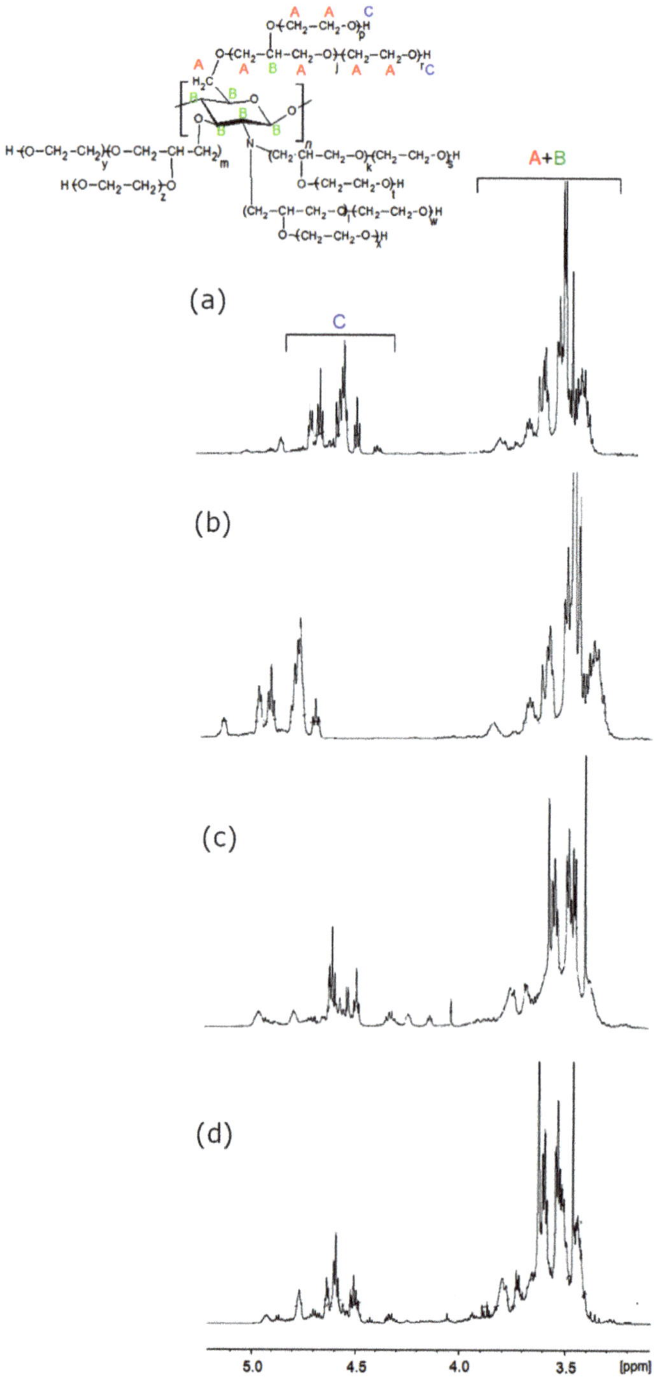

Figure 4. ^{1}H-NMR spectra of polyols (CS + H_2O + GL) + EC (**a**), (CS + GLYC + GL) + EC (**b**), (CS + GL) + EC (**c**), and (CS + GL + EC) (**d**).

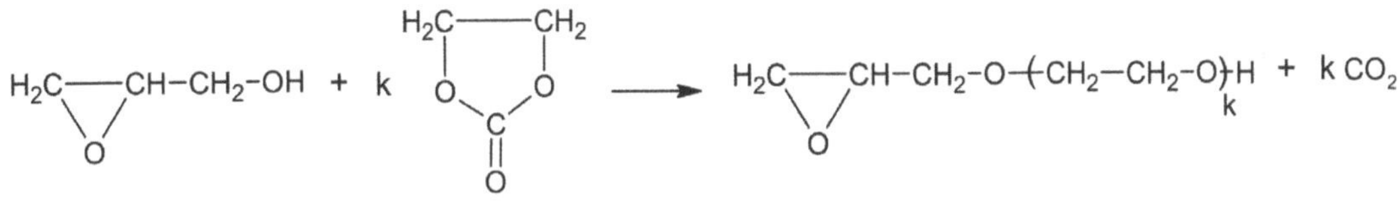

Scheme 7. Hydroxyalkylation of GL with EC.

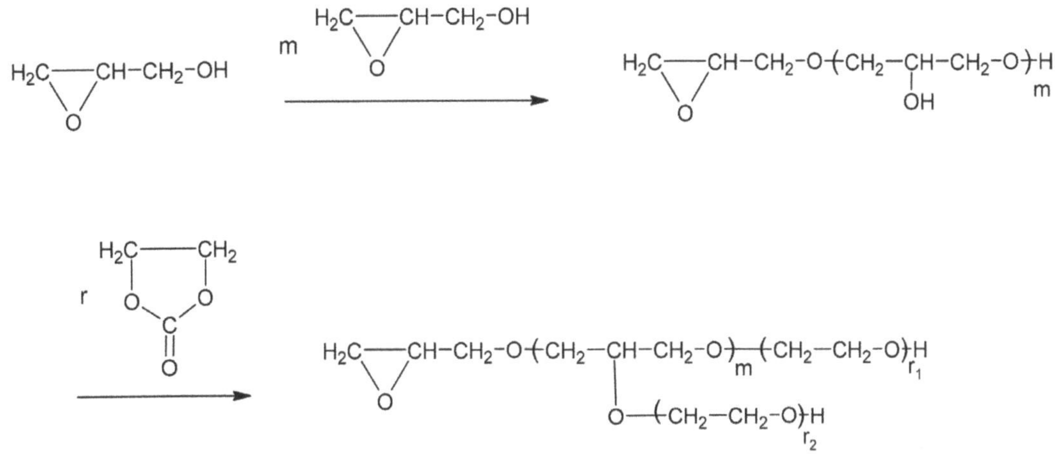

Scheme 8. Oligomerization of GL and hydroxyalkylation of oligomers with EC.

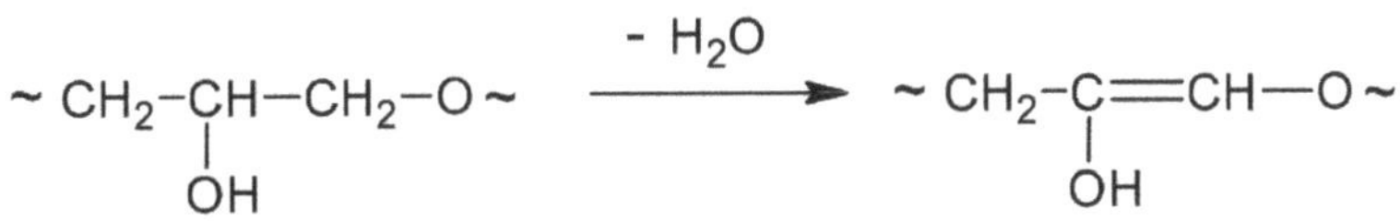

Scheme 9. Dehydration of products at a high temperature.

3.3. Preparation of Polyurethane Foams

The attempts to use the obtained polyols as substrates for the synthesis of PUFs were performed on a laboratory scale in order to select and optimize the kind of polyol, the amount of diisocyanate (pMDI), a catalyst, a surfactant, and a foaming agent. We aimed at a rigid PUF with small pores (Table 5). We concluded that the most promising PUFs were obtained when the amount of pMDI corresponding to the molar ratio of the isocyanate group to the hydroxyl group (isocyanate index, II) was within 1.1–1.3. The exemption was PUF obtained from the (CS + GL + EC) polyol, which was obtained with II 1.5. The relatively high value of II can be attributed to a higher share of chitosan units in the polyol structure, and thus a higher number of amino groups. Amine groups can catalyze trimerization of isocyanates to isocyanuric rings, and this side-reaction may result in isocyanate consumption. Thus, the high thermal resistance of PUF obtained from this polyol can be caused by the high thermal resistance of isocyanuric rings [62]. The optimized surfactant Silicon L-6900 amount was 3.0 or 3.9 g per 100 g of polyol. Two PUFs were obtained from every polyol by using a variable amount of the foaming agent (water), i.e., 2% and 3% related to the polyol mass. The PUFs obtained with less than 2% of water were under-foamed, while those obtained with more than 3% of water were semi-rigid with large pores. The amount of TEA used was variable within 0.5–2.7 g/100 g of polyol. The optimized amount of the catalyst depended on the water content and decreased when the water was increased. The lowest amount of catalyst was successfully used for compositions obtained from (CS + GL + GLYC) + EC and (CS + GL) + EC polyol. When less than the optimized amount of TEA was used, the PUF had irregular, large pores, and was under-crosslinked, while when more than the optimal amount of TEA was used, fast growth and a limited size of PUF was observed, which finally led to a lower foaming degree and an increase of the apparent density of PUF. Creaming, rising, and tack-free times were observed during foaming. They depended on the kind of polyol and the amount of catalyst. The creaming time for optimized compositions was within 21–72 s, while the rising time was short (22–59 s). The longest rising time was observed for the composition with (CS + GL) + EC polyol and 3% water. A typical increase of the creaming time and the rising time of the compositions due to a decreased amount of the catalyst was noticed. The tack-free time was very short (below 6 s) for compositions which used polyols synthesized with water and GLYC, and a longer time (above 10 s) for other PUFs.

3.4. Properties of Polyurethane Foams

The following properties of obtained PUFs were determined: apparent density, water absorption (by volume), dimensional stability, thermal conductivity coefficient, thermal resistance, compressive strength, and glass transition temperature. The apparent density of the obtained PUFs is illustrated in Figure 6. The PUFs obtained with a lower amount of the foaming agent had a higher apparent density due to lower foaming. Thus, the largest density was observed for PUFs obtained from compositions with 2% water (70–73 kg/m^3).

Table 2. Interpretation of MALDI-TOF spectrum of polyol (CS + H_2O +GL) + EC.

Entry	Signal Position, m/z	Relative Intensity of Signal (%)	Molecular Ion Structure	Calc. Molecular Weight (g/mol)
1	96.912	13.60	GL + Na^+	97.027
2	101.080	6.77	GL + OE-H_2O + H^+	101.060
3	108.899	13.69	GL + CH_3OH + H^+	107.071
4	112.900	26.55	GL + K^+	113.000
5	114.877	15.43	GLYC + Na^+	115.037
6	144.010	7.41	GL + 2OE − H_2O	144.079
7	156.861	9.66	GL + OE + K^+	157.027
8	186.206	11.18	2GL + K^+	187.037
9	196.971	35.24	GLYC + GL + CH_3OH	198.110
10	205.068	29.28	3GL-H_2O + H^+	205.108
11	233.106	32.64	GLYC + GL + OE + Na^+	233.100
12	249.080	100.00	3GL + OE − H_2O + H^+	249.133
13	263.077	20.49	GLYC + 2GL + Na^+	263.111
14	279.067	25.71	GLYC + 2GL + K^+	279.0845
15	293.107	99.16	GLYC + GL + 2OE + K^+	293.100
16	307.140	25.74	GLYC + 2GL + OE + Na^+	307.137
17	323.114	40.40	GLYC + 2GL + OE + K^+	323.111
18	351.163	16.91	GLYC + 2GL + 2OE + Na^+	351.163
19	367.139	30.35	4GL + 2OE−H_2O + H^+	367.197
20	397.183	14.63	GLYC + 3GL + OE + K^+	397.148
21	411.169	21.71	GLYC + 4GL + Na^+	411.184
22	425.201	15.63	GLYC + 3GL + 2OE + Na^+	425.200
23	441.154	13.31	GLYC + 3GL + 2OE + K^+	441.174
24	455.203	12.60	GLYC + 4GL + OE + Na^+	455.210
25	471.164	9.66	6GL + OE−H_2O + H^+	471.244
26	485.200	11.57	GLYC + 5GL + Na^+	485.221
27	515.211	12.60	GLYC + 4GL + 2OE + K^+	515.211
28	521.314	11.08	GLYC + 4GL + 3OE + H^+	521.281
29	590.906	20.07	GLYC + 5GL + 2OE + K^+	589.247

Description of abbreviations used in the table: GL—glycidol, GLYC—glycerol, OE—oxyethylene group from EC, H_2O—water, K^+—potassium ion from the catalyst K_2CO_3, Na^+—sodium ion, CH_3OH—methanol.

Table 3. Interpretation of MALDI-TOF spectrum of polyol (CS + GLYC + GL) + EC.

Entry	Signal Position m/z	Relative Intensity of Signal (%)	Molecular Ion Structure	Calc. Molecular Weight (g/mol)
1	96.930	6.08	GL + Na^+	97.026
2	101.063	14.40	GL + OE − H_2O + H^+	101.060
3	112.901	76.62	GL + K^+	113.000
4	114.902	24.69	GLYC + Na^+	115.037
5	120.936	15.39	GL + OE + H^+	119.071
6	140.902	14.34	GL + OE + Na^+	141.053
7	156.907	25.77	GL + OE + K^+	157.027
8	175.069	8.56	2GL + OE-H_2O + H^+	175.097
9	196.881	33.95	GLYC + GL + CH_3OH	198.110
10	205.073	39.21	3GL − H_2O + H^+	205.108
11	219.069	11.52	2GL + 2OE − H_2O + H^+	219.123
12	249.074	96.62	3GL + OE − H_2O + H^+	249.134
13	263.063	6.15	GLYC + 2GL + Na^+	263.111
14	279.087	49.41	GLYC + 2GL + K^+	279.085
15	293.093	41.94	GLYC + GL + 2OE + K^+	293.100
16	323.097	52.76	GLYC + 2GL + OE + K^+	323.111
17	335.114	8.19	GLYC + 2GL + 2O−H_2O + Na^+	333.153
18	353.106	20.29	5GL − H_2O + H^+	353.181
19	367.121	13.63	4GL + 2OE − H_2O + H^+	367.197
20	393.975	100.00	GLYC + 2GL + 3OE + Na^+	395.189
21	397.159	15.38	GLYC + 3GL + OE + K^+	397.148
22	411.143	8.06	GLYC + 4GL + Na^+	411.184
23	427.144	10.43	5GL + OE − H_2O + CH_3OH	428.226
24	441.156	7.13	GLYC + 3GL + 2OE + K^+	441.174
25	471.167	8.45	6GL + OE−H_2O + H^+	471.244
26	515.197	7.26	GLYC + 4GL + 2OE + K^+	515.211
27	534.289	16.09	3GL + 7OE − H_2O + Na^+	535.273
28	575.851	9.48	7GL + OE − H_2O + CH_3OH	576.299
29	590.888	58.08	7GL + 2OE − H_2O + H^+	589.307

Description of abbreviations used in the table: GL—glycidol, GLYC—glycerol, OE—oxyethylene group from EC, H_2O—water, K^+—potassium ion from the catalyst K_2CO_3, Na^+—sodium ion, CH_3OH—methanol.

Table 4. Physical properties and hydroxyl number of polyols at the temperature of 20 °C.

Polyol	Density (g/cm^3)	Viscosity (mPa·s)	Surface Tension (mN/m)	Hydroxyl Number (mg KOH/g)
(CS + GL + H_2O) + EC	1.2789	6847	41.0	654
(CS + GL + GLYC) + EC	1.2702	1475	37.9	560
(CS + GL) + EC	1.3150	88,612	44.2	409
(CS + GL + EC)	1.3208	52,176	41.0	493

Water absorption in the 24 h test was low (usually below 3%), suggesting that the closed pores dominated in the obtained PUFs (Figure 7a,b). This low water absorption is caused by the presence of open cells on the cut of PUF. In order to visualize the pores, the images of cross-sections of PUFs were taken.

Statistical analysis of images enabled to determine an average pore size and thickness of the pore walls (Figure 8). It can be concluded that oval pores of various sizes were present (Table 6). Since it was an ellipsoidal shape, two diameters were calculated. The average longer diameter was within 169–321 μm, while the shorter one was within 84–152 μm. Longer pores were observed in PUFs obtained from (CS + GL + H_2O) + EC and (CS + GL) + EC polyols if 3% water was present in the foaming compositions (Table 6). This led to the release of more CO_2 and also caused a decrease of thickness of the pore walls. Small pores in PUF obtained from (CS + GL + GLYC) + EC polyol and 2% water/100 g of polyol led to the highest compressive strength of all the obtained PUFs. The larger pores enabled to absorb water, and thus the PUFs with large pores showed higher water absorption, especially in cases of PUFs obtained from (CS + GL + H_2O) + EC and (CS + GL + GLYC) + EC polyols.

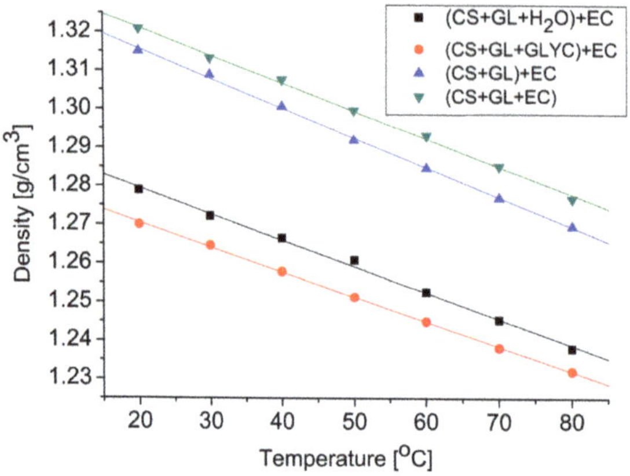

(**a**)

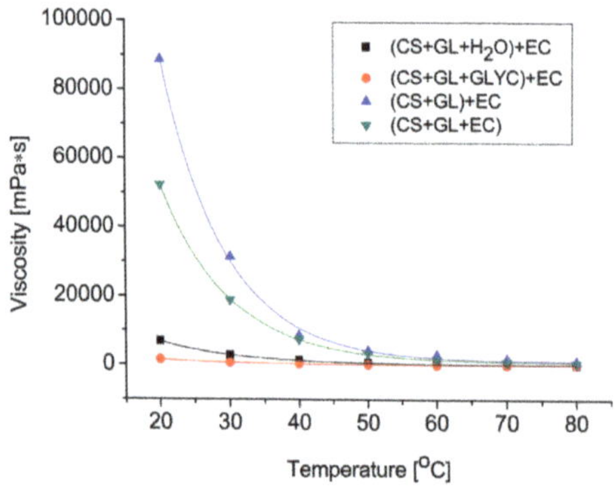

(**b**)

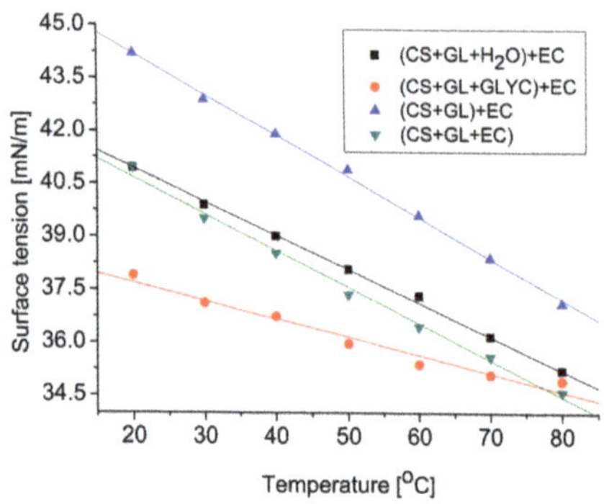

(**c**)

Figure 5. Changes of density (**a**), viscosity (**b**), and surface tension (**c**) of polyol as a function of temperature.

Table 5. The influence of composition on the foaming process.

Polyol	Composition (g/100 g of Polyol)				Isocyanate	Foaming Process		
	pMDI	Water	Silicon L-6900	TEA	Index	Cream Time (s)	Rise Time (s)	Tack-Free Time (s)
(CS + H_2O + GL) + EC	180	2	3.0	2.0	1.1	28	22	5
	190	3	3.9	1.7	1.2	33	46	4
(CS + GL + GLYC) + EC	178	2	3.0	0.8	1.3	55	54	5
	180	3	3.0	0.8	1.3	59	56	5
(CS + GL) + EC	110	2	3.9	0.9	1.1	40	43	14
	120	3	3.9	0.5	1.1	72	59	26
(CS + GL + EC)	150	2	3.9	2.7	1.2	24	20	11
	185	3	3.9	2.2	1.5	21	30	14

Cream time—the time elapsed from the moment of mixing to the start of volume expansion. Rise time—the time from the start of expansion to the moment of reaching the sample's final volume. Tack free time—the time from reaching the sample's final volume to the moment of losing its surface adhesion to powdered substances.

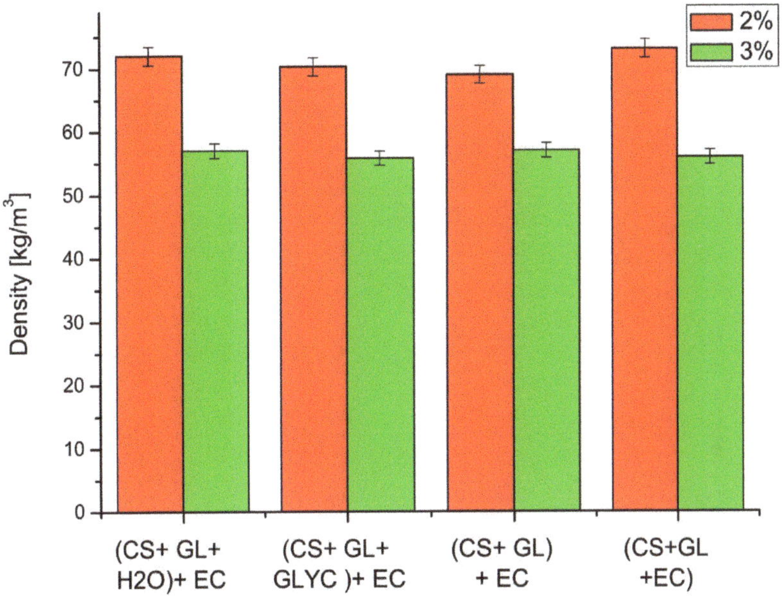

Figure 6. The density of the obtained polyurethane foams (determined with ±2.0% accuracy). The water content in the foam composition per 100 g of polyol is shown in the insert.

Table 6. Results of micrograph analyses.

Foam Obtained from Polyol	Amount of Water/100g of Polyol	Larger Diameter (μm)	Smaller Diameter (μm)	Thickness of Cell Wall (μm)
(CS + GL + H_2O) + EC	2	149 ± 16	129 ± 23	7 ± 1
	3	237 ± 35	85 ± 35	15 ± 2
(CS + GL + GLYC) + EC	2	169 ± 17	80 ± 12	11 ± 1
	3	261 ± 36	149 ± 12	26 ± 7
(CS + GL) + EC	2	209 ± 17	152 ± 23	20 ± 2
	3	321 ± 55	96 ± 15	9 ± 2
(CS + GL + EC)	2	305 ± 65	84 ± 18	12 ± 2
	3	230 ± 24	146 ± 30	10 ± 1

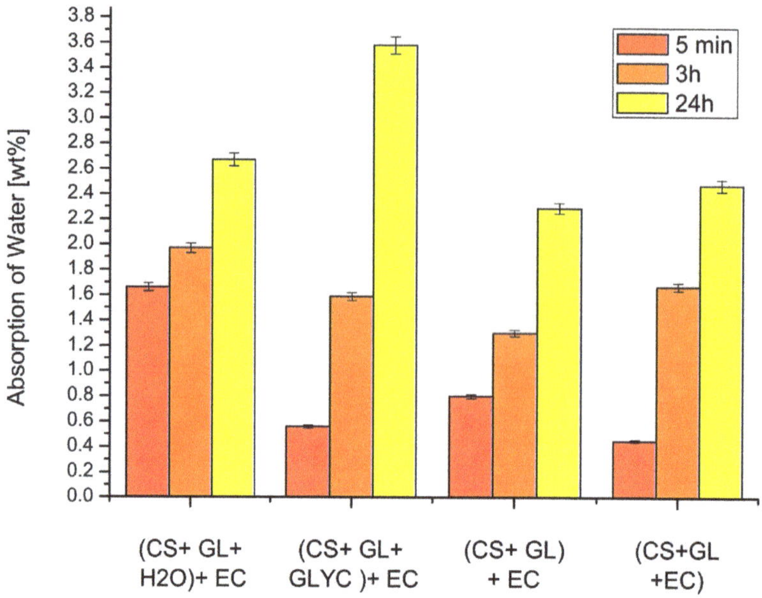

(a)

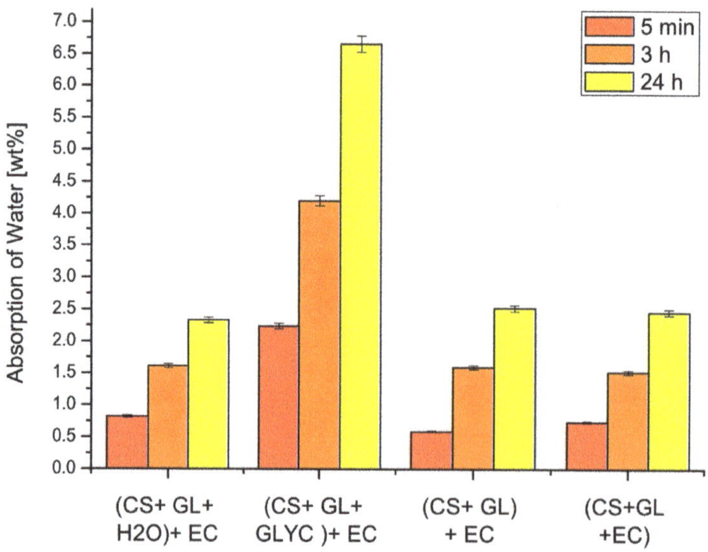

(b)

Figure 7. Absorption of water of the obtained polyurethane foams with 2% (**a**) and 3% (**b**) foaming agent (determined with ±1.9% accuracy). The appropriate shade of color indicates the time the foam stayed in the water (shown in the insert).

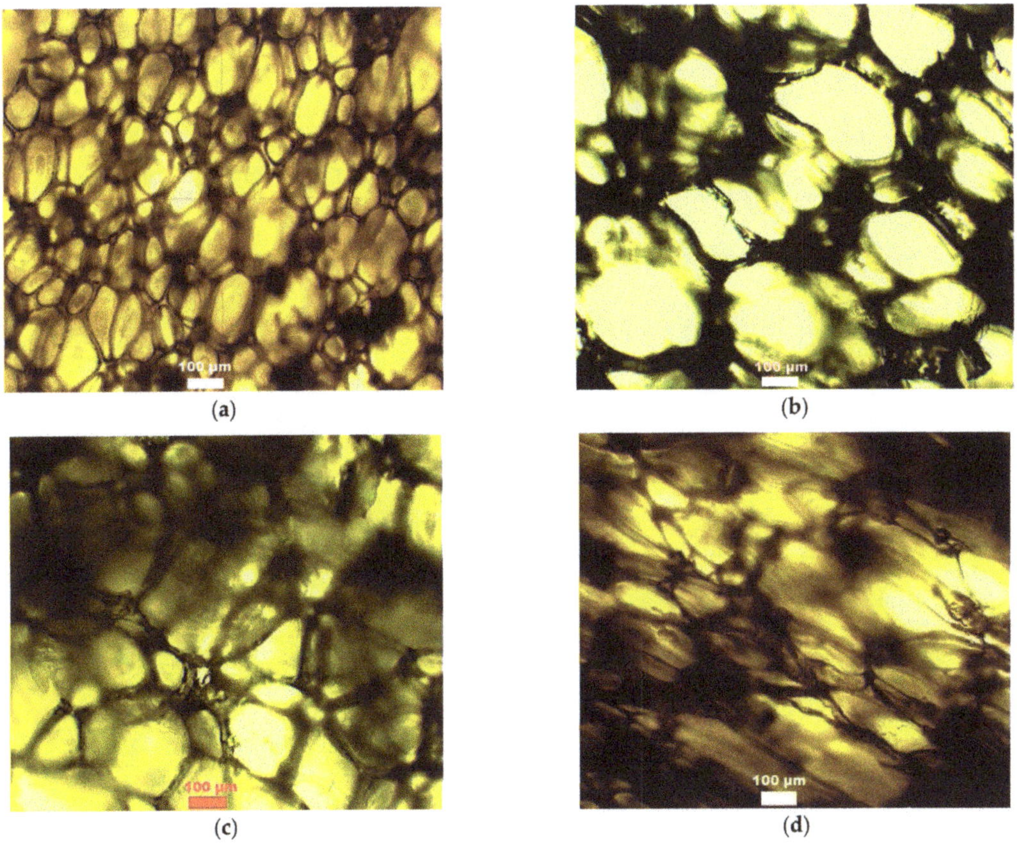

Figure 8. Optical microscopy images of foam obtained from polyols (CS + GL + GLYC) + EC (**a**,**b**) and (CS + GL) + EC (**c**,**d**),with 2% (**a**,**c**) and 3% (**b**,**d**) water content/100g of polyol.

Thermal conductivity coefficient values of the obtained PUFs (Figure 9) in the presence of 2% water were similar to those found in typical rigid PUFs (0.0260 W/m·K) [61].

The obtained PUFs had a good dimensional stability at elevated temperatures, which fell to −2.80% to 3.0% after a 40 h exposure at 150 °C (Table 7). In some cases, elongation in one and shortening in another dimension of exposed PUFs has been observed.

Table 7. Dimensional stability at the temperature of 150 °C (determined with ±1.5% accuracy).

Foam Obtained From	Water (%)	Dimensional Stability (%) at 150 °C					
		Length Change After		Width Change After		Height Change After	
		20 h	40 h	20 h	40 h	20 h	40 h
(CS + GL + H_2O) + EC	2	−2.27	−2.82	−1.14	−0.97	+2.95	+1.22
	3	−0.80	−0.67	−0.88	−1.38	−4.25	−2.30
(CS + GL + GLYC) + EC	2	−0.28	−0.61	−0.92	−3.66	−0.19	−1.57
	3	−1.16	−1.08	+0.75	+0.56	−0.11	−2.67
(CS + GL) + EC	2	+1.20	+0.29	+2.04	+1.47	+2.45	+1.76
	3	−1.66	−1.33	+1.38	−0.81	−2.41	−2.78
(CS + GL + EC)	2	−1.00	−1.09	−1.09	−0.65	−1.18	−1.64
	3	−0.62	−0.62	−0.51	−0.55	−1.35	−0.90

The water content at 2% or 3% of the foam composition is based on 100 g of polyol.

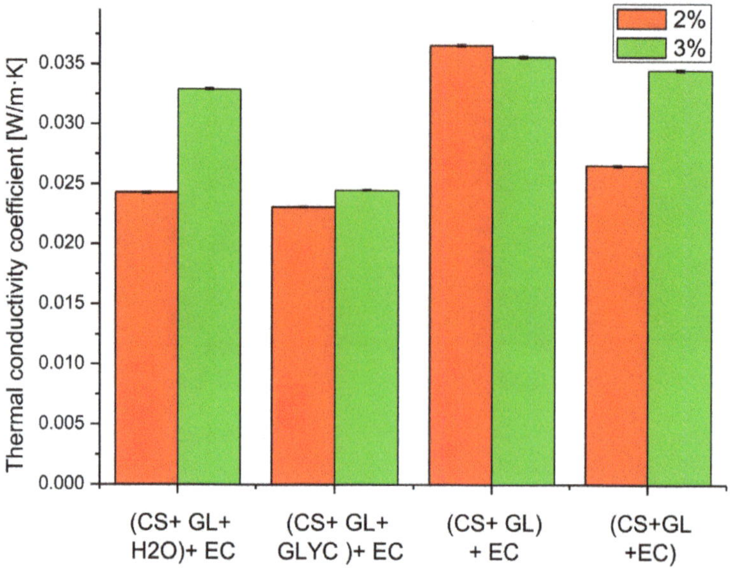

Figure 9. Thermal conductivity coefficients of obtained polyurethane foams (determined with ±0.3% accuracy). The appropriate shade of color indicates the water content in the foam composition per 100 g of polyol (in the insert).

Thermal resistance of PUFs for one month of exposure at 150 °C and 175 °C was determined by the mass loss. The mass loss was due to physical changes, such as water diffusion and evaporation and chemical conversions. The PUFs were not resistant to thermal exposure at 200 °C already in the first days of exposure. The mass loss upon thermal exposure is illustrated in Figure 10a,b. The largest mass loss was observed in the first days of thermal exposure because of the initial water and catalyst (TEA) removal. Thus, after one month of exposure at 150 °C, the mass loss was within 10.1–17.4% (Table 8). After one month at this temperature, the lowest mass loss was observed for the PUFs obtained from (CS + GL + H2O) + EC polyol and the polyol from one-pot synthesis, namely (CS + GL + EC).

Table 8. Mass loss and compressive strength after exposure for one month at 150 and 175 °C.

Foam Obtained From	Water (%)	Mass Loss in %wt. after Exposure for 1 Month		Compressive Strength (MPa)		
				Before Exposure	After Exposure	
		150 °C	175 °C		150 °C	175 °C
(CS + GL + H_2O) + EC	2	10.4	29.1	0.353	0.649	0.791
	3	11.1	27.9	0.337	0.510	0.283
(CS + GL + GLYC) + EC	2	17.4	34.6	0.513	0.694	0.791
	3	10.4	24.2	0.337	0.510	0.238
(CS + GL) + EC	2	14.9	27.7	0.119	0.271	0.373
	3	14.1	28.5	0.203	0.235	0.219
(CS + GL + EC)	2	10.4	24.8	0.238	0.336	0.328
	3	10.1	26.4	0.227	0.249	0.248

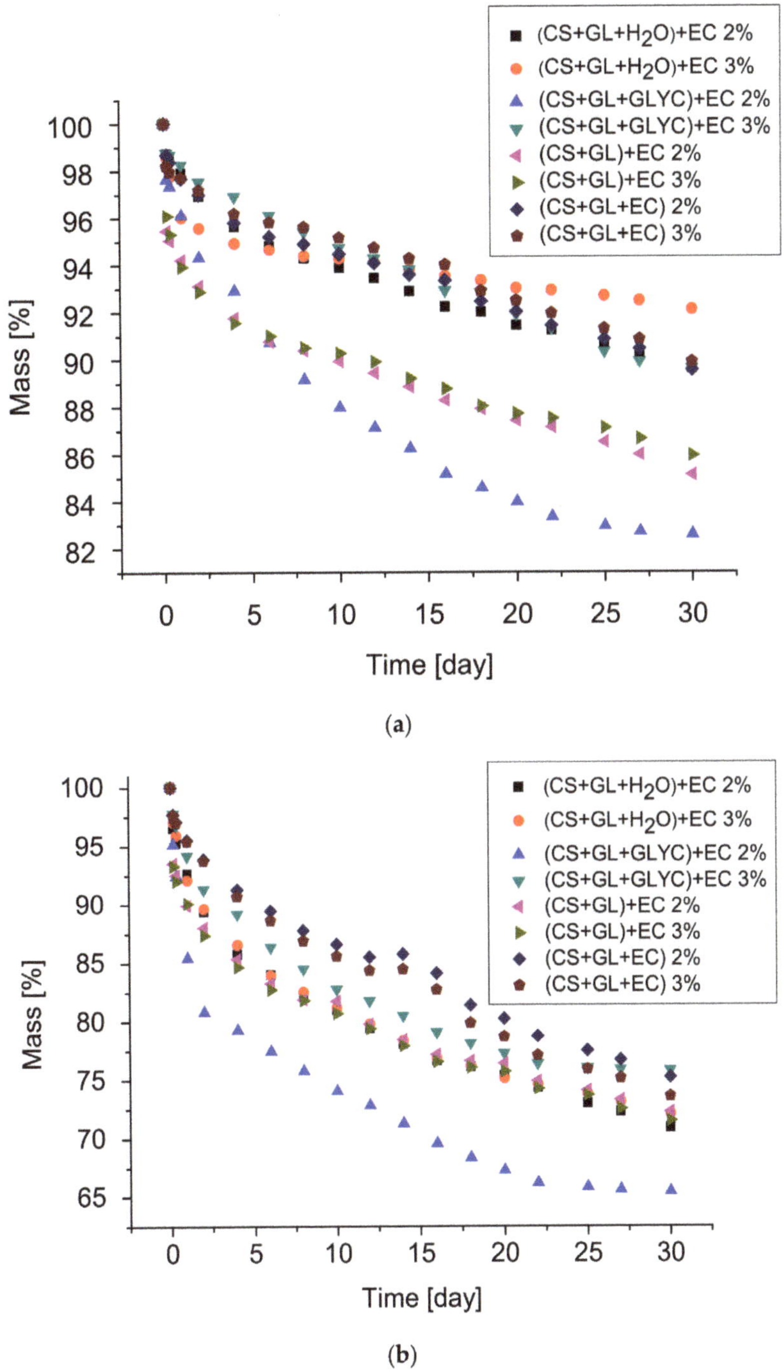

Figure 10. Thermal stability of the polyurethane foam as the mass loss after heating at 150 °C (**a**) and 175 °C (**b**) (the water content of 2% or 3% of the foam composition is based on 100 g of polyol).

These two PUFs also had high thermal resistance at 175 °C. High thermal resistance in the first case is related to high crosslinking of PUF obtained from polyol (CS + GL + H_2O) + EC, which showed a high value of the hydroxyl number as well as the highest percentage of thermally resistant chitosan units in this polyol (5.1%), in comparison to polyols from syntheses 2(3.6%) and 3 (3.7%). Chitosan decomposes at ca. 280 °C [63]. Another polyol containing a slightly lower proportion of chitosan units (4.8%) is (CS + GL + EC). The highest mass loss was found for PUFs obtained from (CS + GLYC + GL) + EC polyol and the 2% foaming agent (17.4% and 34.6% at 150 °C and 175 °C, respectively). The PUFs at ambient temperature were rigid and thermal exposure at 150 °C and 175 °C did not change this.

The obtained PUFs are characterized by compressive strength typical of classic, rigid polyurethane foams (Table 8). The highest compressive strength was for PUFs obtained from (CS + GL + H_2O) + EC and (CS + GL + GLYC) + EC polyols. This is related to the highest functionality of theses polyols, determined by hydroxyl numbers. These polyols provided the best conditions for effective crosslinking because the presence of GLYC resulted in increasing functionalization upon consecutive reactions with GL.

For example, the stress–strain relationship for PUFs obtained from polyol (CS + GL + EC) is shown in the Figure 11. The relationship is typical of rigid PUFs. Initially, the distortion was in line with the compressive strength until 3%. At this point, the PUF lost the ability to transfer the load and the distortion rapidly increased up to 10%, and at this point, the readout of compressive strength of PUF was executed. The PUFs obtained from the composition with 2% water related to the mass of polyol required higher strains forload transfer, which corroborated well with their higher apparent density (compare to Figure 6).

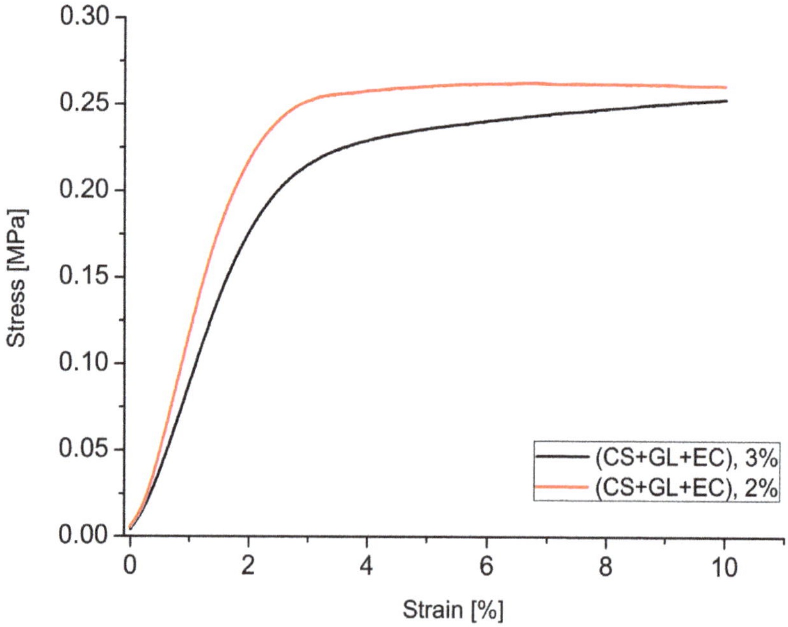

Figure 11. Stress–strain relationship of PUFs obtained from polyol (CS + GL + EC).

It is worth noticing that after annealing at 150 °C and in some cases at 175 °C, some PUFs showed an increase of compressive strength, presumably due to additional crosslinking upon heating. This was especially well-recognized in PUFs obtained from (CS + GL + H_2O) + EC and (CS + GL + GLYC) + EC, which had the largest hydroxyl numbers. The PUF obtained from (CS + GL + H_2O) + EC polyol and 2% water showed 84%

and 124% increases of compressive strength after thermal exposure at 150 and 175 °C. In the case of PUFs from the composition with 3% water, a decrease in compressive strength after annealing at 175 °C was observed in relation to the strength measured at 150 °C, but it was often still greater than that determined before the exposure. Nevertheless, in all the cases, compressive strength increased upon annealing, in comparison with the not-annealed PUFs. Generally, two factors influenced the compressive strength changes upon annealing, namely additional crosslinking and degradation, which contributed to increase and decrease the compressive strength, respectively.

The dynamic thermal resistance of PUFs was evaluated by the thermogravimetric method (Figure 12, Table 9). The TG curves clearly demonstrated that PUFs obtained from (CS + GL + H_2O) + EC and (CS + GL + EC) polyols had the highest thermal resistance. The temperatures of 5% mass loss were 186 °C and 174 °C, or 210 °C and 184 °C, for PUFs obtained from compositions with 2% and 3% water, respectively (Table 9). From these measurements, it has been additionally found that the initial decomposition temperature was lower in case of PUFs obtained with 3% water. The reason for this might be the smaller thickness of the cell walls and the larger pores (see Table 6). Three endothermic peaks were observed at dm/dT vs. temperature curves at 190, 280, and 400 °C. The first peak was attributed to the thermal break of urethane and urea bonds [62], the second one was related to the chitosan ring break [63], and the third to decomposition of polyurethane to amines and carbon dioxide [62]. Decomposition of PUFs was completed at 600 °C. DSC measurements of the obtained PUFs indicated that all samples showed mass loss within 35–110 °C within the first heating/cooling cycle. This endothermic process was due to the presence of TEA and absorbed water. In the second heating cycle, the endothermic process was absent. Therefore, the glass transition temperature could be determined for the PUFs (Figure 12). The glass transition temperature was within 82–148 °C, which allowed the tested materials to be classified as rigid foams (Figure 13). The PUFs obtained from (CS + GL + EC) polyol showed a glass transition temperature beyond this specific range (−40 to 200 °C).

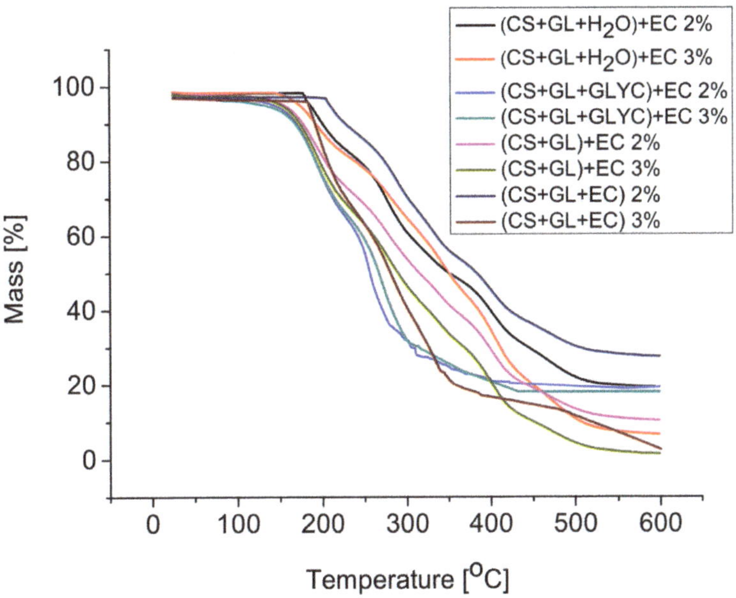

(a)

Figure 12. *Cont.*

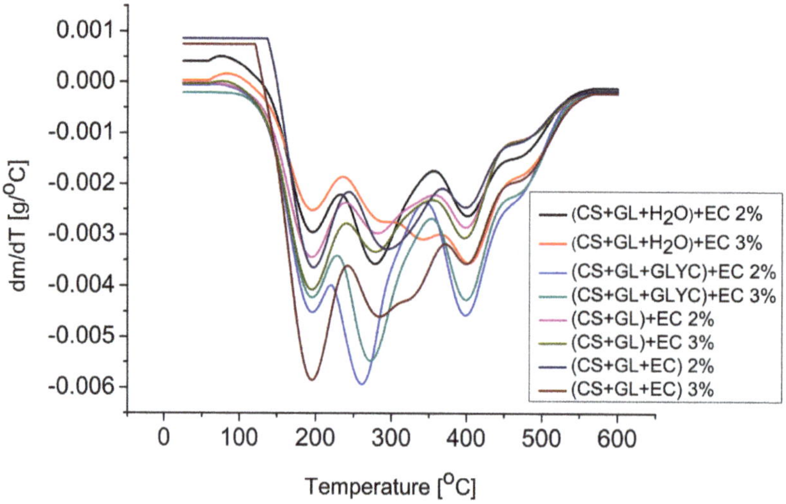

(**b**)

Figure 12. Thermal analysis of foams: mass change as a function of temperature (**a**) and differential mass change as a function of temperature (**b**) (the water content of 2% or 3% of the foam composition is based on 100 g of polyol).

Table 9. Thermal analysis of foam determined by the dynamic method.

Foam Obtained From	Water (%)	$T_{5\%}$ (°C)	$T_{10\%}$ (°C)	$T_{25\%}$ (°C)	$T_{50\%}$ (°C)	Tg (°C)
(CS + H_2O + GL) + EC	2	186	200	263	349	143
	3	174	193	264	350	148
(CS + GLYC + GL) + EC	2	144	171	202	256	111
	3	131	168	203	266	111
(CS + GL) + EC	2	156	178	220	315	91
	3	152	175	208	289	82
(CS + GL + EC)	2	210	229	288	382	-
	3	184	192	216	280	-

The water content of 2% or 3% of the foam composition is based on 100 g of polyol.

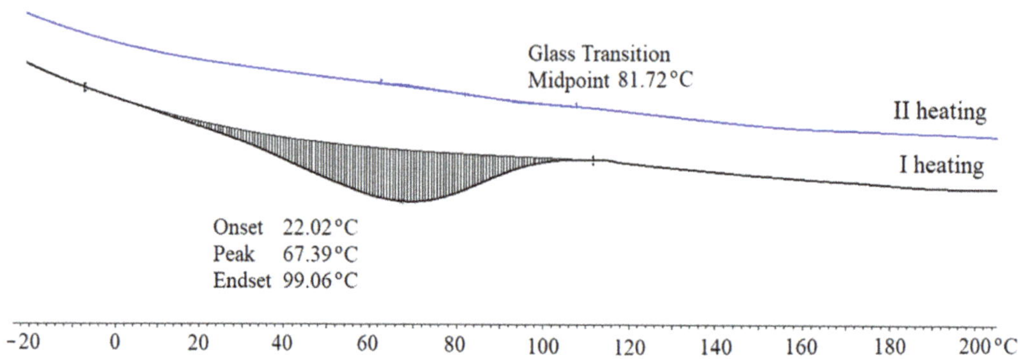

Figure 13. DSC thermogram of foam obtained from polyol (CS + GL) + EC (3% H_2O/100 g of polyol).

Generally, the chitosan-derived polyols as substrates for the formation of PUFs leads toa new possibility to use them as biodegradable materials. In order to determine the biodegradation of such materials, we have tested their biodegradability. Elemental analyses of the tested polyol and PUF are presented in Table 10. Based on these data, the theoretical oxygen demand (*TOD*) was estimated.

Table 10. Mass shares of individual elements in polyol (CS + GL + EC) and the obtained polyurethane foam.

Sample	Element			
	C	H	O	N
Polyol	0.4693	0.0802	0.4505	0
Polyurethane foam	0.6102	0.0577	0.2636	0.0685

BOD_{soil} = 8.5 mg/dm^3.

The degree of degradation (Dt) of the studied samples was determined on the basis of BOD after 28 days and the estimated TOD. The 28-day biodegradation test in soil conditions indicated that the degradation percentage of PUF was 52%. This corroborates very well the presence of biodegradable polyol (Table 11). These are very promising results considering that obtained chitosan-based PUFs have good mechanical properties, improved thermal resistance, and a high degree of biodegradation.

Table 11. Results of susceptibility to biodegradation of polyol and foam.

Sample	BODx (mg/dm^3)	BOD_{28} (mg/dm^3)	Sample Mass (g)	TOD Counts	TOD (mg/dm^3)	Dt (%)
Polyol	116	107.5	0.19	8.4512	44.48	100
Foam	45.1	36.6	0.20	14.1264	70.63	51.82

The properties of the best PUFs obtained from chitosan can be compared to those of PUFs obtained from polyols based on starch or cellulose (Table 12) [10,11,31,32]. Polyols used to obtain those PUFs were received in similar conditions as described here, namely in aqueous solutions via a reaction with GL and/or alkylene carbonates as hydroxyalkylating agents. Cellulose- and starch-based PUFs had a generally lower apparent density in comparison to these obtained from CS (Table 12). This resulted in a lower compressive strength before thermal workup and a higher mass loss upon thermal exposure. The large mass loss also resulted in a lower compressive strength of the annealed PUFs based on cellulose or starch. Better functional properties of foams based on chitosan also resulted from the comparison of water absorption, which was definitely lower for foams with hydroxyalkylated chitosan units. The PUFs based on CS and obtained from the composition with a lower percentage of water (2%) may find applications as an isolating material because they show a lower thermal conductivity coefficient than those based on cellulose and starch.

Table 12. Comparison of properties of PUFs based on natural polymers.

Foam Obtained from Polyol	Water in Polyol (%)	Density (kg/m^3)	Absorption of Water (wt.%)	Thermal Conductivity Coefficient (W/m·K)	Mass Loss after Exposure for 1 Month (wt%)		Compressive Strength (MPa)		
					150 °C	175 °C	Before Exposure	After Exposure 150 °C	After Exposure 175 °C
(CS + GL + H_2O) + EC	2	72.0	2.67	0.0243	10.4	29.1	0.353	0.649	0.791
	3	57.0	2.33	0.0329	11.0	27.1	0.337	0.510	0.283
(CS + GL + EC)	2	73.0	2.47	0.0266	10.4	24.8	0.238	0.336	0.328
	3	56.0	2.46	0.0345	10.1	26.4	0.227	0.249	0.248
S + H_2O + EC [10]	2	59.5	9.03	0.0367	15.2	-	0.223	0.924	-
	3	61.9	7.24	0.0362	14.3	-	0.138	0.934	-
S + H_2O + PC [10]	2	43.2	7.07	0.0459	20.0	38.7	0.121	0.676	0.596
	3	34.9	5.42	0.0561	16.4	34.6	0.120	0.247	0.375
S + GL + EC [32]	4	37.3	4.75	0.0367	6.8	23.9	0.078	0.129	0.159
S + GL + PC [32]	4	26.0	5.00	0.0376	7.6	22.1	0.101	0.127	0.108
(CEL + GL + H_2O) + EC [11]	2	60.5	6.56	0.0338	13.0	30.7	0.234	0.284	0.379
$CEL_{hydrolysate}$ + GL + EC [31]	2	72.9	4.94	0.0364	7.6	21.0	0.112	0.325	0.410

The water content of 2% or 3% of the foam composition is based on 100 g of polyol. —PUF underwent deformation.

4. Summary and Conclusions

1. Four new methods of the synthesis of chitosan-based polyols were elaborated using water-soluble chitosan, glycidol, and ethylene carbonate, with variable environments.
2. The chitosan-derived polyols can be obtained in water in the presence of glycerol or no-solvent conditions. These polyols are suitable to obtain polyurethane foams.
3. The polyurethane foams obtained from these polyols have properties analogous for typical rigid PUFs, except for increased thermal resistance in comparison with classic ones. They can withstand long-term thermal exposure at 175 °C. Additionally, with thermal exposure of the obtained PUFs at 150 °C, the compressive strength of the annealed PUF considerably increased.
4. The chitosan-based polyols and polyurethane foams obtained from them were biodegradable: the polyol was completely biodegraded, while the PUF obtained thereof was 52% biodegradable within 28 days in the soil biodegradation oxygen demand test.
5. Polyurethane foams obtained from polyols based on chitosan converted by the reaction with glycidol and ethylene carbonate in water or in glycerol have useful thermal conductivity, dimensional stability, compressive strength, and low water absorption. Their high thermal resistance renders them as promising candidates to use as thermal insulating materials.

Author Contributions: A.M.S., investigation, data curation, validation, formal analysis, writing—original draft preparation, visualization. J.L., conceptualization, methodology, supervision, writing—review and editing. All authors have read and agreed to the published version of the manuscript.

Funding: This research received no external funding.

Institutional Review Board Statement: Not applicable.

Informed Consent Statement: Not applicable.

Data Availability Statement: Not applicable.

Acknowledgments: We kindly acknowledge Joanna Paciorek-Sadowska and Marcin Borowicz from Kazimierz Wielki University in Bydgoszcz for assistance with the biodegradation studies.

Conflicts of Interest: The authors have declared no conflicts of interest.

References

1. Swinarew, B. Polyurethanes—Modern versatile materials. Part 2, Polyurethane foam. *Przetwórstwo Tworzyw* **2015**, *21*, 428–434.
2. Datta, J.; Włoch, M. *Elastomer Engineering*; Gdańsk University of Technology: Gdańsk, Poland, 2017.
3. Pan, X.; Saddler, J. Effect or replacing polyol by organosolv and kraft lignin on the property and structure of rigid polyurethane foam. *Biotechnol. Biofuels* **2013**, *6*, 1–10. [CrossRef] [PubMed]
4. Prociak, A.; Rokicki, G.; Ryszkowska, J. *Polyurethane Materials*; PWN: Warsaw, Poland, 2014.
5. Noreen, A.; Zia, K.N.; Zuber, M.; Tabasum, S.; Zahoor, A.S. Bio-based polyurethane: An efficient and environment friendly coating systems: A review. *Prog. Org. Coat.* **2016**, *91*, 25–32. [CrossRef]
6. Mosiewicki, M.; Rojek, P.; Michałowski, S.; Aranguren, M.; Prociak, A. Rapeseed Oil—Based Polyurethane Foams Modified with Glycerol and Cellulose Micro/Nanocrystals. *J. Appl. Polym. Sci.* **2015**, *132*, 41602. [CrossRef]
7. Zlatanić, A.; Lava, C.; Zhang, W.; Petrović, Z. Effect of structure on properties of polyols and polyurethanes based on different vegetable oils. *J. Appl. Polym. Sci.* **2004**, *42*, 809–819. [CrossRef]
8. Borowicz, M.; Paciorek-Sadowska, J.; Lubczak, J.; Czupryński, B. Biodegradable, Flame-retardant and Bio-Based rigid Polyurethane/Polyisocyanurate Foams for Thermal Insulation Application. *Polymers* **2019**, *11*, 1816. [CrossRef] [PubMed]
9. Hatakeyama, H.; Hatakeyama, T. Environmentally Compatible Hybrid-Type Polyurethane Foams Containing Saccharide and Lignin Components. *Macromol. Symp.* **2005**, *24*, 219–226. [CrossRef]
10. Lubczak, R.; Szczęch, D.; Broda, D.; Wojnarowska-Nowak, R.; Kus-Liśkiewicz, M.; Dębska, B.; Lubczak, J. Polyetherols and polyurethane foams from starch. *Polym. Test.* **2021**, *93*, 106884. [CrossRef]
11. Szpiłyk, M.; Lubczak, R.; Lubczak, J. The biodegradable cellulose-derived polyol and polyurethane foam. *Polym. Test.* **2021**, *100*, 107250. [CrossRef]
12. Wang, W.; Xue, C.; Mao, X. Chitosan: Structural modification, biological activity and application. *Int. J. Biol. Macromol.* **2020**, *164*, 4532–4546. [CrossRef]

13. Huq, T.; Khan, A.; Brown, D.; Dhayagude, N.; He, Z.; Ni, Y. Sources, production and commercial applications of fungal chitosan: A review. *J. Bioresour. Bioprod.* **2022**, *7*, 85–98. [CrossRef]
14. Rinaudo, M. Chitin and chitosan: Properties and applications. *Prog. Polym. Sci.* **2006**, *31*, 603–632. [CrossRef]
15. Kean, T.; Thanou, M. Biodegradation, biodistribution and toxicity of chitosan. *Adv. Drug. Deliv. Rev.* **2010**, *62*, 3–11. [CrossRef]
16. Pakdel, P.M.; Peighambardoust, S.J. Review on recent progress in chitosan-based hydrogels for waste water treatment application. *Carbohydr. Polym.* **2018**, *201*, 264–279. [CrossRef] [PubMed]
17. Peers, S.; Montembault, A.; Ladavière, C. Chitosan hydrogels for sustained drug delivery. *J. Control. Release* **2020**, *326*, 150–163. [CrossRef]
18. Vunain, E.; Mishra, A.K.; Mamba, B.B. *1-Fundamentals of Chitosan for Biomedical Applications, Chitosan Based Biomaterials*; Woodhead Publishing: Sawston, UK, 2017; Volume 1.
19. Morin-Crini, N.; Lichtfouse, E.; Torri, G.; Crini, G. Applications of chitosan in food, pharmaceuticals, medicine, cosmetics, agriculture, textiles, pulp and paper, biotechnology, and environmental chemistry. *Environ. Chem. Lett.* **2019**, *17*, 1667–1692. [CrossRef]
20. Crini, G. Chitin and chitosan: Production, properties, and applications. In *Chitin and Chitosan. Discoveries and Applications for Sustainability*; Elsevier: Amsterdam, The Netherlands, 2022.
21. Singha, N.R.; Deb, M.; Chattopadhyay, P.K. Chitin and chitosan-based blends and composites. In *Biodegradable Polymers, Blends and Composites*; Elsevier: Amsterdam, The Netherlands, 2022.
22. Azmana, M.; Mahmood, S.; Hilles, A.R.; Rhman, A. Review of chitosan composite as a heavy metal adsorbent: Material preparation and properties. *Carbohydr. Polym.* **2021**, *259*, 117613. [CrossRef]
23. Wang, W.; Meng, Q.; Li, Q.; Liu, J.; Zhou, M.; Jin, Z.; Zhao, K. Chitosan Derivatives and Their Application in Biomedicine. *Int. J. Mol. Sci.* **2020**, *21*, 487. [CrossRef]
24. Sharifi, F.; Atyabi, S.M.; Norouzian, D.; Zandi, M.; Bakhshi, H.; Irani, S. Polycaprolactone/carboxymethyl chitosan nanofibrous scaffolds for bone tissue engineering application. *Int. J. Biol. Macromol.* **2018**, *115*, 243–248. [CrossRef]
25. Borsagli, F.G.L.M.; Carvalho, I.C.; Mansur, H.S. Amino acid-grafted and N-acylated chitosan thiomers: Construction of 3D bio-scaffolds for potential cartilage repair applications. *Int. J. Biol. Macromol.* **2018**, *114*, 270–282. [CrossRef]
26. Shariatinia, Z. Carboxymethyl chitosan: Properties and biomedical applications. *Int. J. Biol. Macromol.* **2018**, *120*, 1406–1419. [CrossRef]
27. Mourya, V.K.; Inamdar, N.N. Chitosan-modifications and applications: Opportunities galore. *React. Funct. Polym.* **2008**, *68*, 1013–1051. [CrossRef]
28. Madni, A.; Kousar, R.; Naeem, N.; Wahid, F. Recent advancements in applications of chitosan-based biomaterials for skin tissue engineering. *J. Bioresour. Bioprod.* **2021**, *6*, 11–25. [CrossRef]
29. Desbriéres, J.; Guibal, E. Chitosan for waste water treatment. *Polym. Int.* **2018**, *67*, 7–14. [CrossRef]
30. Aboelfetoh, E.F.; Abouburaka, A.E.; Ebeid, E.M. Binary coagulation system (graphene oxide/chitosan) for polluted surface water treatment. *J. Environ. Manag.* **2021**, *288*, 112481. [CrossRef] [PubMed]
31. Szpiłyk, M.; Lubczak, R.; Walczak, M.; Lubczak, J. Polyol and polyurethane foam from cellulose hydrolysate. *J. Chem. Technol. Biotechnol.* **2021**, *96*, 881–889. [CrossRef]
32. Lubczak, R.; Szczęch, D.; Lubczak, J. From starch to oligoetherols and polyurethane foams. *Polym. Bull.* **2020**, *77*, 5725–5751. [CrossRef]
33. Fernandes, S.; Freire, C.; Neto, C.P.; Gandini, A. The bulk oxypropylation of chitin and chitosan and the characterization of the ensuing polyol. *Green Chem.* **2008**, *10*, 93–97. [CrossRef]
34. Jain, A.; Gulbake, A.; Shilpi, S.; Jain, A.; Hurkat, P.; Sanjay, K.J. A New Horizon in Modifications of Chitosan: Syntheses and Applications. *Crit. Rev. Ther. Drug Carr. Syst.* **2013**, *30*, 91–181. [CrossRef]
35. Zhu, C.; Zou, S.; Rao, Z.; Min, L.; Liu, M.; Liu, L.; Fan, L. Preparation and characterization of hydroxypropyl chitosan modified with nisin. *Int. J. Biol. Macromol.* **2017**, *105*, 1017–1024. [CrossRef]
36. Fan, L.; Zou, S.; Ge, H.; Xiao, Y.; Wen, H.; Feng, H.; Liu, M.; Nie, M. Preparation and characterization of hydroxypropyl chitosan modified with collagen peptide. *Int. J. Biol. Macromol.* **2016**, *93*, 636–643. [CrossRef] [PubMed]
37. Pedro, R.; Pereira, S.; Goycoolea, F.M.; Schmitt, C.C.; Neumann, M.G. Self-aggregated nanoparticles of N-dodecyl, N′-glycidyl(chitosan) as pH-responsive drug delivery systems for quercetin. *J. Appl. Polym. Sci.* **2018**, *135*, 45678. [CrossRef]
38. Park, J.H.; Kwon, S.; Lee, M.; Chung, H.; Kim, J.H.; Kim, Y.S.; Park, R.W.; Kim, I.S.; Seo, S.B.; Kwon, I.C.; et al. Self-assembled nanoparticles based on glycol chitosan bearing hydrophobic moieties as carriers for doxorubicin: In vivo biodistribution and anti-tumor activity. *Biomaterials* **2006**, *27*, 119–126. [CrossRef] [PubMed]
39. Lee, E.S.; Park, K.-H.; Park, I.S.; Na, K. Glycol chitosan as a stabilizer for protein en-capsulated into poly(lactide-co-glycolide) microparticle. *Int. J. Pharm.* **2007**, *338*, 310–316. [CrossRef]
40. Uscátegui, Y.L.; Arévalo-Alquichire, S.J.; Gómez-Tejedor, J.A.; Vallés-Lluch, A.; Diaz, L.E.; Valero, M.F. Polyurethane-based bio-adhesive synthesized from polyols derived from castor oil (*Ricinuscommunis*) and low concentration of chitosan. *J. Mater. Res.* **2017**, *32*, 3699–3711. [CrossRef]
41. Zuo, D.-Y.; Tao, Y.-Z.; Chen, Y.-B.; Xu, W.-L. Preparation and characterization of blend membranes of polyurethane and superfine chitosan powder. *Polym. Bull.* **2009**, *62*, 713–725. [CrossRef]

42. Schio, R.; da Rosa, B.C.; Goncalves, J.O.; Pinto, L.A.A.; Mallmann, E.S.; Dotto, G.L. Synthesis of a bio–based polyurethane/chitosan composite foam using ricin oleic acid for the adsorption of Food Red 17 dye. *Int. J. Biol. Macromol.* **2019**, *121*, 373–380. [CrossRef]
43. Silva, S.S.; Menezes, S.M.C.; Garcia, R.B. Synthesis and characterization of polyurethane-g-chitosan. *Eur. Polym. J.* **2003**, *39*, 1515–1519. [CrossRef]
44. Zia, K.M.; Anjum, S.; Zuber, M.; Mujahid, M.; Jamil, T. Synthesis and molecular characterization of chitosan based polyurethane elastomers using aromatic diisocyanate. *Int. J. Biol. Macromol.* **2014**, *66*, 26–32. [CrossRef]
45. Qin, H.; Wang, K. Study on preparation and performance of PEG-based polyurethane foams modified by the chitosan with different molecular weight. *Int. J. Biol. Macromol.* **2019**, *140*, 877–885. [CrossRef]
46. Lee, H.C.; Jeong, Y.G.; Min, B.G.; Lyoo, W.S.; Lee, S.C. Preparation and acid dye adsorption behavior of polyurethane/chitosan composite foams. *Fibers Polym.* **2009**, *10*, 636–642. [CrossRef]
47. Sasidharan, A.P.; Meera, V.; Raphael, V.P. Investigations on characteristics of polyurethane foam impregnated with nanochitosan and nanosilver/silver oxide and its effective nessin phosphate removal. *Environ. Sci. Pollut. Res.* **2021**, *28*, 12980–12992. [CrossRef] [PubMed]
48. Kasaai, M.R.; Arul, J.; Charlet, G. Intrinsic Viscosity-Molecular Weight relationship for Chitosan. *J. Polym. Sci. Part B Polym. Phys.* **2000**, *38*, 2591–2598. [CrossRef]
49. Wang, W.; Bo, S.; Li, S.; Qiin, W. Determination of the Mark-Houwink equation for chitosans with different degrees of deacetylation. *Int. J. Biol. Macromol.* **1991**, *13*, 281–284. [CrossRef] [PubMed]
50. Brojer, Z.; Hertz, Z.; Penczek, P. *Epoxy Resins*; WNT: Warsaw, Poland, 1972.
51. Kijowska, D.; Wołowiec, S.; Lubczak, J. Kinetics and mechanism of initial steps of synthesis of polyetherols from melamine and ethylene carbonate. *J. Appl. Polym. Sci.* **2004**, *93*, 294–300. [CrossRef]
52. *Standards PN-93/C-89052.03*; Polyethers for Polyurethanes. Test Methods. Determination of the Hydroxyl Number. Polish Committee for Standardization: Warsaw, Poland, 1993.
53. Nizioł, J.; Zieliński, Z.; Rode, W.; Ruman, T. Matrix-free laser desorption-ionization with silver nanoparticle enhanced steel targets. *Int. J. Mass Spectrom.* **2013**, *335*, 22–32. [CrossRef]
54. *Polish (European) Standards PN-EN ISO 845-2000*; Cellular Plastics and Rubbers. Determination of Apparent (Bulk) Density. Polish Committee for Standardization: Warsaw, Poland, 2001.
55. *Polish (European) Standards PN-EN ISO 2896-1986*; Cellular Plastics, Rigid. Determination of Water Absorption. Polish Committee for Standardization: Warsaw, Poland, 2001.
56. *Polish (European) Standards PN-EN ISO 2796-1986*; Cellular Plastics, Rigid. Test of Dimensional Stability. Polish Committee for Standardization: Warsaw, Poland, 1986.
57. *Polish (European) Standards PN- EN ISO 844-1978*; Cellular Plastics, Compression Test for Rigid Materials. Polish Committee for Standardization: Warsaw, Poland, 2010.
58. *Standard ISO17556-2019*; Plastics—Determination of the Ultimate Aerobic Biodegradability of Plastic Materials in Soil by Measuring the Oxygen Demand in a Respirometer or the Amount of Carbon Dioxide Evolved. ISO/TC 61 Plastics, Subcommittee SC14 Environmental Aspect. Technical Committee: Warsaw, Poland, 2019.
59. *Standard ISO11274:2019*; Soil Quality—Determination of the Water-Retention Characteristic—Laboratory Methods. Chemical and Physical Characterization. Hydrological Properties of Soils Technical Committee ISO/TC 190/SC 3: Warsaw, Poland, 2020.
60. *Standard ISO 10390-2005*; Soil Quality—Determination of pH. ISO/TC 190/SC 3Chemical Characteristics of Soils. Technical Committee: Geneva, Switzerland, 2005.
61. Czupryński, B. *Questions of Chemistry and Technology of Polyurethanes*; The Publishing House of the Academy of Bydgoszcz: Bydgoszcz, Poland, 2004.
62. Wirpsza, Z. *Polyurethanes: Chemistry, Technology, Application*; WNT: Warsaw, Poland, 1991.
63. Diab, M.A.; El-Sonbati, A.Z.; Bader, D.M.D. Thermal stability and degradation of chitosan modified by benzophenone. *Spectrochim. Acta A Mol. Biomol. Spectrosc.* **2011**, *79*, 1057–1062. [CrossRef]

polymers

MDPI

Review

Chitosan-Based Dressing Materials for Burn Wound Healing

Shiyu Li [1], Wenlong Pan [1], Ming Zhang [1], Kailu Song [1], Ziqian Zhou [1], Qilong Zhao [1], Guang-Zhao Li [2] and Chongyu Zhu [1,*]

[1] College of Chemistry and Chemical Engineering, Donghua University, Shanghai 201620, China
[2] Key Laboratory of Materials and Surface Technology (Ministry of Education), School of Materials Science and Engineering, Xihua University, Chengdu 610039, China
* Correspondence: czhu@dhu.edu.cn

Abstract: The treatment of burn injuries remains a significant global challenge. Although conventional cellulose-based dressings are still the dominant clinical choice, chitosan-based burn wound dressing materials have emerged as a promising alternative due to their unique physicochemical properties and biocompatibility. In this mini-review, we aim to provide a summary of recent advances in chitosan-based dressing materials and highlight their advantages in the treatment of burn wounds. Specifically, we first outline the chemical structure and synthesis methods of chitosan and its derivatives. Subsequently, various forms of chitosan-based dressings are introduced, with a particular focus on hydrogels and micro/nanofibers dressings, along with an overview of their preparation methods. Considering the microenvironment of the burn wound site, we then summarize the design principles and clinical efficacy of chitosan-based dressings with antimicrobial and/or antioxidative activity. Additionally, the applications of chitosan dressings in tissue engineering for burn treatment are also discussed, including growth factor delivery, gene therapy, and stem cell-based treatments. Finally, we examine the main challenges of chitosan-based dressing materials and the potential future directions. Through this mini-review, we expect to provide new perspectives for the development of wound dressings for burn care.

Keywords: chitosan; burn wound; dressing forms; antimicrobial activity; antioxidative activity; tissue engineering

1. Burn Wound and Healing Materials

Burns or burn wounds are injuries to the skin caused by various factors, including heat, electricity, chemicals, and radiation [1]. Depending on the severity of the skin damage, burns are typically classified into four degrees [2]. A first-degree burn affects the superficial layers, primarily involving the epidermis. Second-degree burns involve both the epidermis and dermis. Based on the depth of dermal damage, second-degree burns can be further classified into superficial partial-thickness burns and deep partial-thickness burns. A third-degree burn affects the full thickness of the skin, including the subcutaneous fat. A fourth-degree burn extends beyond the skin into the underlying muscles, tendons, and even bones [3]. While superficial burn wounds that only affect the epidermal layer of the skin often heal rapidly without medical intervention, deep burns or those covering a large total body surface area (TBSA) can pose a significant life-threatening risk and may require whole blood replacement and surgical intervention [4,5]. Consequently, the treatment of burn wounds remains a major concern in modern society.

Burn wound healing typically consists of several overlapping and continuous phases: the hemostasis/inflammatory, cellular proliferation and matrix remodeling. Compared to other types of wounds, burn wounds often cause vascular endothelial hyperpermeability, leading to the extravasation of plasma components (e.g., water, small molecules, and proteins) into interstitial space and the subsequent formation of a large amounts of exudates [6]. The nutrient-laden exudates along with necrotic tissues provide microenvironment for bacterial growth, increasing the risk of infection. Moreover, extensive or deep burns wound healing slowly and are prone to hypertrophic scarring and contractures, which affects the appearance and functions of the burnt area. Therefore, it is crucial to strengthen burn wound management to prevent infection, promote tissue regeneration and vascular reconstruction [7].

To address the challenges in burn wound treatment, the development of effective burn wound dressings is critical. While traditional cellulose-based dressings like gauze [8] have proven successful, other bio-derived polymers like sodium alginate [9], silk fibroin [10], collagen [11], and particularly chitosan and its derivatives, have emerged as highly promising alternatives. Among all, burn wound dressings formulated with chitosan and its derivatives stand out due to their cost-effective raw materials, while offering excellent biocompatibility and biodegradability. These features allow them suitable as burn wound dressings to conform the wound healing process and minimizing the risk of adverse reactions or secondary trauma upon removal [12]. Moreover, they possess intrinsic antimicrobial properties, providing a "clean" wound healing environment against infections [13]. Additionally, their hydrophilic nature allows them to absorb wound exudate and maintain a moist wound bed, which supports optimal healing conditions [14]. These attributes, combined with their potential to stimulate cell growth and tissue regeneration, make chitosan and its derivatives be valuable materials in advanced burn wound dressing formulations [15–17].

In this review, we will provide an overview of the recent advancements in chitosan-based dressing materials for burn wound healing in the past decade. We will first introduce the fundamental of chitosan and its derivatives, including their synthesis and chemical/physical properties. Next, we will explore the two major types of burn wound dressing forms (hydrogels and micro/nanofibers), highlighting their structural features and the advantages of each form. Finally, we will discuss the functional aspects of chitosan-based dressings in burn care, focusing on their ability to prevent infections, reduce oxidative stress, and their applications in tissue engineering. Through this mini-review, we aim to offer a deeper insight into the potential and advantages of utilizing chitosan and its derivatives for the management of burn wounds.

2. Chitosan and Its Derivatives for Burn Wound Dressings

Chitosan is a cationic polysaccharide obtained through the deacetylation of chitin, whose unique biological advantages original from its structural properties. Its parent structure, chitin, is found in crustacean exoskeletons [18], insect cuticles, and fungal cell walls [19]. Chitin possesses great biocompatibility, owning to its structure which is similar to extracellular glycosaminoglycans and its enzymatic degradation into non-toxic components. When chemically converted to chitosan, this biopolymer acquires protonatable amino groups that confer pH-dependent cationic charges at physiological conditions. These charged moieties facilitate electrostatic interactions with anionic cell membrane components, enhancing cellular adhesion and proliferation capabilities.

The functional properties of chitosan are collectively determined by three interrelated parameters, including biotic origin, degree of deacetylation, and molecular weight. First, fungal-derived chitosan is safer and less allergenic than crustacean-derived chitosan due to the absence of tropomyosin [20]. Second, the degree of deacetylation, as a core regulatory factor, simultaneously influences the physicochemical characteristics and biological behavior of material. A higher degree of deacetylation enhances crystallinity [21] and thermal stability [22], as well as increases surface charge density by exposing more amino groups. This allows chitosan to improve mechanical properties [23,24] by reinforcing structural integrity and resisting degradation, meanwhile optimizing biological performance by strengthening electrostatic interactions with cells. Finally, differences in molecular weight affect polymer chain entanglement dynamics and degradation rates, providing additional tunability for specific applications.

Chitosan, as a copolymer, is formed by the linkage of N-acetyl-D-glucosamine and D-glucosamine units through β-(1,4)-glycosidic bonds [25]. Notably, the amino groups on these D-glucosamine units make chitosan a unique bio-derived cationic polysaccharide, thereby endowing it with inherent antimicrobial properties [26]. Meanwhile, due to the abundant intermolecular and intramolecular hydrogen bonds within its molecular structure, chitosan displays limited solubility in water and most organic solvents [27,28], which sometimes causes problems during the preparation and application of burn wound dressings [29,30].

To improve the solubility of chitosan, chemical modification has emerged as an effective strategy [31,32]. Through chemical modification, the hydrophilicity/hydrophobicity of the side chains of chitosan is altered and the chain length of chitosan is often shortened, which weakens its intramolecular and intermolecular hydrogen bonding [33]. This, in turn, reduces its crystallinity and improves its water solubility [34]. Generally, the chemical modification of chitosan targets the two functional groups on its D-glucosamine units: the amino group at the C-2 position and the hydroxyl group at the C-6 position owing to their pronounced chemical reactivity [35] (Table 1). For instance, by introducing hydrophilic groups such as carboxyl groups [36], hydroxyl groups, sulfonic acid groups, quaternary ammonium salts, and phosphorylated groups [37], chitosan derivatives with improved water solubility can be obtained. Such examples including commercially available source like carboxymethyl chitosan, glycol chitosan, and hydroxypropyl chitosan. Among them, carboxymethyl chitosan exhibits good solubility in aqueous solutions across wide pH range and demonstrates excellent processing compatibility with other biocompatible macromolecules such as sodium alginate and dextran, making it highly favored in burn wound dressings [38].

Table 1. The common approaches to chemically modify chitosan for improved solubility.

Chemically Modified Chitosan	Principle of Design	Reagents for Modification	Chemical Formula	Reference
Carboxymethyl chitosan	Introduce hydrophilic groups into chitosan	Chloroacetic acid or its sodium salt	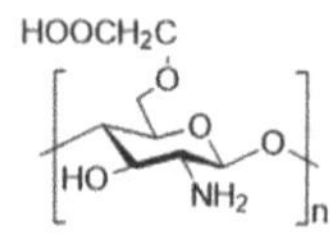 (DS < 1) 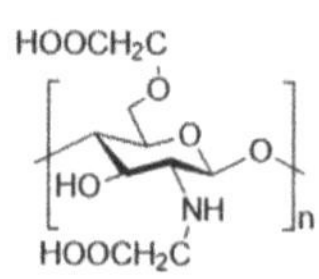 (DS ≥ 1)	[39]
Quaternary amino chitosan		Quaternizing agents, such as methyl iodide and glycidyl trimethylammonium chloride	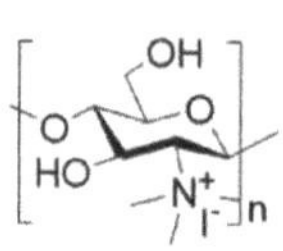 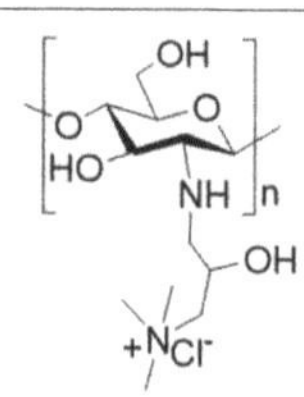	[40,41]

Table 1. *Cont.*

Chemically Modified Chitosan	Principle of Design	Reagents for Modification	Chemical Formula	Reference
Acylated chitosan	Disrupt the intramolecular and intermolecular hydrogen bonds of chitosan	Organic acids, acid anhydrides or acyl chlorides	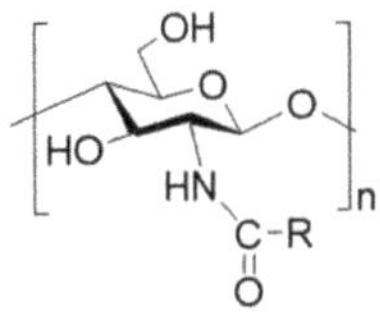	[42]
Alkylated chitosan		Halogenated hydrocarbons or higher aliphatic aldehydes	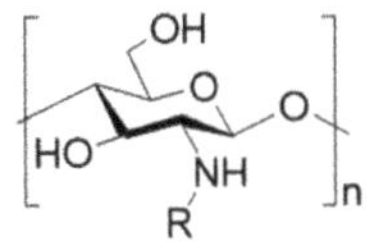	[43–45]

In addition to improving solubility, chemical modification of chitosan, particularly at the amino groups in the C-2 position, can impart additional functions such as antioxidative activity. For instance, chitosan derivatives modified with polyphenol structures such as gallic acid [46] and catechol [47] can be manufactured into burn wound dressings with antibacterial and antioxidant properties, promoting wound healing. Furthermore, these polyphenol structures can further interact with other compounds (i.e., MgO [48], Fe(III) ions [49]) to form additional crosslinking, thereby allowing for the fine-tuning of the mechanical strength of the chitosan-based burn wound dressing.

3. Forms of Chitosan-Based Burn Wound Dressings

To address the diverse needs of burn treatment, various forms of chitosan-based burn wound dressing materials have been developed [50], including films [51], sponges/foams [52], nanoparticles [53], hydrogels [54] and micro/nanofibers [55] (Figure 1). Each form provides different benefits suited to clinical situations. For instance, nanoparticles are typically administered as liquid coatings or through spray-based delivery systems for topical wound application. Alternatively, chitosan-based films function as temporary protective barriers post-debridement, while their sponge/foam variants specialize in exudate absorption through porous architectures. Although all these forms have demonstrated therapeutic potential in burn wound management, chitosan-based hydrogels and micro/nanofibers remain the most extensively researched dressing formats in the past decade [16,56,57]. This is mainly attributed to their superior exudate absorption, improved gas permeability and better mechanical match to the wound areas compared to other forms [58]. More importantly, their porous structures can mimic the natural extracellular matrix, providing a scaffold for cell growth and thereby promoting tissue regeneration [54,55], which is beneficial for deep partial-thickness burn wounds or third-degree burn wounds healing. Therefore, in this section, we will focus on the forms of these two major categories and introduce a brief overview on their preparation methods.

Possessing a three-dimensional network structure, hydrogel exhibits comparable mechanical properties to the skin and holds a large amount of water within its structure [59,60]. Hence, burn wound dressings in the form of hydrogel not only act as a physical barrier for wounds, but also offer effective wound exudate absorption, maintaining a moist environment favorable for wound healing [61]. In addition, hydrogel-based burn wound dressings can offer sustained hydration and localized cooling, reducing pain at the wound area [2,61]. Furthermore, the high porosity of these hydrogels not only facilitates smooth gas exchange between the burn wound and the external environment [62], but also enables them to serve as effective drug carriers [63].

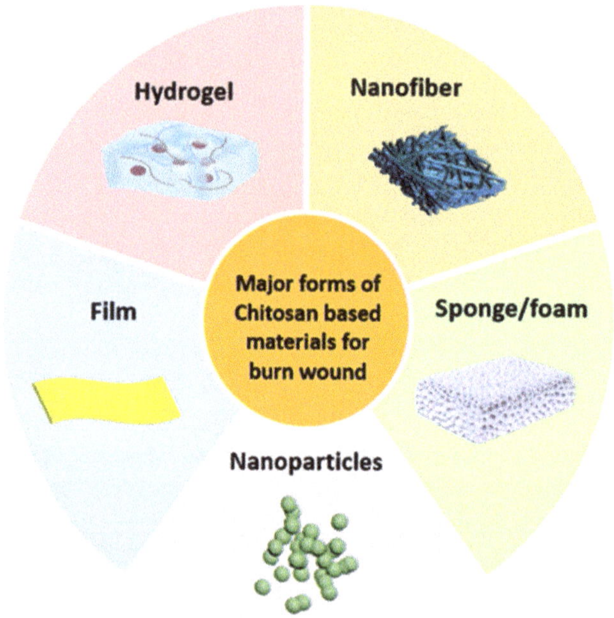

Figure 1. Major forms of chitosan-based wound dressing materials.

The preparation of chitosan-based hydrogel wound dressings typically relies on chemical crosslinking reactions targeting the amino groups in chitosan [64,65]. Common crosslinking agents, such as glyoxal [66] and glutaraldehyde [67], react with amino groups to form stable three-dimensional network structures [68]. Alternatively, the amino groups in chitosan can be further modified with photocrosslinkable groups, such as (meth)acrylic anhydrides, which enables the fabrication of tailored hydrogels via digital light processing (DLP)-based three-dimensional (3D) printing. Wang et al. demonstrated this by developing a methacrylated chitosan hydrogel precursor containing a photoinitiator [69]. By further incorporating therapeutic agents like lidocaine hydrochloride (LIDHCl) and levofloxacin (LVX), they developed drug delivery systems on demand for personalized applications such as pain management or infection-resistant burn wound dressings (Figure 2).

It is noteworthy that the crosslinkers do not necessarily need to form stable crosslinking points to produce hydrogels. Cross-linking can also be formed by dynamic bonds [70] or weaker interactions such as hydrogen bonds, electrostatic attraction, or hydrophobicity [71]. For example, Shen et al. developed a hydrogel by reacting the amino groups of a chitosan derivative (gallic acid-modified chitosan, CSGA) with the aldehyde groups of oxidized dextran via dynamic Schiff base reaction [72] (Figure 3A). Benefiting from the reversibility of Schiff base linkages, this formed hydrogel not only exhibits self-healing properties, but also can be continuously extruded through needles (Figure 3B,C), filling in irregular wound areas and promote in situ regeneration.

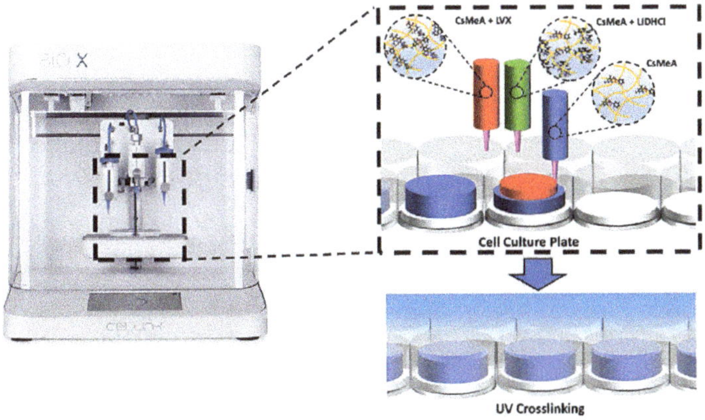

Figure 2. Schematic diagram of manufacturing personalized chitosan methacrylate loaded with lidocaine hydrochloride (LIDHCl) or levofloxacin (LVX) using a multi-nozzle printer. Adapted with permission from Ref. [69]. Copyright © 2021 Wiley-VCH GmbH (Weinheim, Germany).

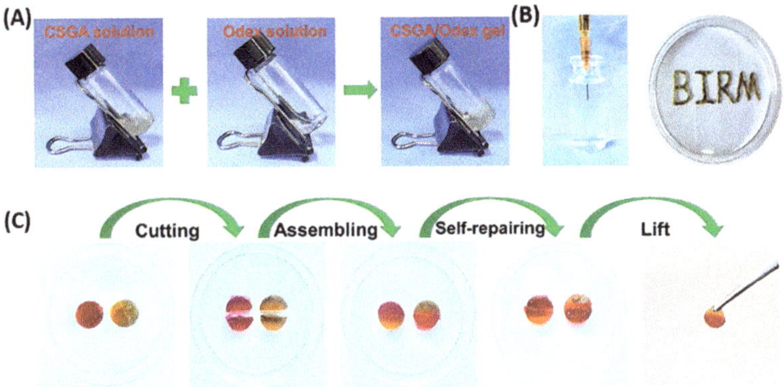

Figure 3. (**A**) The hydrogel by mixing gallic acid-modified chitosan (CSGA) solution and oxidized dextran (ODex) solution. (**B**) CSGA/ODex injectable hydrogel. (**C**) The self-healing property of CSGA/ODex hydrogels. Adapted with permission from Ref. [72]. Copyright © 2023 Published by Elsevier Ltd. (Amsterdam, The Netherlands).

Chitosan-based micro/nanofibers, characterized by their high porosity [55], represent another promising candidate for burn wound treatment due to their excellent gas permeability, effective hemostasis as well as exudate absorption capabilities. Through electrospinning technology, chitosan-based micro/nanofibers can be engineered into core–shell structures [73] or mesoporous structures [74] and subsequently processed into functional dressings. Additionally, this technology facilitates the preparation of chitosan-based dressings with layered structures. A representative study by Dhara et al. demonstrated the electrospinning of bilayered structures with different morphologies using polycaprolactone (PCL)-Chitosan emulsions stirred for different durations [75]. Specifically, the emulsions stirred for 5 min produced nanofibers characterized by a porous and loosely organized architecture. Following functionalization with Type I collagen, this layer served as a substrate that mimicked the dermal matrix. In contrast, the emulsions stirred for 12 h generated densely packed nanofibers that structurally replicated the basement membrane, thereby forming an ultrathin yet continuous top layer (Figure 4).

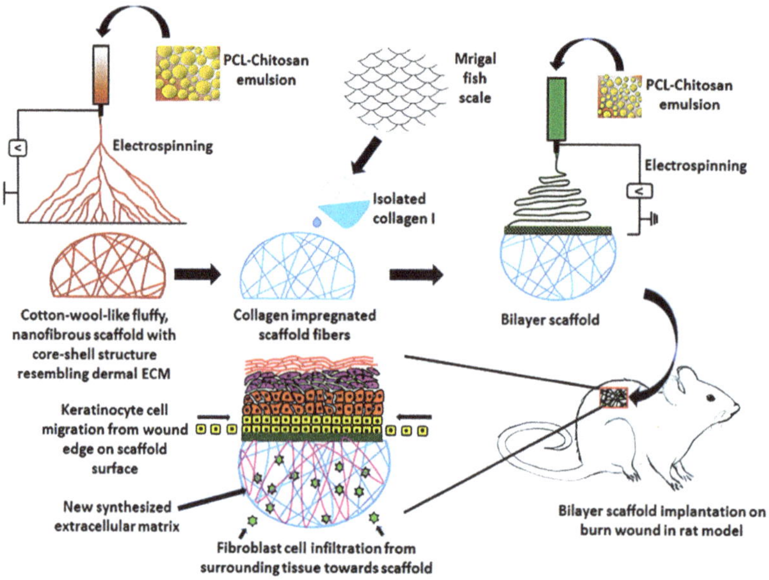

Figure 4. Schematic diagram representing development of a bilayer electrospun dressing and its application to healing burn wound in a rat model. Reprinted with permission from Ref. [75]. Copyright © 2013 Royal Society of Chemistry (London, UK).

4. Functional Chitosan-Based Wound Dressings for Burn Wound Treatment

Second-degree burns damage both the epidermis and part of the dermis, often accompanied by severe pain and a high risk of infection [76,77]. Therefore, it is essential for burn wound dressings to control bacterial growth as well to reduce inflammatory responses [78]. To address this issue, antimicrobial and antioxidant properties have been incorporated into chitosan-based burn wound dressings [79]. For more severe burn wounds, tissue engineering techniques are often required to assist in wound repair. Hence, there is a growing interest in developing new chitosan-based burn wound dressings containing bioactive components to promote tissue regeneration at the wound sites. In this section, we will present an overview of the design and application of functional chitosan-based dressing materials. We will begin by exploring strategies for developing chitosan-based burn wound dressings possessing antimicrobial and/or antioxidant properties, followed by introducing the recent advances in tissue engineering facilitated by chitosan-based dressings.

4.1. Antimicrobial and Antioxidant Chitosan-Based Burn Wound Dressings

Despite the antimicrobial nature of chitosan and its derivatives, additional antimicrobial agents are often incorporated to provide a better protection for the burn wounds against infection. To achieve this goal, the most straightforward approach is to direct mixing antimicrobial agents into the chitosan-based burn wound dressing. For example, Doaa Alshora et al. prepared a burn wound dressing in the film form composed of chitosan and sodium alginate and introduced silver sulfadiazine [80], a small-molecular antimicrobial agent, into this burn wound dressing using the solvent casting method. Compared with non-medicated chitosan-sodium alginate film, the silver sulfadiazine -loaded biofilm exhibited significantly superior antibacterial effects [81].

Antimicrobial agents can also be encapsulated into chitosan-based hydrogels in addition to simple mixing [82]. For instance, colistin is a potent polypeptide antibiotic against various Gram-negative "superbugs" such as Pseudomonas aeruginosa. However, its potential nephrotoxicity limits its application. To tackle this problem, Velkov and Haddleton et al. integrated colistin into a glycol chitosan-based hydrogel by forming imine bonds among colistin, glycol chitosan, and an aldehyde-modified poly (ethylene glycol) crosslinker (DF-PEG) [83], as illustrated in Figure 5A. Within this system, colistin was uniformly distributed in the hydrogel, leading to a high drug loading efficiency. More interestingly, the presence of colistin also accelerated the gelation process, indicating the incorporation of colistin is beyond simple encapsulation. In an in vivo burn infection model, the colistin-loaded hydrogel not only killed colistin-sensitive Pseudomonas aeruginosa strains effectively, but also demonstrated potency against colistin-resistant "superbugs". (Figure 5B,C). Owing to the dynamic feature of imine bonds as well as the biodegradability of glycol chitosan, this burn wound dressing showed "on-wound" degradation over 24 h.

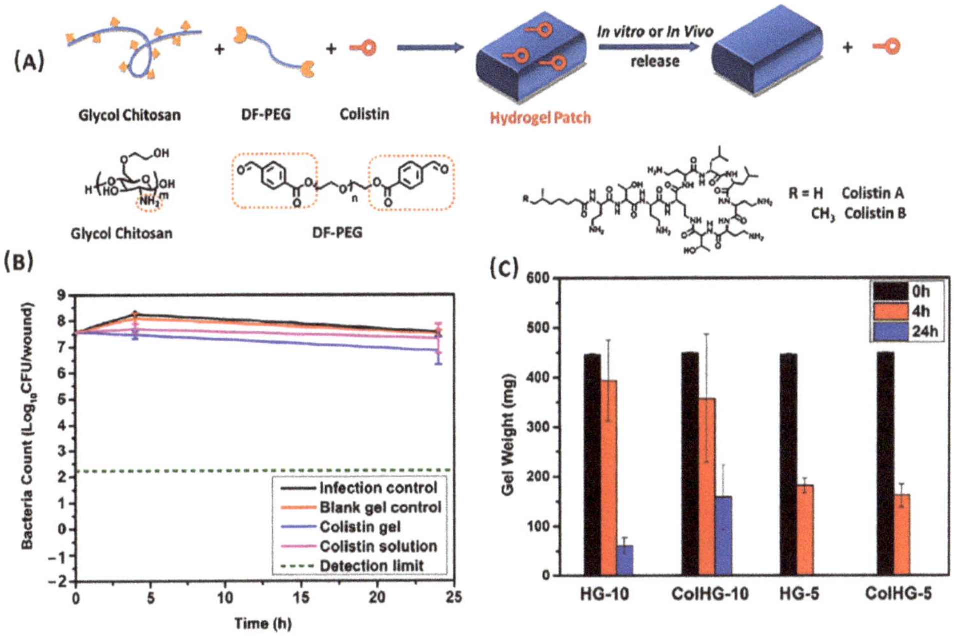

Figure 5. (**A**) Schematic diagram of the synthesis of the colistin-containing hydrogel. (**B**) The animal "burn" infection model test of the colistin-loaded hydrogel against colistin-resistant strain at a low dose (0.3 mg/wound). Black line: blank infection control; red line: blank HG-10 hydrogel; blue line: HG 10 with colistin; pink line: colistin solution. The detection limit is shown in dash line. (**C**) The weight loss of the colistin-loaded hydrogels in vivo over the time. Adapted with permission from Ref. [83]. Copyright © 2016 The WILEY-VCH (Weinheim, Germany).

In addition to infections, when a burn occurs, the skin and surrounding tissues are severely damaged, immediately triggering a series of complex inflammatory responses [84]. During this process, immune cells respond rapidly, releasing a large number of inflammatory mediators, including reactive oxygen species (ROS) and reactive nitrogen species (RNS), which play crucial roles in clearing pathogens and damaged cells. However, excessive ROS and RNS can also lead to oxidative stress, further exacerbating tissue damage and causing severe complications such as systemic inflammatory response syndrome, immunosuppression, infection, sepsis, and multiple organ failure.

Incorporating antioxidants into chitosan-based burn wound dressings is an effective approach to address the abovementioned challenge. Antioxidants can be categorized into non-enzymatic and enzymatic types based on their mechanisms of action [85]. Non-enzymatic antioxidants include low-molecular-weight compounds such as vitamin E, vitamin C, flavonoids, and glutathione, while enzymatic antioxidants include catalase and superoxide dismutase. Additionally, some natural products, such as curcumin, salvianolic acid, crocin, and quercetin, also show strong antioxidative activity. They can protect burn wounds from oxidative stress damage through various mechanisms, such as scavenging free radicals, inhibiting oxidase activity, and regulating the expression of antioxidant enzymes. Through a similar encapsulation strategy as described in antimicrobial burn wound dressing, chitosan-based dressings can also be loaded with antioxidants to further enhance their antioxidant properties.

For example, Salvianolic acid B is an active component extracted from the traditional Chinese herb Salvia miltiorrhiza, known for its excellent antioxidant and anti-inflammatory properties. Incorporating Salvianolic acid B, He and Zhang et al. designed an antioxidant hydrogel dressing for burn treatment. Crosslinked through the dynamic imine bonds between glycol chitosan and tetra-arm polyethylene glycol (terminated with acetaldehyde or benzaldehyde) [1], this hydrogel exhibits shape adaptability, self-healing property, as well as rapid degradation. The encapsulated Salvianolic acid B, on the other hand, can be released to the wound site through a combined mechanism of passive diffusion and hydrogel degradation, showing excellent antioxidant performance in vitro. In a rat model of deep second-degree burn wounds, the Salvianolic acid B-loaded hydrogel rapidly reduced wound temperature, modulated the oxidative microenvironment of the wound, promoted angiogenesis, and eventually accelerated wound healing.

Recently, it has been found that the presence of bacteria often induces oxidative stress in wounds, while the wounds under oxidative stress are more prone to bacterial colonization [86]. Therefore, the combination of antibacterial and antioxidant properties can further enhance the therapeutic effect of dressings on burn wounds. For example, baicalein, which exhibits excellent antimicrobial and antioxidant activities, was used by Wang, Bai, Du, and their research team to develop a dual-functional nanofiber membrane for burn wound treatment [87]. In this study, baicalein was first modified using valeric anhydride and then reacted with the amino groups in chitosan to form baicalein-modified chitosan. This baicalein-modified chitosan was subsequently combined with polyvinyl alcohol and electrospun into a nanofiber membrane. In vitro experiments demonstrated that the baicalein-modified chitosan nanofiber membrane not only effectively neutralizes ROS but also significantly inhibits the growth of both Staphylococcus aureus and Escherichia coli. Furthermore, the nanofiber membrane exhibited outstanding antimicrobial and wound-healing effects in infected wound models.

Apart from encapsulating a single active ingredient, chitosan-based dressings that are loaded with multiple components have the potential to deliver programmable antimicrobial and antioxidant effects, providing an enhanced therapeutic strategy to address the varied requirements of wound healing throughout its different stages. For instance, Nie, Ma, and their team prepared a core-shell structured dual-drug-loaded nanofiber dressing using electrospinning technology for the treatment of deep second-degree burns [88]. The shell, composed of Polycaprolactone and Chitosan, was loaded with Asiaticoside (Asia) to promote collagen deposition and tissue repair. The core, on the other side, encapsulated curcumin using 2-hydroxypropyl-β-cyclodextrin (HP-β-CD) for sustained release (Figure 6A). Chitosan in the shell provided biocompatibility and antimicrobial properties, while the porous structure in the nanofiber dressing ensured uniform drug distribution and release. In vitro, 66.82% of Asia was rapidly released within 0.5 h, while 83.22% of CUR was released over 120 h (Figure 6B). Mass loss ratios of the nanofiber components in PBS buffer confirmed these release rates (Figure 6C). In a rat model, the dressing achieved a 95% wound healing rate within 14 days, with significant collagen deposition and angiogenesis.

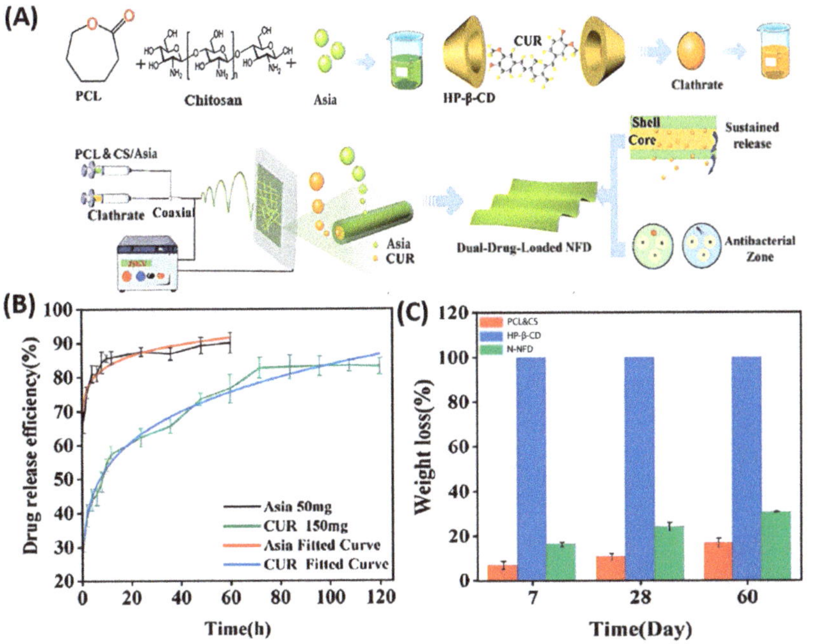

Figure 6. (**A**) Preparation process of the dual-loaded NFD; (**B**) Release curves of Asia and CUR; (**C**) Statistical graph of the degradation of N-NFD, PCL&CS and HP-β-CD at 7, 28 and 60 days in PBS solution. Adapted with permission from Ref. [88]. Copyright © 2024, American Chemical Society (Washington, DC, USA).

4.2. Chitosan-Based Burn Wound Dressings for Tissue Engineering

In addition to providing antibacterial activity and an antioxidant microenvironment, the treatment of deep partial-thickness burns wounds or third-degree burn wounds often requires the integration of tissue engineering strategies [89]. With the aid of bioactive scaffolds and suitable bioactive additives, tissue engineering to activate critical repair mechanisms such as cell migration, angiogenesis, and tissue regeneration. In recent years, chi-tosan-based dressings have shown power in tissue engineering [90]. Owing to the extracellular matrix-mimicking structure of chitosan, chitosan-based dressings are known to promote cellular adhesion and proliferation. Furthermore, by incorporating suitable bioactive components (such as growth factors, genetic materials, stem cells), chitosan-based dressings can be transformed into versatile platforms for treating different burn wounds.

Growth factor therapy delivers growth factors to wound sites, activating the proliferation of surrounding cells to facilitate wound healing. Chitosan-based dressings not only serve as a delivery system for growth factors but also promote tissue regeneration through their inherent biological activity. For instance, Hieu Tran-Van et al. developed a hydrogel film composed of carboxymethyl chitosan (CMCS) and hydroxyethyl cellulose (HEC) for the delivery of fibroblast growth factor (FGF-2), accelerating the repair of burn wounds [91]. This hydrogel film is formed through hydrogen bonding between CMCS and HEC, creating an interpenetrating polymer network with excellent swelling properties and controlled drug release capabilities, while effectively avoiding the potential toxicity issues associated with chemical cross-linking agents (Figure 7A,B). Studies have shown that the CMCS/HEC hydrogel film significantly promotes the proliferation of NIH/3T3 fibroblasts (Figure 7C), protects FGF-2 from protease degradation in vitro, and exhibits remarkable wound healing effects in a burn mouse model (Figure 7D), including accelerated epithelialization, enhanced formation of granulation tissue and blood vessels, and reduced scar formation.

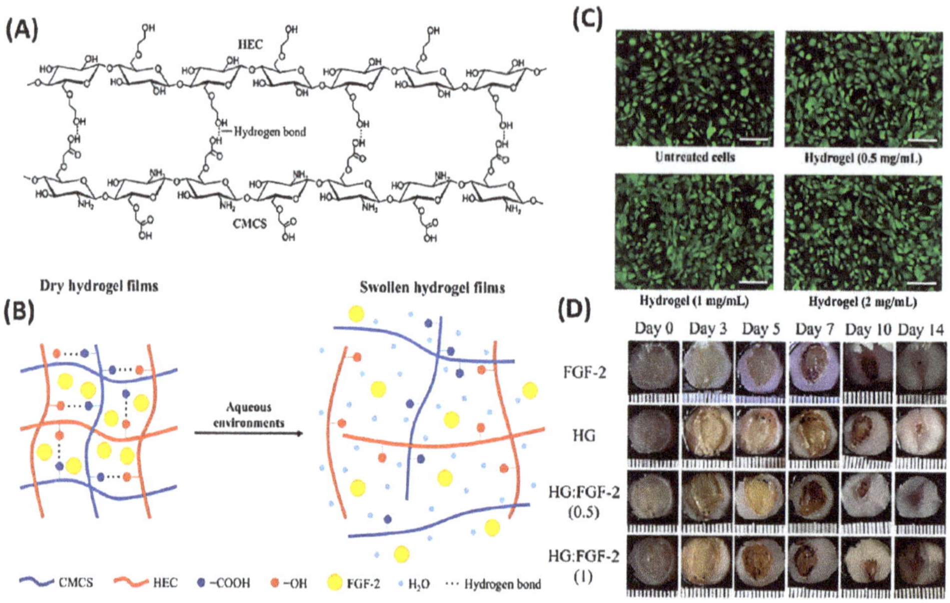

Figure 7. (**A**) Interpolymeric bonding between CMCS and HEC. (**B**) Release of FGF-2 from the swollen hydrogel film in an aqueous environment. (**C**) Effects of hydrogel of different concentrations in the culture medium on the proliferation of NIH/3T3 cells. (**D**) Optical photographs of burn wounds treated with hydrogel films with different FGF-2 release amounts. Adapted with permission from Ref. [91]. Copyright © 2023 Elsevier B.V. (Amsterdam, The Netherlands).

Gene therapy involves delivering genetic material (DNA/RNA) into cells using viral or non-viral vectors to restore or correct cellular functions. Chitosan, positively charged at physiological pH, binds to negatively charged DNA/RNA, forming protective nanoparticles that resist enzymatic degradation. As a result, chitosan-based materials are extensively studied as non-viral gene delivery vectors. However, bio-derived chitosan suffers from poor water solubility and low transfection efficiency. To address these issues, researchers have chemically modified chitosan by incorporating basic amino acids like arginine, histidine, and lysine, enhancing its water solubility and mimicking viral envelope components [92]. Notably, arginine-modified chitosan (Arg-CS) demonstrates high gene transfection efficiency both in vitro and in vivo, showcasing its potential for gene therapy applications.

Inspired by this, Chang et al. utilized Arg-CS as a gene delivery vector and developed a composite hydrogel as gene-activated matrix (GAM) for in situ treatment of deep second-degree burn wounds [93]. They first produced therapeutic plasmids/Arg-CS complexes by complexing Arg-CS with plasmid DNA (pDNA) encoding mVEGF165 and TGF-β1. These complexes were then incorporated into a composite hydrogel based on N-carboxymethyl chitosan and sodium alginate (NS-GAM). This hydrogel, with an average pore size of 100 μm and a porosity of 50.9%, facilitated pDNA release (Figure 8A,B) and supported cell adhesion and growth. In vitro, the hydrogel loaded with gene fragments sustained gene expression for at least 9 days and protected the gene fragments from enzymatic degradation and immune responses during delivery. In a rat burn model, the experimental group treated with this burn wound dressing achieved complete wound healing within 22 days (Figure 8C,D), with significantly increased expression levels of VEGF and TGF-β1 proteins, promoting neovascularization and collagen regeneration.

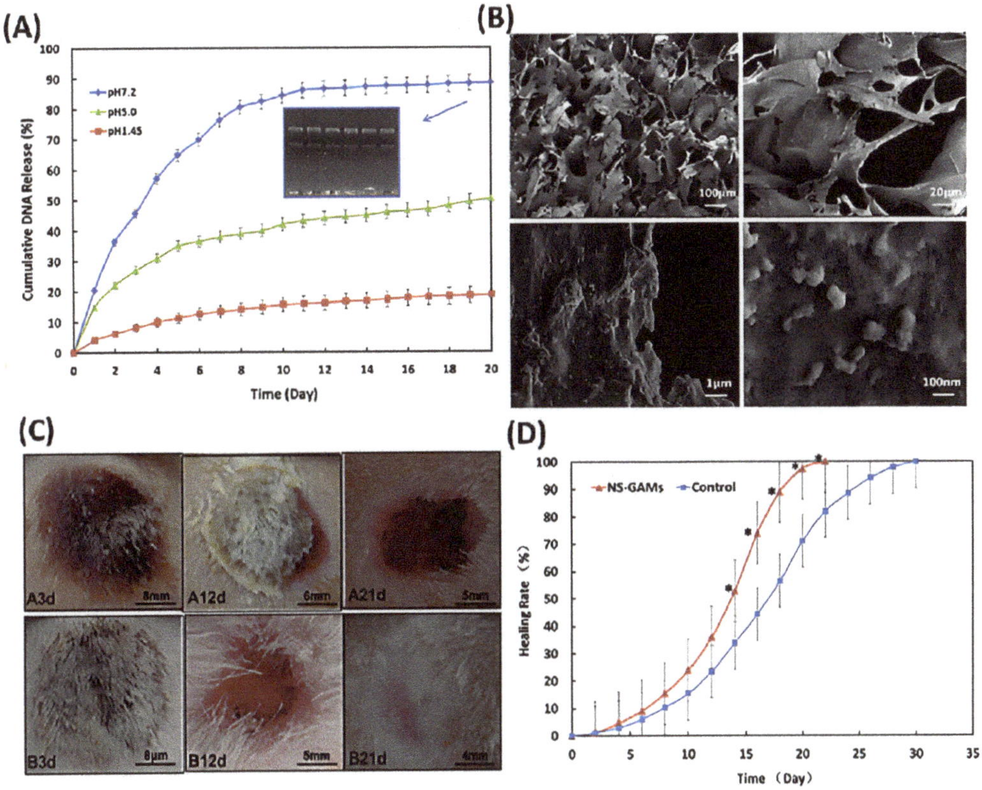

Figure 8. In vitro release of pDNA from NS-GAM. (**A**) Cumulative amount of pDNA released in vitro from NS-GAM and agarose gel electrophoresis of plasmids. (**B**) SEM of the surface of NS-GAM. Gross examination and healing rate. (**C**) Observation of deep 2nd degree burn wound. upper: control group; Below: NS-GAM group. (**D**) The calculated wound size decreases. Adapted with permission from Ref. [93]. Copy © 2021 Elsevier B.V. (Amsterdam, The Netherlands).

In addition to gene therapy, stem cell therapy has also garnered considerable attention in burn wound treatment in recent years. Unlike growth factor or gene therapy, which promotes the cell growth around the wound, stem cell therapy leverages the self-renewal and differentiation of stem cells to direct participate in the remodeling of burn-injured tissues. Combining stem cells with chitosan-based dressings can serve as a carrier material for tissue engineering, providing a suitable microenvironment for the delivery and cultivation of stem cells. For example, Yao and Li et al. developed a zwitterionic polysaccharide hydrogel based on chitosan [4] (Figure 9A). Formed by the dynamic imine bonds between sulfobetaine-modified dextran and carboxymethyl chitosan (Dex-SB-CHO), this hydrogel was injectable with self-healing performance, allowing the delivery of adipose-derived stem cells (ADSCs) to the deep wound site. Moreover, the hydrogel exhibited high nonfouling and antimicrobial properties, creating a clean microenvironment for proliferation and maintaining the stemness of ADSCs. (Figure 9B). Furthermore, this hydrogel evaded immune system recognition, reducing inflammation, thereby promoting collagen deposition as well as angiogenesis. In a mouse burn model, the ADSC-loaded hydrogel minimized the wound area in 28 days (Figure 9C) and offered scarless skin tissue regeneration.

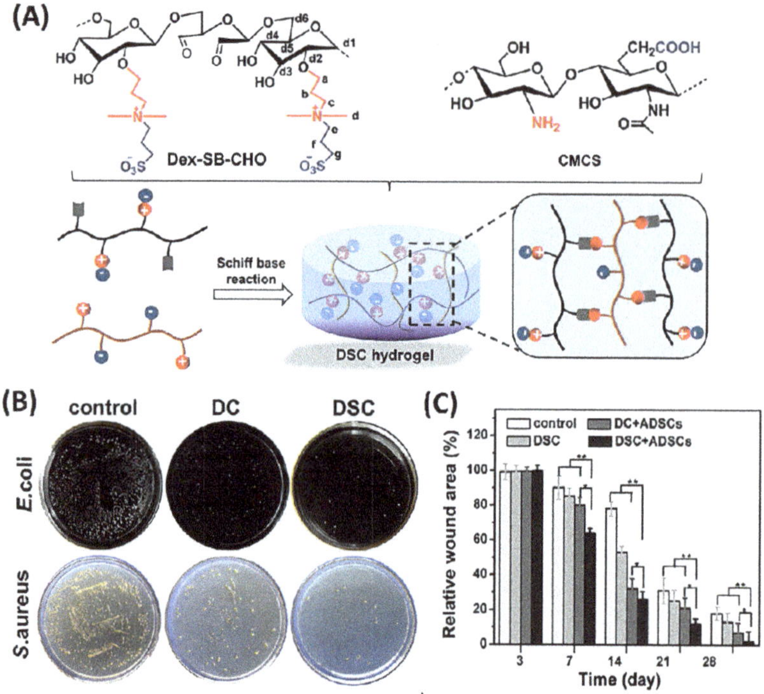

Figure 9. (**A**) Preparation of zwitterionic polysaccharide hydrogel. (**B**) Optical photographs of the relative adsorption of proteins on the surface of DC hydrogel (without zwitterionic modification) and DSC hydrogel. (**C**) The relative wound closure area, all experiments were performed in triplicate and data are reported as mean ± SD (n = 3), ** $p < 0.01$. Adapted with permission from Ref. [4]. Copy © 2022 Wiley-VCH GmbH (Weinheim, Germany).

5. Outlook and Summary

5.1. Outlook

Chitosan and its derivatives have a broader range of raw material sources than other biomacromolecules [94], ranging from traditional crustaceans and fungi to emerging sources such as bee and silkworm exoskeletons [95]. However, this diversity leads to quality variability, posing challenges for medical applications. Variations in molecular weight, degree of deacetylation, processing methods [96], and batch-to-batch differences in chain length and crystallinity result in inconsistent performance [97], while residual impurities may also elevate immunogenicity risks [98]. Furthermore, the lack of standardized detection methods causes discrepancies in the measurement of critical parameters. For instance, when the degree of deacetylation of the same batch of chitosan was tested by infrared spectroscopy versus titration method, the results differed significantly, rendering the data incomparable, further complicating standardization and regulatory [91]. To facilitate clinical use, it is essential to establish unified quality standards, optimize production processes, develop standardized testing methods and material traceability procedures. Concurrently, regulatory oversight should extend beyond chitosan raw materials to encompass various forms of chitosan-based dressing products, alongside the formulation of product-specific regulations. Such as standards for hydrogel water content and permeability, or specifications for nanofiber diameters in burn wound dressings.

Additionally, there is still potential for improvement in the management of burn wound exudate using chitosan-based dressings [99]. At present, these dressings primarily rely on their hydrophilic and porous structures to absorb exudate, which may not be sufficient when dealing with large volumes of exudate [100]. Janus dressings with asymmetric structures and unidirectional biofluid transport properties present a novel strategy for tackling this challenge [101]. Through the combination of advanced techniques like electrospinning and 3D printing, coupled with thoughtful molecular engineering and structural refinement, it is expected that chitosan-based burn wound dressings with unidirectional biofluid transport will prevent exudate reflux and secondary injuries resulting from dressing adherence. Additionally, the incorporation of intelligent monitoring components into chitosan materials paves the way for "smart" wound management [6]. By embedding sensors into these dressings may facilitate the continuous collection and analysis of burn exudate parameters, offering real-time insights to track wound healing progress and assess treatment effectiveness.

Furthermore, to address the complex and dynamic microenvironments encountered during the various stages of burn wound healing, future research should focus on developing chitosan-based burn wound dressings that offer multifunctionality and high tunability. Specifically, drawing on recent advancements in multi-component reactions for polymer modification [102], chemically modified chitosan through these reactions to produce multifunctional burn wound dressings are expected to enhance the efficacy of burn treatment. Simultaneously, integrating highly biocompatible responsive motifs, such as enzyme-responsive modules, into existing chitosan-based burn wound dressings enables them to dynamically respond to changes in the wound microenvironment throughout the healing process, thereby facilitating more efficient wound care. Moreover, it is important to develop chitosan-based burn wound dressings with enhanced biomimetic performance. These dressings should replicate the multilayered structure, mechanical properties, and biological functions of human skin, thereby offering superior protection, supporting tissue regeneration, and effectively minimizing scar formation. In summary, by conducting more in-depth research on chitosan dressings, we can offer personalized and adaptive solutions for burn treatment, tailored to the unique needs of each patient and the specific stages of the healing process.

5.2. Summary

Chitosan-based dressings have emerged as promising biomaterials for burn wound care due to their inherent biocompatibility, antimicrobial activity, and tunable functionalities. In this mini-review, we summarize recent advancements in this field, with a focus on hydrogel and micro/nanofiber formulations. These dressings not only act as physical barriers to protect wound surfaces but also maintain a moist environment conducive to healing. Crucially, they can be tailored to modulate the wound microenvironment based on burn severity, while simultaneously serving as scaffolds to support tissue regeneration in severe injuries. Despite substantial evidence supporting their therapeutic potential, the clinical translation of chitosan dressings faces hurdles such as regulatory approvals and scalable manufacturing. Continuous refinement of fabrication processes, develop standardized testing methods as well to adjust mechanical robustness and functions, is expected to overcome limitations of conventional burn wound dressings, thereby providing more efficient and patient-friendly solutions for burn treatment. Furthermore, the development of next-generation chitosan-based materials with stimuli-responsive and multifunctional capabilities holds promise for advancing wound management and improving rehabilitation outcomes for burn patients.

Author Contributions: Conceptualization, C.Z. and G.-Z.L.; formal analysis, C.Z.; investigation, C.Z., S.L. and M.Z.; resources, C.Z.; data curation, C.Z. and Q.Z.; writing—original draft preparation, C.Z., S.L. and K.S.; writing—review and editing, C.Z., S.L. and W.P.; visualization, Z.Z.; supervision, C.Z.; funding acquisition, C.Z. and G.-Z.L. All authors have read and agreed to the published version of the manuscript.

Funding: The authors gratefully acknowledge the financial support received from the Fundamental Research Funds for the Central Universities (No. 2232024D-13) and the Key Laboratory of Materials and Surface Technology, Ministry of Education (No. xxx-2024-zd003).

Institutional Review Board Statement: Not applicable.

Data Availability Statement: No new data were created.

Conflicts of Interest: The authors declare no conflicts of interest.

References

1. Ren, H.; Zhang, Z.; Lu, K.L.; Shen, Y.Y.; He, C.L.; Chen, X.S. Injectable Chitosan Hydrogels Loaded with Antioxidant Agent as First-Aid Dressings for Second-Degree Burn Wounds. *Sci. China-Technol. Sci.* **2024**, *67*, 891–901. [CrossRef]
2. Surowiecka, A.; Struzyna, J.; Winiarska, A.; Korzeniowski, T. Hydrogels in Burn Wound Management-a Review. *Gels* **2022**, *8*, 122. [CrossRef]
3. Cook, K.A.; Martinez-Lozano, E.; Sheridan, R.; Rodriguez, E.K.; Nazarian, A.; Grinstaff, M.W. Hydrogels for the Management of Second-Degree Burns: Currently Available Options and Future Promise. *Burn. Trauma* **2022**, *10*, tkac047. [CrossRef]
4. Yu, Q.; Sun, H.; Yue, Z.; Yu, C.; Jiang, L.; Dong, X.; Yao, M.; Shi, M.; Liang, L.; Wan, Y.; et al. Zwitterionic Polysaccharide-Based Hydrogel Dressing as a Stem Cell Carrier to Accelerate Burn Wound Healing. *Adv. Health. Mater.* **2022**, *12*, 2202309. [CrossRef]
5. Xu, H.L.; Chen, P.P.; ZhuGe, D.L.; Zhu, Q.Y.; Jin, B.H.; Shen, B.X.; Xiao, J.; Zhao, Y.Z. Liposomes with Silk Fibroin Hydrogel Core to Stabilize Bfgf and Promote the Wound Healing of Mice with Deep Second-Degree Scald. *Adv. Health. Mater.* **2017**, *6*, 1700344. [CrossRef]
6. Widgerow, A.D.; King, K.; Tocco-Tussardi, I.; Banyard, D.A.; Chiang, R.; Awad, A.; Afzel, H.; Bhatnager, S.; Melkumyan, S.; Wirth, G.; et al. The Burn Wound Exudate-an under-Utilized Resource. *Burns* **2015**, *41*, 11–17. [CrossRef]
7. Dewi, I.S.; Handayani, E.; Masithoh, R.F.; Kamal, S.; Nugroho, S.H.P.; Wahyuningtyas, E.S.; Amin, M.K.; Priyo; Priyanto, S. Pain Management with Virtual Reality in Burn Patients: A Literature Review. *Br. J. Community Nurs.* **2024**, *29*, S22–S28. [CrossRef]
8. Craig, C.K.; Williams, J.W.; Carter, J.E.; Holmes, J.H. Bismuth/Petroleum Gauze Plus High Density Polyethylene vs. Bismuth/Petroleum Gauze: A Comparison of Donor Site Healing and Patient Comfort. *Burns* **2022**, *48*, 1917–1921. [CrossRef]
9. Sanchez, M.F.; Guzman, M.L.; Flores-Martín, J.; Del Puerto, M.C.; Laino, C.; Soria, E.A.; Donadio, A.C.; Genti-Raimondi, S.; Olivera, M.E. Ionic Complexation Improves Wound Healing in Deep Second-Degree Burns and Reduces in-Vitro Ciprofloxacin Cytotoxicity in Fibroblasts. *Sci. Rep.* **2022**, *12*, 16035. [CrossRef]
10. Ju, H.W.; Lee, O.J.; Lee, J.M.; Moon, B.M.; Park, H.J.; Park, Y.R.; Lee, M.C.; Kim, S.H.; Chao, J.R.; Ki, C.S.; et al. Wound Healing Effect of Electrospun Silk Fibroin Nanomatrix in Burn-Model. *Int. J. Biol. Macromol.* **2016**, *85*, 29–39. [CrossRef]
11. Song, J.H.; Gu, J.T.; Dang, G.P.; Wan, M.C.; Bai, Y.K.; Bai, Q.; Jiao, K.; Niu, L.N. DNA-Collagen Dressing for Promoting Scarless Healing in Early Burn Wound Management. *Adv. Compos. Hybrid Mater.* **2025**, *8*, 212. [CrossRef]
12. Shah, J.; Patel, D.; Rananavare, D.; Hudson, D.; Tran, M.; Schloss, R.; Langrana, N.; Berthiaume, F.; Kumar, S. Recent Advancements in Chitosan-Based Biomaterials for Wound Healing. *J. Funct. Biomater.* **2025**, *16*, 45. [CrossRef] [PubMed]
13. Ma, L.; Zhu, Y.; Wang, L.; Yang, Z.; Li, X.; Chen, M.; Yuan, M.; Li, W.; Ma, X.; Xiong, H.; et al. Multi-Form Antibacterial Dressings Based on a Deep Eutectic Supramolecular Polymer Promote Healing of Burn Wounds through a Weakly Acidic Microenvironment. *Chem. Eng. J.* **2024**, *494*, 153153. [CrossRef]
14. Zhai, G.T.; Wang, Y.; Han, P.; Xiao, T.; You, J.; Guo, C.L.; Wu, X.C. Drug Loaded Marine Polysaccharides-Based Hydrogel Dressings for Treating Skin Burns. *Int. J. Biol. Macromol.* **2024**, *281*, 135779. [CrossRef]
15. Bano, I.; Arshad, M.; Yasin, T.; Ghauri, M.A. Preparation, Characterization and Evaluation of Glycerol Plasticized Chitosan/Pva Blends for Burn Wounds. *Int. J. Biol. Macromol.* **2019**, *124*, 155–162. [CrossRef]
16. Cherng, J.H.; Lin, C.A.J.; Liu, C.C.; Yeh, J.Z.; Fan, G.Y.; Tsai, H.D.; Chung, C.F.; Hsu, S.D. Hemostasis and Anti-Inflammatory Abilities of Aunps-Coated Chitosan Dressing for Burn Wounds. *J. Pers. Med.* **2022**, *12*, 1089. [CrossRef]

17. Wang, W.Q.; Meng, Q.Y.; Li, Q.; Liu, J.B.; Zhou, M.; Jin, Z.; Zhao, K. Chitosan Derivatives and Their Application in Biomedicine. *Int. J. Mol. Sci.* **2020**, *21*, 487. [CrossRef]
18. Alam Shah, M.K.; Nawaz, A.; Latif, M.S.; Ullah, W.; Ullah, A.; Khan, A.A.; Malik, A.; Kumarasamy, V.; Subramaniyan, V.; Azad, A.K. Chitosan-Based Mupirocin and Alkanna Tinctoria Extract Nanoparticles for the Management of Burn Wound: In Vitro and in Vivo Characterization. *Nanotechnol. Rev.* **2024**, *13*, 20240065. [CrossRef]
19. Huq, T.; Khan, A.; Brown, D.; Dhayagude, N.; He, Z.; Ni, Y. Sources, Production and Commercial Applications of Fungal Chitosan: A Review. *J. Bioresour. Bioprod.* **2022**, *7*, 85–98. [CrossRef]
20. Ayers, L.C.; Johnson, R.M.; Tolfree, A.R.; Al Kharji, N.M.; Senarathna, M.C.; Arora, N.; Smaldone, R.A. Effects of the Chitosan Source on the Properties of 3d Printed Polymer Composites. *ACS Appl. Polym. Mater.* **2025**, *7*, 2644–2651. [CrossRef]
21. Hsu, S.H.; Whu, S.W.; Tsai, C.L.; Wu, Y.H.; Chen, H.W.; Hsieh, K.H. Chitosan as Scaffold Materials: Effects of Molecular Weight and Degree of Deacetylation. *J. Polym. Res.* **2004**, *11*, 141–147. [CrossRef]
22. Zhang, L.H.; Gao, C.C.; Wang, Z.J.; Xie, F.W.; Chen, Y.; Meng, L.H.; Tang, X.Z. Structure and Properties of Thermomechanically Processed Chitosan-Based Biomimetic Composite Materials: Effect of Chitosan Molecular Weight. *ACS Sustain. Chem. Eng.* **2023**, *11*, 708–717. [CrossRef]
23. Tien, N.D.; Geng, T.X.; Heyward, C.A.; Reseland, J.E.; Lyngstadaas, S.P.; Blaker, J.J.; Haugen, H.J. Solution Blow Spinning of Highly Deacetylated Chitosan Nanofiber Scaffolds for Dermal Wound Healing. *Biomater. Adv.* **2022**, *137*, 212871. [CrossRef] [PubMed]
24. Heidari, F.; Razavi, M.; Bahrololoom, M.E.; Tahriri, M.; Tayebi, L. Investigation of the Mechanical Properties and Degradability of a Modified Chitosan-Based Scaffold. *Mater. Chem. Phys.* **2018**, *204*, 187–194. [CrossRef]
25. Xu, Y.; Li, L.; Cao, S.; Zhu, B.; Yao, Z. An Updated Comprehensive Review of Advances on Structural Features, Catalytic Mechanisms, Modification Methods and Applications of Chitosanases. *Process. Biochem.* **2022**, *118*, 263–273. [CrossRef]
26. Zhu, J.Y.; Ma, Z.Y.; Xiao, M.; Wang, L.S.; Jiang, X.K. Molecular Dynamics Simulations of the Interaction between Chitosan and Bacterial Membranes. *Prog. Biochem. Biophys* **2023**, *50*, 1995–2005.
27. Eulálio, H.Y.C.; Vieira, M.; Fideles, T.B.; Tomás, H.; Silva, S.M.L.; Peniche, C.A.; Fook, M.V.L. Physicochemical Properties and Cell Viability of Shrimp Chitosan Films as Affected by Film Casting Solvents. I-Potential Use as Wound Dressing. *Materials* **2020**, *13*, 5005. [CrossRef]
28. Ji, M.C.; Li, F.Y.; Li, J.Y.; Li, J.F.; Wang, X.J.; Zhang, C.W.; Peng, S.X.; Man, J. Physical, Antibacterial, Blood Coagulation, and Healing Promotion Evaluations of Chitosan Derivative-Based Composite Films. *Int. J. Biol. Macromol.* **2024**, *278*, 134714. [CrossRef]
29. Mottaghitalab, F.; Khodadadi Yazdi, M.; Reza Saeb, M.; Bączek, T.; Farokhi, M. Green and Sustainable Hydrogels Based on Quaternized Chitosan to Enhance Wound Healing. *Chem. Eng. J.* **2024**, *492*, 152288. [CrossRef]
30. Chiu, P.-H.; Wu, Z.-Y.; Hsu, C.-C.; Chang, Y.-C.; Huang, C.-M.; Hu, C.-T.; Lin, C.-M.; Chang, S.C.; Hsieh, H.-J.; Dai, C.-A. Enhancement of Antibacterial Activity in Electrospun Fibrous Membranes Based on Quaternized Chitosan with Caffeic Acid and Berberine Chloride for Wound Dressing Applications. *RSC Adv.* **2024**, *14*, 34756–34768. [CrossRef]
31. Panda, P.K.; Dash, P.; Chang, Y.-H.; Yang, J.-M. Improvement of Chitosan Water Solubility by Fumaric Acid Modification. *Mater. Lett.* **2022**, *316*, 132046. [CrossRef]
32. Zhou, C.; Jiang, T.; Liu, S.J.; He, Y.J.; Yang, G.C.; Nie, J.Q.; Wang, F.Y.; Yang, X.F.; Chen, Z.B.; Lu, C.F. Agnps Loaded Adenine-Modified Chitosan Composite Poss-Peg Hybrid Hydrogel with Enhanced Antibacterial and Cell Proliferation Properties for Promotion of Infected Wound Healing. *Int. J. Biol. Macromol.* **2024**, *267*, 131575. [CrossRef] [PubMed]
33. He, X.R.; Zhang, Z.P.; Xuan, X.; Tan, T.Y.; Sun, J.X.; Wang, B.; Tian, Y.; Chen, H. Structure and Properties of Chitosan Plasticized with Hydrophobic Short-Chain Fatty Acid Cosolvent. *Int. J. Biol. Macromol.* **2025**, *300*, 140250. [CrossRef]
34. Chalitangkoon, J.; Ronte, A.; Monvisade, P. Carboxyethylation of Chitosan-Based Polymeric Dyes for Potential Ph-Sensing Applications. *J. Taiwan Inst. Chem. Eng.* **2023**, *149*, 105001. [CrossRef]
35. Xu, R.B.; Aotegen, B.; Zhong, Z.M. Synthesis, Characterization and Biological Activity of C_6-Schiff Bases Derivatives of Chitosan. *Int. J. Biol. Macromol.* **2017**, *105*, 1563–1571. [CrossRef]
36. Pang, G.; Yi, M.; Yin, X.; Wu, W.; Xu, S. Immobilization of Chitosan on Polyether Ether Ketone Surface Modified with Acrylic Acid by Uv-Induced Graft Polymerization. *Iran. Polym. J.* **2022**, *31*, 1399–1407. [CrossRef]
37. Huang, Y.; Zhang, G.; Liu, J.; Wang, Q.; Shen, S.; Liu, D.; Hong, Y.; Sun, S.; Wyman, I. Overview on Modified Membranes by Different Polysaccharides and Their Derivatives: Preparation and Performances. *J. Environ. Chem. Eng.* **2024**, *12*, 111980. [CrossRef]
38. Tan, L.; Sun, B.B.; Luo, W.; Liu, S.T.; Qiu, B.W.; Zhang, H.; Liang, M.; Xia, S.; Zou, H.W. Enhanced Mechanical Property and Water Solubility of Polyvinyl Alcohol Film Via Constructing Hydrogen-Bond Network by Adding Carboxymethyl Chitosan. *J. Macromol. Sci. Part A* **2024**, *61*, 31–39. [CrossRef]
39. Yang, C.; Chen, Y.; Huang, H.; Fan, S.; Yang, C.; Wang, L.; Li, W.; Niu, W.; Liao, J. Ros-Eliminating Carboxymethyl Chitosan Hydrogel to Enhance Burn Wound-Healing Efficacy. *Front. Pharmacol.* **2021**, *12*, 679580. [CrossRef]

40. Huang, Y.; Mu, L.; Zhao, X.; Han, Y.; Guo, B. Bacterial Growth-Induced Tobramycin Smart Release Self-Healing Hydrogel for Pseudomonas Aeruginosa-Infected Burn Wound Healing. *ACS Nano* **2022**, *16*, 13022–13036. [CrossRef]
41. Rúnarsson, Ö.; Holappa, J.; Jónsdóttir, S.; Steinsson, H.; Másson, M. N-Selective 'One Pot' Synthesis of Highly N-Substituted Trimethyl Chitosan (Tmc). *Carbohydr. Polym.* **2008**, *74*, 740–744. [CrossRef]
42. Cai, J.P.; Dang, Q.F.; Liu, C.S.; Fan, B.; Yan, J.Q.; Xu, Y.Y.; Li, J.J. Preparation and Characterization of N-Benzoyl-O-Acetyl-Chitosan. *Int. J. Biol. Macromol.* **2015**, *77*, 52–58. [CrossRef] [PubMed]
43. Palacio, D.A.; Urbano, B.F.; Palencia, M.; Rivas, B.L. Preparation of Alkylated Chitosan-Based Polyelectrolyte Hydrogels: The Effect of Monomer Charge on Polymerization. *Eur. Polym. J.* **2019**, *118*, 551–560. [CrossRef]
44. Zhang, Y.; Guan, J.; Wu, J.; Ding, S.; Yang, J.; Zhang, J.; Dong, A.; Deng, L. N-Alkylated Chitosan/Graphene Oxide Porous Sponge for Rapid and Effective Hemostasis in Emergency Situations. *Carbohydr. Polym.* **2019**, *219*, 405–413. [CrossRef]
45. Chen, Z.; Yao, X.; Liu, L.; Guan, J.; Liu, M.; Li, Z.; Yang, J.; Huang, S.; Wu, J.; Tian, F.; et al. Blood Coagulation Evaluation of N -Alkylated Chitosan. *Carbohydr. Polym.* **2017**, *173*, 259–268. [CrossRef]
46. Zarandona, I.; Puertas, A.I.; Dueñas, M.T.; Guerrero, P.; de la Caba, K. Assessment of Active Chitosan Films Incorporated with Gallic Acid. *Food Hydrocoll.* **2020**, *101*, 105486. [CrossRef]
47. Liu, L.H.; Zheng, J.; Li, S.H.; Deng, Y.Y.; Zhao, S.F.; Tao, N.; Chen, W.S.; Li, J.H.; Liu, Y.N. Nitric Oxide-Releasing Multifunctional Catechol-Modified Chitosan/Oxidized Dextran Hydrogel with Antibacterial, Antioxidant, and Pro-Angiogenic Properties for Mrsa-Infected Diabetic Wound Healing. *Int. J. Biol. Macromol.* **2024**, *263*, 130225. [CrossRef]
48. Sukumaran, S.A.; Kalimuthu, B.; Selvamurugan, N.; Mani, P. Wound Dressings Based on Chitosan/Gelatin/Mgo Composite Films. *Int. J. Polym. Mater.* **2022**, *71*, 1252–1261. [CrossRef]
49. Kaczmarek, B.; Mazur, O.; Milek, O.; Michalska-Sionkowska, M.; Das, A.; Jaiswal, A.; Vishnu, J.; Tiwari, K.; Sionkowska, A.; Osyczka, A.M.; et al. Design, Characterization and in Vitro Evaluation of Thin Films Enriched by Tannic Acid Complexed by Fe(Iii) Ions. *Prog. Biomater.* **2020**, *9*, 249–257. [CrossRef]
50. Jayakumar, R.; Prabaharan, M.; Kumar, P.T.S.; Nair, S.V.; Tamura, H. Biomaterials Based on Chitin and Chitosan in Wound Dressing Applications. *Biotechnol. Adv.* **2011**, *29*, 322–337. [CrossRef]
51. Jiang, Z.W.; Wang, Y.T.; Li, L.L.; Hu, H.W.; Wang, S.; Zou, M.Y.; Liu, W.S.; Han, B.Q. Preparation, Characterization, and Biological Evaluation of Transparent Thin Carboxymethyl-Chitosan/Oxidized Carboxymethyl Cellulose Films as New Wound Dressings. *Macromol. Biosci.* **2022**, *22*, 2100308. [CrossRef] [PubMed]
52. Bianchi, E.; Ruggeri, M.; Vigani, B.; Fila, C.T.; Cornaglia, A.I.; Boselli, C.; Viseras, C.; Rossi, S.; Sandri, G. Gas Foamed Scaffolds as Smart 3d Structures in Skin Tissue Engineering. *J. Drug Delivery Sci. Technol.* **2024**, *95*, 105541. [CrossRef]
53. El-Feky, G.S.; Sharaf, S.S.; El Shafei, A.; Hegazy, A.A. Using Chitosan Nanoparticles as Drug Carriers for the Development of a Silver Sulfadiazine Wound Dressing. *Carbohydr. Polym.* **2017**, *158*, 11–19. [CrossRef]
54. Stoica, A.E.; Chircov, C.; Grumezescu, A.M. Hydrogel Dressings for the Treatment of Burn Wounds: An up-to-Date Overview. *Materials* **2020**, *13*, 2853. [CrossRef]
55. Zhang, S.W.; Yang, W.; Gong, W.J.; Lu, Y.H.; Yu, D.G.; Liu, P. Recent Progress of Electrospun Nanofibers as Burning Dressings. *RSC Adv.* **2024**, *14*, 14374–14391. [CrossRef]
56. Sharifi, A.; Mashjoor, S.; Makhmalzadeh, B.S.; Khorsandi, L.; Shushizadeh, M.R. Baicalin-Loaded Proline and Hydroxy Proline Functionalized Chitosan Derivative Nanofiber Composite as Burning Wound Dressings. *Appl. Mater. Today* **2024**, *41*, 102519. [CrossRef]
57. Nematollahi, S.; Maghsoudian, S.; Motasadizadeh, H.; Nouri, Z.; Azad, K.; Fatahi, Y.; Samadi, N.; Mahmoudieh, M.; Shaabani, A.; Dinarvand, R. Polyhexamethylene Biguanidine Coated Silver Nanoparticles Embedded into Chitosan Thiourea/Pva Nanofibers as Wound Healing Mats: In Vitro and in Vivo Studies. *Carbohydr. Polym.* **2025**, *347*, 122704. [CrossRef]
58. Wang, J.; Duan, X.X.; Zhong, D.H.; Zhang, M.Q.; Li, J.Y.; Hu, Z.J.; Han, F. Pharmaceutical Applications of Chitosan in Skin Regeneration: A Review. *Int. J. Biol. Macromol.* **2024**, *261*, 129064. [CrossRef]
59. Zahra, D.; Shokat, Z.; Ahmad, A.; Javaid, A.; Khurshid, M.; Ashfaq, U.A.; Nashwan, A.J. Exploring the Recent Developments of Alginate Silk Fibroin Material for Hydrogel Wound Dressing: A Review. *Int. J. Biol. Macromol.* **2023**, *248*, 125989. [CrossRef]
60. Huang, W.J.; Wang, Y.X.; Huang, Z.Q.; Wang, X.L.; Chen, L.Y.; Zhang, Y.; Zhang, L.N. On-Demand Dissolvable Self-Healing Hydrogel Based on Carboxymethyl Chitosan and Cellulose Nanocrystal for Deep Partial Thickness Burn Wound Healing. *ACS Appl. Mater.* **2018**, *10*, 41076–41088. [CrossRef]
61. Goh, M.; Du, M.; Peng, W.R.; Saw, P.E.; Chen, Z.Y. Advancing Burn Wound Treatment: Exploring Hydrogel as a Transdermal Drug Delivery System. *J. Drug Deliv.* **2024**, *31*, 2300945. [CrossRef] [PubMed]
62. Ding, Q.T.; Liu, X.L.; Liu, X.X.; Chai, G.D.; Wang, N.; Ma, S.; Zhang, L.F.; Zhang, S.; Yang, J.L.; Wang, Y.J.; et al. Polyvinyl Alcohol/Carboxymethyl Chitosan-Based Hydrogels Loaded with Taxifolin Liposomes Promote Diabetic Wound Healing by Inhibiting Inflammation and Regulating Autophagy. *Int. J. Biol. Macromol.* **2024**, *263*, 130226. [CrossRef] [PubMed]
63. Tian, B.; Hua, S.; Tian, Y.; Liu, J. Chemical and Physical Chitosan Hydrogels as Prospective Carriers for Drug Delivery: A Review. *J. Mater. Chem. B* **2020**, *8*, 10050–10064. [CrossRef] [PubMed]

64. Iqbal, Y.; Ahmed, I.; Irfan, M.F.; Chatha, S.A.S.; Zubair, M.; Ullah, A. Recent Advances in Chitosan-Based Materials; the Synthesis, Modifications and Biomedical Applications. *Carbohydr. Polym.* **2023**, *321*, 121318. [CrossRef]
65. Lu, J.W.; Chen, Y.; Ding, M.; Fan, X.K.; Hu, J.W.; Chen, Y.H.; Li, J.; Li, Z.H.; Liu, W.Y. A 4arm-Peg Macromolecule Crosslinked Chitosan Hydrogels as Antibacterial Wound Dressing. *Carbohydr. Polym.* **2022**, *277*, 118871. [CrossRef]
66. Tsai, C.C.; Young, T.H.; Chen, G.S.; Cheng, N.C. Developing a Glyoxal-Crosslinked Chitosan/Gelatin Hydrogel for Sustained Release of Human Platelet Lysate to Promote Tissue Regeneration. *Int. J. Mol. Sci.* **2021**, *22*, 6451. [CrossRef]
67. Keshavarz, H.; Khavandi, A.; Alamolhoda, S.; Naimi-Jamal, M.R. Ph-Sensitive Magnetite Mesoporous Silica Nanocomposites for Controlled Drug Delivery and Hyperthermia. *RSC Adv.* **2020**, *10*, 39008–39016. [CrossRef]
68. Thirupathi, K.; Raorane, C.J.; Ramkumar, V.; Ulagesan, S.; Santhamoorthy, M.; Raj, V.; Krishnakumar, G.S.; Phan, T.T.V.; Kim, S.C. Update on Chitosan-Based Hydrogels: Preparation, Characterization, and Its Antimicrobial and Antibiofilm Applications. *Gels* **2023**, *9*, 35. [CrossRef]
69. Teoh, J.H.; Mozhi, A.; Sunil, V.; Tay, S.M.; Fuh, J.; Wang, C.-H. 3D Printing Personalized, Photocrosslinkable Hydrogel Wound Dressings for the Treatment of Thermal Burns. *Adv. Funct. Mater.* **2021**, *31*, 2105932.
70. Salagierski, S.; Dziadek, M.; Zagrajczuk, B.; Domalik-Pyzik, P.; Serafim, A.; Stancu, I.-C.; Cholewa-Kowalska, K. When Dynamic Covalent Chemistry Meets Bioactive Glasses—Novel Multi-Level Crosslinked, Injectable, Self-Healing Chitosan-Based Hydrogels. *Sustain. Mater. Technol.* **2024**, *39*, e00850. [CrossRef]
71. Pita-López, M.L.; Fletes-Vargas, G.; Espinosa-Andrews, H.; Rodríguez-Rodríguez, R. Physically Cross-Linked Chitosan-Based Hydrogels for Tissue Engineering Applications: A State-of-the-Art Review. *Eur. Polym. J.* **2021**, *145*, 110176. [CrossRef]
72. Shen, J.; Jiao, W.; Chen, Z.; Wang, C.; Song, X.; Ma, L.; Tang, Z.; Yan, W.; Xie, H.; Yuan, B.; et al. Injectable Multifunctional Chitosan/Dextran-Based Hydrogel Accelerates Wound Healing in Combined Radiation and Burn Injury. *Carbohydr. Polym.* **2023**, *316*, 121024. [CrossRef] [PubMed]
73. Monavari, M.; Sohrabi, R.; Motasadizadeh, H.; Monavari, M.; Fatahi, Y.; Ejarestaghi, N.M.; Fuentes-Chandia, M.; Leal-Egaña, A.; Akrami, M.; Homaeigohar, S. Levofloxacin Loaded Poly (Ethylene Oxide)-Chitosan/Quercetin Loaded Poly (D,L-Lactide-Co-Glycolide) Core-Shell Electrospun Nanofibers for Burn Wound Healing. *Front. Bioeng. Biotechnol.* **2024**, *12*, 1352717. [CrossRef]
74. Ailincai, D.; Cibotaru, S.; Anisiei, A.; Coman, C.G.; Pasca, A.S.; Rosca, I.; Sandu, A.I.; Mititelu-Tartau, L.; Marin, L. Mesoporous Chitosan Nanofibers Loaded with Norfloxacin and Coated with Phenylboronic Acid Perform as Bioabsorbable Active Dressings to Accelerate the Healing of Burn Wounds. *Carbohydr. Polym.* **2023**, *318*, 121135. [CrossRef]
75. Pal, P.; Dadhich, P.; Srivas, P.K.; Das, B.; Maulik, D.; Dhara, S. Bilayered Nanofibrous 3d Hierarchy as Skin Rudiment by Emulsion Electrospinning for Burn Wound Management. *Biomater. Sci.* **2017**, *5*, 1786–1799. [CrossRef]
76. Hu, J.L.; Lin, Y.S.; Cui, C.S.; Zhang, F.F.; Su, T.T.; Guo, K.Y.; Chen, T.N. Clinical Efficacy of Wet Dressing Combined with Chitosan Wound Dressing in the Treatment of Deep Second-Degree Burn Wounds: A Prospective, Randomised, Single-Blind, Positive Control Clinical Trial. *Int. Wound J.* **2023**, *20*, 699–705. [CrossRef]
77. Cibotaru, S.; Anisiei, A.; Platon, V.-M.; Rosca, I.; Sandu, I.-A.; Coman, C.-G.; Mititelu-Tartau, L.; Andreica, B.-I.; Marin, L. Imino-Quaternized Chitosan/Chitosan Nanofibers Loaded with Norfloxacin as Potential Bandages for Wound Healing. *Int. J. Biol. Macromol.* **2025**, *314*, 144304. [CrossRef]
78. Pasaribu, K.M.; Ilyas, S.; Tamrin, T.; Radecka, I.; Swingler, S.; Gupta, A.; Stamboulis, A.G.; Gea, S. Bioactive Bacterial Cellulose Wound Dressings for Burns with Collagen in-Situ and Chitosan Ex-Situ Impregnation. *Int. J. Biol. Macromol.* **2023**, *230*, 123118. [CrossRef]
79. Dragostin, O.M.; Samal, S.K.; Dash, M.; Lupascu, F.; Pânzariu, A.; Tuchilus, C.; Ghetu, N.; Danciu, M.; Dubruel, P.; Pieptu, D.; et al. New Antimicrobial Chitosan Derivatives for Wound Dressing Applications. *Carbohydr. Polym.* **2016**, *141*, 28–40. [CrossRef]
80. Alshora, D.; Ashri, L.; Alfaraj, R.; Alhusaini, A.; Mohammad, R.; Alanaze, N.; Ibrahim, M.; Badran, M.M.; Bekhit, M.; Alsaif, S.; et al. Formulation and in Vivo Evaluation of Biofilm Loaded with Silver Sulfadiazine for Burn Healing. *Gels* **2023**, *9*, 855. [CrossRef]
81. Zhao, Y.Y.; Dai, Z.B.; Huang, H.M.; Tian, J.; Cai, H.B. Injectable Silver Nanoparticle-Based Hydrogel Dressings with Rapid Shape Adaptability and Antimicrobial Activity. *Appl. Biochem. Biotechnol.* **2025**, *197*, 821–836. [CrossRef] [PubMed]
82. Zhang, X.; Qin, M.; Xu, M.; Miao, F.; Merzougui, C.; Zhang, X.; Wei, Y.; Chen, W.; Huang, D. The Fabrication of Antibacterial Hydrogels for Wound Healing. *Eur. Polym. J.* **2021**, *146*, 110268. [CrossRef]
83. Zhu, C.; Zhao, J.; Kempe, K.; Wilson, P.; Wang, J.; Velkov, T.; Li, J.; Davis, T.P.; Whittaker, M.R.; Haddleton, D.M. A Hydrogel-Based Localized Release of Colistin for Antimicrobial Treatment of Burn Wound Infection. *Macromol. Biosci.* **2016**, *17*, 1600320. [CrossRef] [PubMed]
84. Shi, S.Y.; Ou, X.H.; Long, J.F.; Lu, X.Q.; Xu, S.Q.; Zhang, L. Nanoparticle-Based Therapeutics for Enhanced Burn Wound Healing: A Comprehensive Review. *Int. J. Nanomed.* **2024**, *19*, 11213–11233. [CrossRef]
85. Comino-Sanz, I.M.; López-Franco, M.D.; Castro, B.; Pancorbo-Hidalgo, P.L. The Role of Antioxidants on Wound Healing: A Review of the Current Evidence. *J. Clin. Med.* **2021**, *10*, 3558. [CrossRef]

86. Salze, M.; Giard, J.-C.; Riboulet-Bisson, E.; Hain, T.; Rincé, A.; Muller, C. Identification of the General Stress Stimulon Related to Colonization in Enterococcus Faecalis. *Arch. Microbiol.* **2019**, *202*, 233–246. [CrossRef]
87. Song, Y.; Han, N.; Guo, Z.S.; Li, H.H.; Guo, M.X.; Dou, M.H.; Ye, J.H.; Peng, Z.W.; Lu, X.Y.; Li, M.H.; et al. Baicalein-Modified Chitosan Nanofiber Membranes with Antioxidant and Antibacterial Activities for Chronic Wound Healing. *Int. J. Biol. Macromol.* **2024**, *279*, 134902. [CrossRef]
88. Han, W.; Wang, L.; Sun, J.; Shi, Y.; Cui, S.; Yang, D.; Nie, J.; Ma, G. Dual-Drug-Loaded Core–Shell Electrospun Nanofiber Dressing for Deep Burns. *ACS Appl. Bio Mater.* **2024**, *7*, 1179–1190. [CrossRef]
89. Owida, H.A.; Al-Nabulsi, J.I.; Alnaimat, F.; Al-Ayyad, M.; Turab, N.M.; Al Sharah, A.; Shakur, M.; Merodio, J. Recent Applications of Electrospun Nanofibrous Scaffold in Tissue Engineering. *Appl. Bionics Biomech.* **2022**, *2022*, 1953861. [CrossRef]
90. Liu, S.N.; Wei, L.; Huang, J.Q.; Luo, J.Y.; Weng, Y.J.; Chen, J.Y. Chitosan/Alginate-Based Hydrogel Loaded with Ve-Cadherin/Fgf as Scaffolds for Wound Repair in Different Degrees of Skin Burns. *J. Biomed. Mater. Res. Part B-Appl. Biomater.* **2025**, *113*, e35533. [CrossRef]
91. Le, K.T.; Nguyen, C.T.; Lac, T.D.; Nguyen, L.G.T.; Tran, T.L.; Tran-Van, H. Facilely Preparing Carboxymethyl Chitosan/Hydroxyethyl Cellulose Hydrogel Films for Protective and Sustained Release of Fibroblast Growth Factor 2 to Accelerate Dermal Tissue Repair. *J. Drug Deliv. Sci. Technol.* **2023**, *82*, 104318. [CrossRef]
92. Chang, J.; Xu, X.; Li, H.; Jian, Y.; Wang, G.; He, B.; Gu, Z. Components Simulation of Viral Envelope Via Amino Acid Modified Chitosans for Efficient Nucleic Acid Delivery: In Vitro and in Vivo Study. *Adv. Funct. Mater.* **2012**, *23*, 2691–2699. [CrossRef]
93. Wang, L.; Sun, L.; Gu, Z.; Li, W.; Guo, L.; Ma, S.; Guo, L.; Zhang, W.; Han, B.; Chang, J. N-Carboxymethyl Chitosan/Sodium Alginate Composite Hydrogel Loading Plasmid DNA as a Promising Gene Activated Matrix for in-Situ Burn Wound Treatment. *Bioact. Mater.* **2022**, *15*, 330–342. [CrossRef]
94. Kou, S.; Peters, L.M.; Mucalo, M.R. Chitosan: A Review of Sources and Preparation Methods. *Int. J. Biol. Macromol.* **2021**, *169*, 85–94. [CrossRef]
95. Zhu, L.; Zou, D.Q.; Fan, Z.Q.; Wang, N.; Bo, Y.Y.; Zhang, Y.Q.; Guo, G. Properties of a Novel Carboxymethyl Chitosan Derived from Silkworm Pupa. *Arch. Insect Biochem. Physiol.* **2018**, *99*, 21499. [CrossRef]
96. Pellis, A.; Guebitz, G.M.; Nyanhongo, G.S. Chitosan: Sources, Processing and Modification Techniques. *Gels* **2022**, *8*, 393. [CrossRef]
97. Poznanski, P.; Hameed, A.; Orczyk, W. Chitosan and Chitosan Nanoparticles: Parameters Enhancing Antifungal Activity. *Molecules* **2023**, *28*, 2996. [CrossRef]
98. Cord-Landwehr, S.; Moerschbacher, B.M. Deciphering the Chitocode: Fungal Chitins and Chitosans as Functional Biopolymers. *Fungal Biol. Biotechnol.* **2021**, *8*, 19. [CrossRef]
99. Gong, X.; Wang, F.P.; Yang, J.L.; Du, H.; Jiang, M.Y.; Tan, M.M.; Chen, G.B.; Chen, Z.M. Engineered Composite Dressing with Exudate Management Capabilities for the Process of Entire Wound Healing. *Mater. Today Commun.* **2024**, *39*, 108557. [CrossRef]
100. Feng, F.; Zhao, Z.H.; Li, J.W.; Huang, Y.S.; Chen, W.C. Multifunctional Dressings for Wound Exudate Management. *Prog. Mater. Sci.* **2024**, *146*, 101328. [CrossRef]
101. Wang, X.; Zhang, D.; Wang, J.; Tang, R.P.; Wei, B.; Jiang, Q. Succinyl Pullulan-Crosslinked Carboxymethyl Chitosan Sponges for Potential Wound Dressing. *Int. J. Polym. Mater.* **2017**, *66*, 61–70. [CrossRef]
102. Mao, T.F.; Yang, L.; Liu, G.Q.; Wei, Y.; Gou, Y.Z.; Wang, J.; Tao, L. Ferrocene-Containing Polymer Via the Biginelli Reaction for in Vivo Treatment of Oxidative Stress Damage. *ACS Macro Lett.* **2019**, *8*, 639–645. [CrossRef] [PubMed]

MDPI AG
Grosspeteranlage 5
4052 Basel
Switzerland
Tel.: +41 61 683 77 34

Polymers Editorial Office
E-mail: polymers@mdpi.com
www.mdpi.com/journal/polymers

www.ingramcontent.com/pod-product-compliance
Lightning Source LLC
Chambersburg PA
CBHW061436210926
51850CB00008B/1528
9783725854530